Decoding Aristarchus

Alberto Gomez Gomez

Decoding Aristarchus

An investigation into the life and work of
Aristarchus of Samos, The Mathematician

Bibliographic Information published by the Deutsche Nationalbibliothek
The Deutsche Nationalbibliothek lists this publication in the Deutsche Nationalbibliografie; detailed bibliographic data is available in the internet at http://dnb.d-nb.de.

Library of Congress Cataloging-in-Publication Data
A CIP catalog record for this book has been applied for at the Library of Congress.

Cover illustration courtesy of Prabhu Astrophotography

ISBN 978-3-631-89261-9 (Print)
E-ISBN 978-3-631-89294-7 (E-PDF)
E-ISBN 978-3-631-89295-4 (E-PUB)
10.3726/b20352

© Peter Lang GmbH
Internationaler Verlag der Wissenschaften
Berlin 2023
All rights reserved.
Peter Lang – Berlin · Bruxelles · Lausanne · New York · Oxford

All parts of this publication are protected by copyright. Any utilisation outside the strict limits of the copyright law, without the permission of the publisher, is forbidden and liable to prosecution. This applies in particular to reproductions, translations, microfilming, and storage and processing in electronic retrieval systems.
This publication has been peer reviewed.
www.peterlang.com

To my parents, Manuel and Carmen, whose love has made this book possible, and to my friends John Westfall and William Sheehan, whose influence can be felt all over it.

Table of Contents

Acknowledgements 9
Acronyms, conventions, and symbols 11
Introduction 13

1 The sources 15
    1.1   The main sources 15
    1.2   The genealogy of *On Sizes* 17
    1.3   The genealogy of *The Sand Reckoner* 18
    1.4   Aetius and other sources 20
    1.5   Forerunners 23
    1.6   The authenticity of *On Sizes* 39
    1.7   The authenticity of *The Sand Reckoner* 43

2 *On Sizes* 47
    2.1   Hypotheses, conclusions, and the Euclid connection 47
    2.2   The angular resolution of the human eye 49
    2.3   Estimating sizes and distances 58
    2.4   Conclusion 77

3 Measuring the Sun's distance 79
    3.1   Introduction 79
    3.2   Thomas Harriot 79
    3.3   Johannes Kepler 86
    3.4   Godfrey Wendelin 87
    3.5   Arthur Allen Hoag 88
    3.6   A lucky conjunction 92
        3.6.1   The Aristarchus Procedure Series 96
        3.6.2   The author's observations 103
        3.6.3   Westfall's *dichotometer* 113
        3.6.4   The *cosmometer* 122

	3.7	An overall assessment of the half-Moon method	124
	3.8	Conclusion	126
4	*The Sand Reckoner*		129
	4.1	Aristarchus' model of the universe	129
	4.2	Archimedes' assumptions	142
	4.3	Archimedes' telltale ratio	146
	4.4	Archimedes' big numbers	177
	4.5	Conclusion	181
5	On shadows, time, and light		183
	5.1	The Moon's distance	183
	5.2	The length of the year	196
	5.3	The irradiation illusion	229
	5.4	Conclusion	235
A	The length of the Moon's shadow		237
B	Trigonometry for Proposition 13		239
C	The eclipse method		247
	C.1	The lunar eclipse method	247
	C.2	The solar eclipse method	253
	C.3	Linear approximations	256
	C.4	The eclipse of March 14, 190 BC	257
	C.5	Summary	260

Glossary 261
References 265
List of Figures 279
List of Tables 280
Index 281

Acknowledgements

This book started as a joint project of three people who, in all justice, should have ended up as coauthors. Unfortunately, the untimely death of one of them left our team badly maimed, and it was only after several years of frequently interrupted work that the author's many commitments finally allowed him to complete it—if such a thing can ever really be done. The other member of the original team eventually opted out of the authors' list on the grounds that 'the book represents mostly *your* [the present author's] hard work and insight' (as he humbly said in a private communication). But the truth is that without Dr William Sheehan, this book would not exist, so it is only fitting that it should be dedicated to him and to the memory of Dr John Westfall, both of whom enthusiastically joined the author in the effort to unravel the mystery surrounding the birth of heliocentrism (even though it cannot be assumed that they would agree to everything that has been written here).

Several other people helped the author along the way. Among them, his own father and sister, Manuel and Begoña, who patiently assisted him in the experiments made; Drs Angelo Gioè, David Runia, and Jaap Mansfeld, whose mastery of the Classics and unrivalled linguistic knowledge of the original texts, proved an invaluable aid to understanding critical issues; Drs Alexander Jones and Fabio Acerbi, who generously shared their knowledge of ancient manuscripts; Dr Kurt Guckelsberger, with whom the author used to muse about the nature of the universe, and whose realistic comments helped the author to set his feet on the ground, dreamy as they usually are; Dr Stella Bradbury, who helped the author understand the surprising behaviour of the Moon's shadow as it enters the Earth's atmosphere; Dr Martin Beech, who kindly provided details of his trigonometric work on lunar eclipses; Dr Matthias Schemmel, whose stimulating conversations helped the author deepen his understanding of some of Thomas Harriot's astronomical achievements; the staff of the Bodleian and Vatican Libraries, who kindly provided access to relevant documents; Dr Wayne Orchiston and at least two other anonymous reviewers, whose sharp eyes went over the manuscript at several stages, eventually prompting the full rewriting of it in what is hopefully a more rigorous form. Many thanks are heartily given to all these people for kindly contributing to anything good this book may contain; any errors are, of course, the author's sole responsibility.

Acronyms, conventions, and symbols

Acronyms

IAU International Astronomical Union
AU   Astronomical Unit

Conventions

thou used as a gender-fair third-person singular pronoun

Symbols

Astronomical symbols

⊕ Earth
☾ Moon
☉ Sun
☿ Mercury
♀ Venus
♂ Mars
♃ Jupiter
♄ Saturn

Greek symbols

ε elongation of the Moon from the Sun
ζ half the angular size of an object as seen from the tip of the shadow it casts
μ parallax
$\pi \simeq 3.14...$
ρ angular radius
ϕ latitude
φ angular radius of the Earth's shadow at the distance of the Moon
ψ phase angle of the Moon

Roman symbols

e radius of the Earth
F radius of the Sphere of the Fixed Stars (or firmament)
L length of the Moon's shadow
M distance between the Earth and the Moon

- m radius of the Moon
- S distance between the Earth and the Sun
- s radius of the Sun
- t time
- U radius of the Earth's shadow (measured in Earth radii)
- V volume; length of the Earth's shadow minus the distance between the Earth and the Moon
- W length of the Earth's shadow

Introduction

Aristarchus of Samos, The Mathematician, is best known for his discovery that the Earth moves around the Sun. This simple idea led to a radical change in our understanding of the cosmos and eventually sparked the greatest upheaval in human thought ever. Yet, the details of how this idea came to life have always been shrouded in mystery. His only surviving work (namely, *On the Sizes and Distances of the Sun and Moon*, often abbreviated to just *On Sizes*) was once a must-read among astronomers, but today it is not even clear whether it belongs to the field of Astronomy, at least, not as much as it used to be when this science was one with Geometry and Harmony. Along with (and perhaps because of) the modern redefinition of Astronomy as a separate science, have come accusations of Aristarchus being 'obscure' (starting with Voltaire, 1775:230), but this does not mean that he really is, or that his book will forever defy classification.

In fact, the present book hopes to show that Aristarchus' extant work can be classified into a well defined niche of modern science (namely, Applied Mathematics), and that his style is remarkably clear for someone who was born at a time when the language of science was much, much younger than it is now. The only thing needed to understand his delightfully primal words is to pay attention, and this is what the present book intends to do.

First, we shall see what the ancient sources teach us about him. Then, we shall examine his extant book in detail and design experiments to get a hint of what he might have done. Then, we shall examine another famous ancient book (namely, *The Sand Reckoner*) in which Archimedes describes Aristarchus' heliocentric model of the cosmos. Finally, we shall analyse the scant evidence that remains on his measurement of the length of the year. But before we start, let us know the man and his work a little better.

His name is made of two Greek words (namely, ἄριστος and ἀρχός) meaning 'best' and 'leader'. Today, more than a dozen historical figures bearing it are still remembered who once came from the most varied corners of the Hellenistic World (Smith, 1849:290). The one in this book, named after his birthplace, Samos, is also known by the epithet 'The Mathematician', a word which originally meant something like 'learned', 'erudite', 'polymath', or 'all-rounder'.

His only surviving work was translated several times into Arabic: (at least) twice in the ninth century (by Qusta ibn Luqa and Thabit ibn Qurra) and once in the thirteenth century (by Nasir al-Din al-Tusi) (Berggren and Sidoli, 2007:235). Later, Latin translations were made by Giorgio Valla (1488) and Federico Commandino (1572). In 1913, it was translated into English by Sir Thomas Heath, whose work is a cornerstone in the study of this ancient treatise. He (1921*b*:1) started the second volume of his monumental *History of Greek Mathematics* complaining that scholars had paid little attention to Aristarchus. The cause, he said, was that historians of mathematics had little interest in someone who was widely regarded as an astronomer. However, this does not explain why historians of astronomy, who should certainly have been more interested, have also neglected his study. This suggests that something else is at work here. Perhaps the ultimate cause of this neglect is the sheer immensity of a task that requires those seeking to understand The All-Rounder to be all-rounders themselves, and like Heath, be good at both Mathematics, Astronomy, History, Classics, Palaeography, and even also the Psychology of Perception.

Each of these disciplines provides a different point of view from which to attack the problem. Mathematics alone affords absolute certainty, but not so History, where such a thing is unattainable. Our knowledge of the past is continuously being refined and reshaped in an attempt to make sense of the new pieces of evidence that stubbornly keep turning up. Thus, no historical investigation is ever complete: it is forever an ongoing business. This is why the present book can claim to be nothing more than a mere attempt to bridge the scholarly gap exposed by Heath.

For the same reason, the call for objectivity is a hopeless dream (outside the realm of Mathematics). It is, however, possible to approach it ever more closely, and the author hopes to have made every effort in this direction, especially knowing that Aristarchus' discovery polarized the world of science for over two thousand years in a debate that saw more than strong words before it was finally settled in his favour.

These things said, the stage is set for our investigation to begin. The goal: trying to gather as much information about him as possible from the extant sources and understand how he made his great contribution to science.

1 The sources

Abstract: In this chapter, the extant sources on Aristarchus will be visited; their authenticity, assessed; and questions about his background, discoveries, and methods, addressed.

1.1 The main sources

The main sources of information on Aristarchus are two books, namely, *On the Sizes and Distances of the Sun and Moon*, supposedly by Aristarchus himself, and *The Sand Reckoner*, supposedly by Archimedes, one of the greatest mathematicians of all time.

The authorship of these books has been contested on some occasions. The first to contest that of *On Sizes* (as we shall refer to the first of these books, for short) was the French philosopher Voltaire (1775:410), and more recently, the American astronomer and historian Dennis Rawlins (2008:19). The first to contest the authorship of *The Sand Reckoner* were the German American philologists Erhardt and Erhardt-Siebold (1942). The arguments these authors advanced will be discussed in Sections 1.6 and 1.7.

Aristarchus' theory that the Earth goes round the Sun eventually became known as *heliocentrism*, as opposed to *geocentrism*, a pair of terms (coined from the Greek words ἥλιος, γῆ, and κέντρον for 'Sun', 'Earth', and 'centre') which appeared for the first time in history, as adjectives, in a book by the German astronomer Johann Jacob Zimmermann (1679:28, 33). Aristarchus and Archimedes themselves used neither.

In his book, Archimedes described Aristarchus' model of the universe as the largest then conceived, and estimated the number of grains of sand needed to fill it in order to illustrate the ease with which mathematics can handle huge numbers.

Despite being well acquainted with Aristarchus' model, Archimedes did not openly embrace it. This prudent move was indeed a blessing in disguise as long as it favoured the providential preservation of so many of his works, *The Sand Reckoner* among them. By contrast, Aristarchus' works have all vanished, except for the one mentioned above.

All we know about him comes from these two books and from several brief references by some other authors. Happily, much can be inferred from these sources, scant as they are. For example, the very fact that his theory was rejected by most people and (at best) secretly acknowledged by the likes of Archimedes, yet turned out to be true, says something about his intelligence quotient—that of

Archimedes having been rated (by Thims, 2016:217) at about a staggering 190 (if such a thing can ever be fathomed).

One of these men once boasted of what he could do should he be given a lever long enough and a place to stand (Diodorus Siculus, 1957:194; Pappus, 1878:1060), but it was the other man who actually shook the foundations of the Earth, setting it in motion around the Sun for the first time in human awareness (Plutarch, 1957:54; Goodwin, 1878b:240).

Among the stars of human intellect, he is what might be called an *outlier* (to use a statistical term), in the sense that his discovery took over eighteen centuries to start dawning on the rest of humanity. Yet, his early book *On Sizes* formed part of a small set of recommended readings (known as *Little Astronomy*) previous to the study of Ptolemy's *Almagest*.[1] It contains no trace of heliocentrism. The one that does is *The Sand Reckoner*, and awareness of the Earth's motion started to bloom in the West soon after the arrival of this book as a refugee from the besieged Byzantium.

But the new philosophy did not have an easy start, and eventually, was described as 'foolish' and 'formally heretical' by Saint Robert Bellarmine, then Cardinal Inquisitor, on February 24, 1616, during Galileo's trial (Brodrick, 1961:372; Finocchiaro, 1989:146). Thankfully, Galileo's name was formally and publicly cleared of heresy by Pope John Paul II in his speech to the members of the Pontifical Academy of Sciences delivered on October 31, 1992 (John Paul II, 2003:336). It was then officially and unequivocally acknowledged that Galileo had been right all along, even when he said that Scripture needed to be reinterpreted in light of the heliocentric theory.

Heliocentrism has ever since been held to be true by the Catholic Church, and since 'truth cannot contradict truth' (Leo XIII, 1893:23), those passages of Scripture which seem to support geocentrism if interpreted literally (such as Joshua 10:12, 1 Chronicles 16:30, Psalm 93:1, Psalm 96:10, Psalm 104:5, Ecclesiastes 1:5, Job 26:7, ...) need to be reexamined if their proper sense is to be determined (John Paul II, 2003:341, 371). But this is a task better left to theologians. Let the present book be concerned solely with the scientific side of Aristarchus' amazing breakthrough.

---

1 In the 4th century, Pappus of Alexandria counted Aristarchus' geocentric work among the *Little Astronomy* books, which included Euclid's *Optics* and *Phaenomena*, Autolycus' *On the Moving Sphere* and *On Risings and Settings*, Theodosius' *Sphaerica*, *On Habitations*, and *On Days and Nights*, Aristarchus of Samos' *On the Sizes and Distances of the Sun and Moon*, and Hypsicles' *Risings* (Hultsch, 1877:554; Heath, 1913:317).

Before we start, however, it must be said that the long centuries that elapsed between Galileo's conviction as a heretic and his full rehabilitation left their mark: the word 'foolish' once used to describe heliocentrism was often extended to its father, Aristarchus, and repeated long after the Church had ceased hostilities, as can be read in the language of some scholars, who created an air of 'clumsiness' around the Samian mathematician, dismissing his discoveries as mere good luck, and saying these were based on little or no observation. As an example of this kind of literature, we may take the Austrian-American mathematician and historian of science Otto Eduard Neugebauer (1975:4), to whom Aristarchus' work was 'pure mathematical pedantry, [having] little to do with practical astronomy' (1975:643). Such words are a clear reminder that it was not only religion that once opposed Aristarchus' theory, but also a large part of the scientific community, including some of the most brilliant and talented people ever. However, as mathematicians should know better than anyone else, science is built on proof, rather than personal opinion and hearsay. This is why only the most careful scrutiny of all the extant evidence at our disposal can reveal whether such claims as Neugebauer's have any substance.

1.2 The genealogy of *On Sizes*

The extant copies of the book *On Sizes* have all been studied in depth by the Italian philologist Angelo Gioè (2007), who concludes the language in it is typically that of the third-century BC Greek academic world. The oldest and best preserved of these copies—and seemingly 'the ultimate source of all the others' (Heath, 1913:325)—is the *Codex Vaticanus Graecus* 204 (folios 108v to 117v), which is believed to have been copied from one or two Byzantine prototypes in the ninth or tenth century (Berggren and Sidoli, 2007:236).

There are also two Medieval Arabic versions of this book: one by Thabit ibn Qurra and another by Nasir al-Din al-Tusi. The former predates the oldest extant Greek copy. The colophon of this text states (in folio 133r) that it is a 'revision' by Thabit ibn Qurra (Berggren and Sidoli, 2007:235). At present, this manuscript is privately owned and is usually called the Kraus Manuscript, after its former owner (Sidoli and Kusuba, 2014:156). The book *On Sizes* can be found there in folios 124r to 133r. Tusi's version can be found at Tabriz National Library, as MS 3484, 184. A copy of it dating from 1319 (Uri, 1787:187) is housed in the Bodleian Library, under the shelfmark MS *Archivum Seldenianum* A.45 (folios 142v to 150r).

The book *On Sizes* can be traced still further back in time, when the Greek mathematician Pappus of Alexandria (c. AD 320) quoted 'word for word' some passages from it, from which it can be deduced that the version he had access to (if indeed quoted verbatim) was slightly different from the one we have. The oldest extant copy of Pappus' work, *Synagoge*, can be found in the tenth century *Codex Vaticanus Graecus* 218 (folios 8r to 205v), and his comments on the book *On Sizes* are in Book VI (folios 87v to 118r) of this codex (Treweek, 1957:196; Hultsch, 1877:554; Heath, 1913:412).

1.3 The genealogy of *The Sand Reckoner*

As we saw in Section 1.1, Aristarchus is exceptional in many ways. The Roman writer Marcus Vitruvius Pollio regarded him as one of the most learned men he knew, placing him first in his short, non-chronological list in *On Architecture* 1.1.17, reproduced below.

> **Passage 1** Those to whom nature has granted such wits, acuity, and good memory that they are fully skilled in geometry, astronomy, music and related disciplines, pass beyond the business of architects and are turned into mathematicians. As a result they can easily hold their own in such fields of study, because they have a well-stocked scholarly arsenal with missiles from several disciplines. Nonetheless, such people are seldom to be found, such as once were Aristarchus of Samos, Philolaos and Archytas of Tarentum, Apollonius of Perge, Eratosthenes of Cyrene, Archimedes, and Scopinas of Syracuse, who made all manner of discoveries on measurement through mathematics and natural philosophy, and left treatises on these subjects for subsequent generations (Rowland and Howe, 2001:24; see also Granger, 1931:23; Morgan, 1914:12).

Vitruvius' words in Passage 1 suggested to Gwilt (1874:9) that Aristarchus 'wrote on all the sciences', but the extant sources do not confirm this, apart from a single reference in *Sand Reckoner* 1.4, according to which, Aristarchus 'published in writing' the hypotheses leading to his vast model of the world.

> **Passage 2** Aristarchus, the Samian, published in writing certain hypotheses, in which it follows from the suppositions that the cosmos must be many times greater than the one mentioned before (Erhardt and Erhardt-Siebold, 1942:579; see also Heiberg, 1881:244).

It is tempting to think that this 'published writing' took the form of a book or treatise. However, Erhardt and Erhardt-Siebold (1942:579) pointed out that the Greek text can also be taken to mean that Aristarchus lectured on his model by means of 'graphical representations'. It is, however, unlikely that Aristarchus, literate as he was, would limit himself to just lectures or drawings. His book *On Sizes*, which seems to be an early version of the book Archimedes actually read, contains both text and illustrations, and Archimedes' description of Aristarchus' model is detailed enough to suggest he had access to more than just drawings. So, in the present work, we shall assume that Aristarchus did write a book expounding his ideas, and that this book (now lost) inspired Archimedes to write *The Sand Reckoner*, which is the main source of information we have on Aristarchus' theory.

Archimedes' book was dedicated to King Gelo II of Syracuse, who reigned from about 266 to 216 BC, so it must have been written between these years. After the wanton destruction that followed the fall of Rome, knowledge of Archimedes (and most classics) dwindled to almost nothing in the western part of the Roman Empire. The eastern part, however, was spared and preserved the knowledge of the past. As far as we can trace, a first attempt at compiling Archimedes' works seems to have been made there by Isidore of Miletus (AD 442 – 537), one of the architects of Hagia Sophia; another was made by Leon, The Mathematician (c. 790 – after 869), archbishop of Thessalonica and head of the Magnaura School of philosophy in Constantinople (Netz, 2004:14). The latter compilation consisted of (at least) three codices, designated by the Danish philologist Johan Ludvig Heiberg (1972:iii) as A, 𝔅, and C. Codex C was destined to become the famous *Archimedes Palimpsest*, which never made it to the West during the Middle Ages. Codices A and 𝔅 did eventually arrive in the papal court in Viterbo, near Rome. In 1269, the Flemish Dominican William of Moerbeke translated into Latin all the books in these codices, except for *The Sand Reckoner* and Eutocius' *Commentary on the Measurement of the Circle* (both in Codex A). His translation is preserved in the Vatican Library in a manuscript on parchment called *Codex Ottobonianus Latinus* 1850.

Codices A and 𝔅 were both listed (as items 612 and 608) in an inventory of papal manuscripts made in 1311 by order of Pope Clement V (Ehrle, 1890:96), but after this, Codex 𝔅 disappeared and was never seen again (Jones, 1986:19). Only Codex A remained when Pope Nicholas V commissioned a new Latin translation of Archimedes' works to be made in 1450 by the Italian humanist Iacopo da San Cassiano (Iacobus Cremonensis), who, this time, translated the whole of it, including *The Sand Reckoner* and Eutocius' *Commentary on the Measurement of the Circle* (James of Cremona, 1544:120, 155).

Eventually, Codex A became the property of the Italian humanist Giorgio Valla, a great scholar and collector of Greek manuscripts, who allowed a copy of it to be made in 1491 by his friend Angelo Poliziano at the request of Lorenzo de Medici. This copy is now housed in the Laurentian Library in Florence, under the shelfmark *Pluteus* 28.4. When Valla died, Alberto III Pio, Prince of Carpi, acquired his library, which in time was inherited by his nephew, Cardinal Rodolfo Pio da Carpi. When the latter died, in 1564, an inventory was made of his manuscripts, and Codex A was listed (as item 161) in it (Mercati, 1938:233), and that was the last time it was heard of (Clagett, 1964:57). This makes the copy in the Laurentian Library—denoted Codex D by Heiberg—one of the oldest and most authoritative ones, as it was made in imitation of the hand and page layout of Codex A (Napolitani, 2013:111). *The Sand Reckoner* can be found there (between folios 104r and 111r).

1.4 Aetius and other sources

There are other sources from which we learn about Aristarchus. Special mention is due to Aetius, a shadowy author of the first to second century AD, whose name is known only from two references in the book *A Cure for Pagan Maladies* (Gaisford et al., 1839:167, 195) by Theodoret, Bishop of Cyrus, and whose works (now lost) can be partially reconstructed from two surviving anthologies: namely, *Opinions of the Philosophers*, attributed to Plutarch, but probably written in the second century AD by someone else (often called pseudo-Plutarch), and *Selections on Natural Philosophy*, written by the Macedonian anthologist Johannes Stobaeus (John of Stobi) in the fifth century AD (McKirahan, 2011:4). Of these sources for Aetius, only Stobaeus mentions Aristarchus, and he does so four times (Mansfeld and Runia, 2018:355). We will now see three of these mentions, reserving the fourth for a later occasion (in Chapter 4). The first of them (*Selections* 1.15.5) runs as follows.

> **Passage 3** Aristarchus of Samos, the astronomer, a pupil of Strato, [says] that light is the colour which falls on the underlying objects (Fortenbaugh, 2017:41, 307; see also Heath, 1913:300; Diels, 1879:313; Wachsmuth and Hense, 1884:149).

Aetius says here that Aristarchus studied under the Peripatetic philosopher Strato of Lampsacus (c. 335 – c. 269 BC), who, according to the biographer Diogenes Laertius (1925a:509), spent the best part of his life in Alexandria, tutoring Prince Ptolemy Philadelphus, son of Ptolemy I Soter and Berenice I, the Macedonian couple who founded the Ptolemaic Kingdom. The former general of Alexander the Great made every effort to attract brilliant minds to Alexandria and even move the grounds of Aristotle's Lyceum to the City of the Great Library. Inviting Strato was part of this ambitious plan, which might have succeeded had Strato not moved to Athens in 287 BC to succeed Theophrastus of Eresus as head of the Lyceum.[2]

---

2 Aristotle's school was called the *Peripatos* ('walks') and was located on the grounds of the Lyceum, a temple dedicated to Apollo Lyceus ('Apollo the Wolf-God') outside the walls of Athens.

Passage 4 [Theophrastus'] successor in the school was Strato, the son of Arcesilaus, a native of Lampsacus, whom he mentioned in his will; a distinguished man who is generally known as 'The Physicist', because more than anyone else he devoted himself to the most careful study of nature. Moreover, he taught Ptolemy Philadelphus and received, it is said, eighty talents from him. According to Apollodorus in his *Chronicle* [of which only fragments now remain] he became head of the school in the hundred and twenty-third Olympiad [288 to 284 BC], and continued to preside over it for eighteen years (Laertius, 1925a:509).

Combining Passages 3 and 4, we learn that Aristarchus studied under one of the best tutors available at the time, possibly in one of the most intellectually stimulating environments ever. Namely, Alexandria (and, less likely, perhaps also Athens). Being a Samian (and therefore, likely to have strong ties with Macedon), he surely had unrestricted access to the Library and to all the knowledge the Macedonians had brought from Babylon. Aetius' comments also show that Aristarchus applied himself to the study of light, colour, and vision, articulating his definition of *light* in a way (Aetius deemed) worthy of note.

Another of Aetius' comments (in *Selections* 1.52.3-4) runs as follows.

Passage 5 Strato says that colours travel from bodies and give their colour to the intervening air [that is, the air between the thing seen and the person seeing it] (Fortenbaugh, 2017:151). Aristarchus [says] that in some way, shapes give their form to the air (Mansfeld and Runia, 2009:188; see also Diels, 1879:403; Wachsmuth and Hense, 1884:483).

Here Aetius says that both Strato and Aristarchus supported the Aristotelian theory of vision according to which, 'the object that is seen sends something from itself to us', rather than the alternative theory of Empedocles of Akragas (c. 494 – c. 434 BC) by which 'the object that is seen waits for some sensory power to come to it from us'. According to Mansfeld and Runia (2009:188), it follows from the cited fragment that Aristarchus thought that what travelled from the seen to the seer was some kind of 'force', rather than 'quality', 'part', or 'image' of the external body, as supported, respectively, by Strato (in *Selections* 1.52.3), Empedocles and Timagoras (in *Selections* 1.52.2 and 1.52.12), and Empedocles and the atomists (in *Selections* 1.52.1 and 1.52.12).

There is yet another fragment (in *Selections* 1.15.9) in which Aetius tells us that Epicurus and Aristarchus remarked on the inability of the eye to see colours in the dark.

Passage 6 Epicurus and Aristarchus [say that] bodies in the dark do not have colour (Mansfeld and Runia, 2018:420; see also Diels, 1879:314; Wachsmuth and Hense, 1884:149).

They were simply echoing Aristotle's observation that 'colour is not visible without light' (Hicks, 1907:77). This means that, apart from correctly noting that 'things are colourless in the dark' (Huby and Gutas, 1999:66), these scientists were all aware that day and night vision are quite different things.

It is to be noted that Aetius' entries (in both pseudo-Plutarch and Stobaeus) are arranged thematically, not chronologically, but when several names appear in the same entry, they are usually arranged in chronological order. So, in Passage 6, Aetius is telling us that Epicurus (341 – 270 BC) was older than Aristarchus. In fact, since Strato (c. 335 – c. 269 BC) tutored both Ptolemy Philadelphus (309 – 246 BC) and Aristarchus, it is likely that Aristarchus was many years Epicurus' junior.

Now, according to Laertius (1925*b*:528), Epicurus was born in Samos to Athenian parents, and at the age of eighteen, went to Athens to do his military service. Then, Alexander died (in 323 BC) and his loyal general Perdiccas carried out his will to expel the Athenians from Samos to make room for Macedonian settlers and Samian exiles. So, after his two years in Athens, Epicurus rejoined his parents in Colophon, where they had moved. He lived in a few more places on the western coast of Anatolia before he finally settled in Athens in 306 BC.

He may never have met Aristarchus in Samos, especially if the latter, as it is most likely the case (since he was a pupil of Strato), was born after the expulsion of the Athenians (in 321 BC). The extant sources do not tell us when Aristarchus was born or who his parents were, but the evidence so far shows that Aristarchus received a thorough Aristotelian education and agreed—at least in the part that has survived—with the Aristotelian theory of vision. Both Aristotle, Epicurus, and Strato were committed empiricists (or hands-on scientists) and keen observers of nature. So Aristarchus is likely to have followed suit. After all, the very nature of his discovery strongly suggests that he excelled in the art of observation.

But these are all possibilities. Remember, we can never be sure of anything that happened in the past. Our witnesses might be misinformed or lying. Even assuming they are neither, there is still the question of where did Strato meet Aristarchus. Was it in Alexandria, in Athens, in both, or elsewhere? The scant evidence we have, together with the fact that mathematicians tend to make their greatest contributions when young (Guterman, 2000:18-20), points to Alexandria as the most likely candidate.

It is also possible that Aristarchus, who was destined to excel in mathematics, would not be confined to what the Macedonian philosopher Aristotle had to offer. The more mathematically inclined Pythagoras of Samos might have exerted an alluring pull on his imagination too. After all, it was in the Pythagorean school that the novel idea of a moving Earth was first conceived.

> **A summary of the evidence for Aristarchus being a committed empiricist**
> - Aristarchus was a pupil of Strato (Passage 3), the third headmaster of the Lyceum. So he is likely to have received a thorough Aristotelian education.
> - Aristotle was a Macedonian, and Ptolemy Philadelphus, the only other pupil of Strato known by name (Passage 4), was of pure Macedonian descent.
> - Being a student of Strato, Aristarchus is likely to have been born after Perdiccas expelled the Athenians from Samos to give the island back to the Samians (Laertius, 1925b:528). If so, Aristarchus is also likely to have had strong ties with Macedon.
> - Strato taught Ptolemy Philadelphus in Alexandria, the city named after Aristotle's most famous pupil. So Aristarchus is likely to have studied there.
> - Aristarchus agreed with Aristotle's theory of vision (as Passages 3, 5 and 6 show).
> - Aristotelians were committed empiricists (Passage 4), and most likely so was Aristarchus.
> - Mathematicians tend to make their discoveries when young (Guterman, 2000:18-20), so Aristarchus is likely to have been young when he studied under Strato.
> - Aristarchus' method of proving the huge distance (and size) of the Sun is crucially based on the *observation* of lunar terminators.

1.5 Forerunners

Laertius (1925b:398) says that the first to conceive of a moving Earth were either Philolaus of Croton (c. 470 – c. 385 BC) or Hicetas of Syracuse (c. 400 – c. 335 BC), two of the main exponents of Pythagoreanism.

> Passage 7 [Philolaus] was the first to declare that the Earth moves in a circle, though some say that it was Hicetas of Syracuse (Laertius, 1925b:398; see also Diels, 1906:233).

This was an important achievement. For the first time in history, someone had displaced the Earth from the centre of the universe, setting it in motion around something else.

Aetius mentions Hicetas only once (in pseudo-Plutarch's *Opinions* 3.9.2), saying he believed there was an extra heavenly body called the *Antichthon*, meaning 'Counter-Earth', or, as Graham (2015:229) prefers, 'Other-Earth'.[3]

---

3   The word ἀντί means 'opposite to', and the word χθών means 'earth', 'underground', or 'underworld'. Their combined meaning (in this context) is something like 'object facing the Earth's underside'.

Passage 8 Thales and his successors [say] there is [only] one Earth. Hicetas the Pythagorean [says that] there are two, this one and the Counter-Earth (Mansfeld and Runia, 2020:2115; see also Goodwin, 1878a:154; Diels, 1906:265, 1879:376).

But this is just part of a bigger theory which Aetius ascribes twice to Philolaus. On one of these occasions (in pseudo-Plutarch's *Opinions* 3.13.1-2), he says

Passage 9 The rest [say] the Earth stands still, but Philolaus the Pythagorean says it travels in a circle around the [central] fire in an oblique orbit like those of the Sun and Moon (Graham, 2010:503; see also Diels, 1906:237, 1879:378).

On another occasion (in Stobaeus' *Selections* 1.22.1), he mentions Philolaus again as the sole author of a *pyrocentric* model of the universe which he describes in the following terms.[4]

Passage 10 Philolaus [declares that] there is fire in the middle around the centre, which he calls the universe's hearth and Zeus' house and the gods' mother, altar and maintenance and measure of nature. And again there is another highest fire, that which surrounds [the universe]. The centre is first by nature, and around this, ten divine bodies dance: the heaven, the five planets, after them the Sun, under it the Moon, under it the Earth, under it the Counter-Earth, and after all of them there is fire, which has the position of the hearth in relation to the centres. Moreover, he calls the highest part of the surrounding [region] **Olympus**, in which he says the purity of the elements exists, while the [region] under the orbit of Olympus, in which the five planets together with the Sun and the Moon are positioned, [he calls] **Cosmos**. The sublunary and earthly part below these, in which the [realm] of change-loving generation [is located], [he calls] **Heaven**. He also declares that wisdom arises concerning what is ordered in the regions on high, whereas excellence arises concerning the disorder of what comes into being, and the former is complete, but the latter incomplete (Mansfeld and Runia, 2009:679, 2020:2094; see also Huffman, 1993:396; Barnes, 2001:179; Kirk et al., 1957:260; Diels, 1906:237, 1879:336; Wachsmuth and Hense, 1884:196).

So, Philolaus' universe has a centre called *Hestia* (Hearth or Central Fire), around which ten heavenly bodies move in circles, whose names (in expanding order) are: Antichthon, Earth, Moon, Sun, the five planets, and the sky. Beyond the sky, there is more fire. Note that the sky is said to be moving too (presumably, at a slower rate than that of the outermost planet). Whether it did so or not is something we cannot tell, because there is no further background to compare it with. So, for all intents and purposes, the starry sky of Philolaus' cosmos can be considered to be a fixed frame of reference. (See Figure 1.1, where the planets are arranged as in the Ptolemaic system for reasons that will soon be revealed.)

---

4 The terms *pyrocentric* and *hestiocentric* were coined, respectively, by Maniatis (2009:401) and Graham (2015:219) in order to describe Philolaus' cosmos.

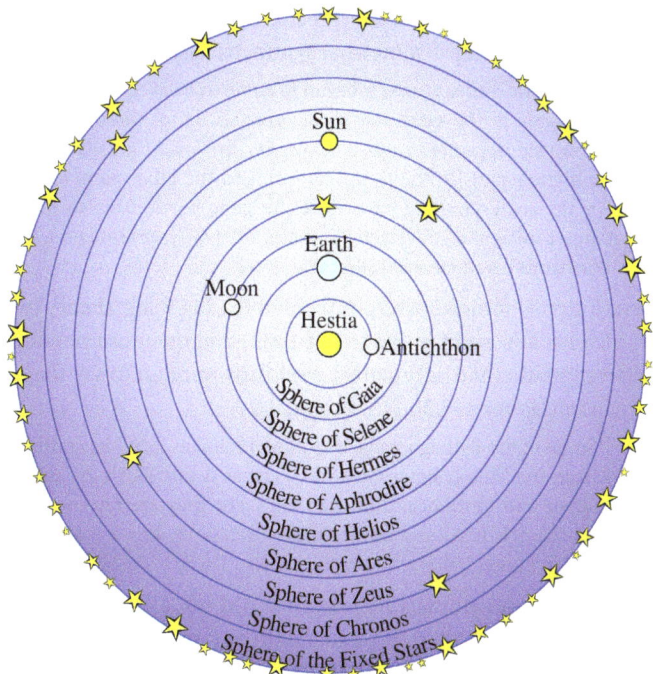

Figure 1.1 Philolaus' *hestiocentric* cosmos, conceived in the fifth century BC, consists of ten heavenly bodies orbiting a central fire that cannot be seen from the far side of the Earth, where we live. In his scheme of things, we are closest to the Sun at noon and furthest from it at midnight.

Aetius' words in another passage (Stobaeus, *Selections* 1.21.8) reveal that Philolaus' cosmos was spherical in shape and had a beginning in space and time.

Passage 11 The first thing fitted together, the one in the centre of the sphere, is called the hearth (Huffman, 1993:227; see also Diels, 1906:242; Wachsmuth and Hense, 1884:189).

He says (in *Selections* 1.21.15) that it all started when the 'All-Maker' fitted the Hearth together.[5]

Passage 12 [Philolaus locates] what is controlling in the Central Fire, which the demiurgic god set down under [the sphere of] the whole like a keel (Huffman, 1993:400; see also Diels, 1906:237, 1879:332; Wachsmuth and Hense, 1884:186).

---

5 Note that Huffman (1993:400) warns that the presence of the stoic word ἡγεμονικὸν ('hegemon') and the Platonic word δημιουργὸς ('demiurge') in this fragment indicate it may not be authentic.

Then, the universe expanded in all directions, forming a sphere whose upper and lower halves are arranged somewhat symmetrically (*Selections* 1.15.7).

> **Passage 13** The world-order is one. It began to come to be at the middle, and from the middle upwards in the same way as downwards, [and] the things above the middle are arranged opposite to those below. For the lowest part is to the things below as the highest part [to the things above], and the other parts are arranged similarly. For the corresponding parts have the same relationship to the middle, except that their positions are reversed (Graham, 2010:497; see also Diels, 1906:246, 1879:377; Wachsmuth and Hense, 1884:148).

This description is somewhat reminiscent of the Big Bang Theory (which also starts the universe at a point in space and time) and quantum physics (with its duality between matter and antimatter). Philolaus went on to say that there was life on the Moon (*Opinions* 2.30.1, *Selections* 1.26.4).

> **Passage 14** Some of the Pythagoreans, including Philolaus, [say] its earthy appearance arises from the Moon's being inhabited, just as is our Earth, by animals and plants, but larger and more beautiful than ours. For they are fifteen times more powerful than animals here, and do not make excretions, and a day there is that much longer than here (Graham, 2010:501; see also Diels, 1906:237, 1879:361).

The lunar plants and animals he imagined did not excrete and were fifteen times bigger and stronger than those on Earth and more beautiful. Aetius also says that Philolaus thought a lunar day is about fifteen times longer than an earthly day. This may be connected to the fact that fifteen days are about half a month, which Philolaus took to be 29½ earthly days, as can be deduced from two little fragments in *The Birthday Book*, written (in AD 238) by the Roman grammarian Censorinus. The first fragment (*Birthday Book* 18.8) runs as follows.

> **Passage 15** There is also a [great] year of Philolaus the Pythagorean consisting of fifty-nine years, in which there are twenty-one intercalary months (Huffman, 1993:276; see also Hultsch, 1867:38).

This passage says that Philolaus took a **great year** (that is, a cycle involving whole numbers of both lunar months and solar years) to be 59 years long. This information can be used to show that Philolaus equated 59 solar years (of 364½ days each) to 729 lunar months (of 29½ days each), since $59 \times 364.5 = 729 \times 29.5$. This is confirmed by the second fragment (*Birthday Book* 19.2), which runs as follows.

> **Passage 16** Philolaus proposed that the natural year has 364½ days (Huffman, 1993:277; see also Hultsch, 1867:40).

Note that the length of Philolaus' year is rather short, but in his Pythagorean mind, this must have been amply compensated by the 'fascinating' fact that the number 729 can be expressed both as a square, a cube, and a sixth power.

Philolaus must also have observed that the Moon always shows the same face to us. 'So', he must have reasoned, 'those living on the far side of the Moon can never see the Earth'. In the same way, the Earth always shows the same face to the Hearth and to the Antichthon, which is why we, who live on the far side of the Earth, can see neither. Aetius describes this phenomenon (in *Opinions* 3.11.3) in the following terms.

> **Passage 17** Philolaus the Pythagorean says that fire is central (for this is the hearth of the universe), the Counter-Earth second, and third the Earth we inhabit, which is located and orbits opposite the Counter-Earth (that is why the people on that earth are not seen by those on this one) (Barnes, 2001:220; see also Huffman, 1993:238; Diels, 1906:237, 1879:377).

Thus, Philolaus developed the first geokinetic model of the world we know of. It is fascinating in many ways and shows a bold imagination. For example, his thinking up the possibility of life on the Moon means that he thought of the Moon as a place that could be inhabited, possibly with mountains, valleys, rivers, lakes, and seas.[6] It was a giant leap for mankind which paved the way for further thinking.

One of the most striking features of this model is that bodies travel faster the closer they are to the Central Fire (Maniatis, 2009:404). The Earth, for example, completes a revolution in just one day. The Moon, which is further, does so in 29½ earthly days, and the Sun, which is even further, in 364½ earthly days. This immediately opens up the possibility of speculating on the relative distances of these bodies.

One way to proceed is to pierce a hole in the bottom of a tank of water and follow the movement of floating particles in the whirl that forms as the water leeks out. This can be used as a rough approximation to the motion of the planets in Philolaus' cosmos. If we carry out this experiment, we should find that the Moon is $\sqrt{29.5} \approx 5$ times as far away from the Central Fire as the Earth is, and the Sun, $\sqrt{364.5} \approx 19$ times.[7] This model is very rough because the solar system

---

[6] Laertius (1925a:137) says that Anaxagoras of Clazomenae, who was about forty years older than Philolaus, also declared that there were dwellings, hills, and ravines on the Moon.

[7] This calculation is based on Equations 1.3 and 1.4 and Solution 1.2(a) in Unit 5 of MST326 (Open University, 2009:16, 33, 58), according to which, the velocity field **u** for a two-dimensional flow is given (in SI units) by the time derivative of the particle's position **r**. In the case of a vortex at the origin, **u** is also given by the formula in Table 3.1 of the cited unit. Equating these, we have

$$\mathbf{u} = \frac{d\mathbf{r}}{dt} = \frac{dr}{dt}\mathbf{e}_r + r\frac{d\theta}{dt}\mathbf{e}_\theta = \frac{k}{r}\mathbf{e}_\theta,$$

is not a vortex, though it looks like one. It does, however, correctly predict that particles in a whirlpool, as well as planets, move more slowly the further they are from the centre of motion.

Another way to proceed is to use a simple rule of three and reason as follows: if the Moon takes 29½ days to orbit the centre of the universe (be that what it may), and the Sun takes 364½ days to do this, then the Sun moves $364.5/29.5 \approx$ 12 times more slowly than the Moon, and since both luminaries seem to cut the same figure in the sky, this ratio might also be an approximation to their relative sizes and distances. This procedure is still rough, but less so than the whirlpool analogy, and might explain the solar distance Archimedes attributed to his father, Phidias, in *Sand Reckoner* 1.9 (Heiberg, 1881:248; Vardi, 1997:2). (Archimedes also reports there that Eudoxus of Cnidus (c. 400 – c. 350 BC), the first astronomer known to have developed a theory to account for the retrograde motion of the planets, took the Sun to be nine times farther away than the Moon.)

As for the planets, there is no extant record of Philolaus having arranged them in any particular order, but this does not mean he did not. In fact, the Byzantine philosopher Simplicius of Cilicia (c. AD 490 – c. 560) quotes the Aristotelian Philosopher Eudemus of Rhodes (c. 370 – c. 300 BC) as saying that the Pythagoreans were the first to find the correct order of the planets.

Passage 18 Anaximander was the first to find an account of the sizes and distances [of the planets], as Eudemus says, adding that the Pythagoreans were the first who found the order of their position (Zhmud, 2008:246; see also Diels, 1906:15; Heiberg, 1894:471).

Eudemus (as quoted by Simplicius) did not mention Philolaus by name, but he surely meant him (Burkert, 1972:313). Neither did he mention the actual ordering, which might (or might not) be that which Aetius had in mind when he wrote Passage 10 (where the five planets were placed above the Sun). This might also be the order that appears implicitly stated in Plato's *Timaeus* (1888:122) and

---

where e_r and e_θ are the radial and tangential unit vectors corresponding to the particle's plane polar coordinates r and θ (measured in metres and radians, each), t is the time (measured in seconds), and k is a positive constant. Assuming there is no displacement in the transverse direction, the above equation reduces to

$$\frac{d\theta}{dt} = \frac{k}{r^2}.$$

This can be integrated with respect to time to obtain the position θ of the particle as a function of t and r, which in turn can be used to model the flow of particles in a vortex, as in the interactive illustration of Philolaus' cosmos which the reader may find by googling the words 'Vortical model of Philolaus' cosmos'.

explicitly stated in his pupil Philip of Opus' *Epinomis* (1854:27). Namely, Earth, Moon, Sun, Venus, Mercury, Mars, Jupiter, Saturn, and Sky. But this arrangement is in conflict with other sources. For example, the Alexandrian astronomer Claudius Ptolemy (c. AD 100 – c. 170) says (in *Almagest* 9.1) that the 'foremost astronomers' placed the spheres of Venus and Mercury below the Sun's.

> **Passage 19** Almost all the foremost astronomers agree that all the spheres are closer to the Earth than that of the fixed stars, and farther from the Earth than that of the Moon, and that those of the three [outer planets] are farther from the Earth than those of the other [two] and the Sun, Saturn's being greatest, Jupiter's the next in order towards the Earth, and Mars' below that. But concerning the spheres of Venus and Mercury, we see that they are placed below the Sun's by the more ancient astronomers, but by some of their successors, these too are placed above [the Sun's], for the reason that the Sun has never been obscured by [Venus and Mercury] either (Toomer, 1984:419).

Also, in *Refutation of all Heresies* 4.1.9, the Christian theologian Hippolytus of Rome (c. 170 – c. 235) identifies Archimedes as one of those adopting this planetary arrangement.

> **Passage 20** Archimedes understands the circumference of the zodiacal circle to be 447,310,000 stadia. Consequently, a radius drawn from the centre to the outer surface is a sixth of the above-mentioned number [namely, 74,551,666.6̇]. Moreover, a straight line from the surface of the Earth on which we walk as far as the zodiacal circle is the just mentioned sixth of the total, minus 40,000 stadia (the distance from the centre of the Earth to its surface). Yet he says that the distance from the circle of Saturn to the Earth is 222,692,711 stadia, and from the circle of Jupiter to the Earth 202,720,646 stadia, and from the circle of Mars to the Earth 132,418,581 stadia, and from the Sun to the Earth 121,604,454 stadia, and from the circle of Mercury to the Earth 52,688,259 stadia, and from the circle of Venus to the Earth 50,815,160 stadia. The distance of the Moon [to the Earth] was given before [namely, 5,544,130] (Hippolytus, 2016:116).

Also, in *Commentary on the Dream of Scipio* 1.19.2 and 2.3.13, the 4th-century AD Roman scholar Macrobius Ambrosius Theodosius counted both the Chaldeans, Archimedes, and Cicero among those who used this planetary order.

> **Passage 21** We must say a few things about the order of the spheres, a matter in which it is possible to find Cicero differing with Plato, in that he speaks of the sphere of the Sun as the fourth of seven, occupying the middle position, whereas Plato says that it is just above the Moon, that is, holding the sixth place from the top among the seven spheres. Cicero is in agreement with Archimedes and the Chaldean system; Plato followed the Egyptians, the authors of all branches of philosophy, who preferred to have the Sun located between the Moon and Mercury even though they

discovered and made known the reason why others believed that the Sun was above Mercury and Venus (Macrobius, 1963:73; 1952:162,196).

So, as Osborne (1983:235) says, there were two main schools of thought on the order of the planets: the earliest one (used by the Egyptians and Platonists) put all the planets above the Sun. The other one (used by the Chaldeans, the Pythagoreans, Archimedes, Cicero, and Ptolemy) put Mercury and Venus below the Sun.

But the ancient testimony must be taken with a pinch of salt here, because, as Kugler (1907:13), Parker (1974:60), and Neugebauer (1969:175, 1975:690) noted, the planets in Chaldean and Egyptian monuments are not arranged as the Greeks said. Also, if Eudemus is correct and the Pythagoreans were the first to find the correct order of the planets, then they must be the 'foremost astronomers' referred to by Ptolemy (in Passage 19) and whose planetary order he adopted (immediately after Passage 19). Ptolemy's adoption of this order makes sense from both a geocentric and hestiocentric point of view, because in both systems, the planets moved more slowly the further they were from the centre of the universe.[8] Thus, as seen from Earth, the sidereal cycles of each of these bodies are (approximately) as follows. Moon 27⅓ days, Mercury 87 days, Venus 225 days, Sun 365¼ days, Mars 690 days, Jupiter 12 years, and Saturn 30 years. This is exactly the order Ptolemy ascribes to, and adopts from the 'foremost astronomers'. Since this is not the Chaldeans' order, we will assume in this book that he meant the Pythagoreans'. In fact, this assumption is supported by the Greek mathematician Theon of Smyrna (c. AD 70 – c. 135) when he says (in his *Useful Mathematics* 3.15),

> **Passage 22** Here are the opinions of certain Pythagoreans relative to the position and the order of the spheres or circles on which the planets are moving. The circle of the Moon is closest to the Earth, that of Hermes is second above, then comes that of Venus, that of the Sun is fourth, next comes those of Mars and Jupiter, and that of Saturn is last and closest to that of the distant stars (Lawlor and Lawlor, 1979:91; see also Dupuis, 1892:226).

(See Table 1.1 for a comparison of the above-mentioned planetary arrangements.)

Now, we may ask, what could have made Philolaus think that the Earth is not the centre of the universe? Aristotle gives two answers to this question in *On the Heavens* (1922:293^b), where he explains that for the Pythagoreans,

---

8 Note that the geocentric and hestiocentric systems are not as different as one might think, since the former can be obtained from the latter by filling up the sphere in which the Earth moves.

Table 1.1: Ancient planetary arrangements

originators	from c.	0	1	2	3	4	5	6	7
Chaldeans	700 BC	⊕	☾	☉	♂	☿	♄	♀	♃
Chaldeans	400 BC	⊕	☾	☉	♂	♄	☿	♀	♃
Egyptians	1500 BC	⊕	☾	☉	♀	☿	♂	♄	♃
Platonists	400 BC	⊕	☾	☉	♀	☿	♂	♃	♄
Pythagoreans	450 BC	⊕	☾	☿	♀	☉	♂	♃	♄

- 'fire is more precious than earth', and therefore, fire (rather than earth) must occupy the most important place in the world: the centre;
- the presence of the Antichthon (and perhaps more like objects) explains why 'eclipses of the Moon are more frequent than eclipses of the Sun'.

The latter explanation was also given by Philip of Opus, as Aetius reports (in *Opinions* 2.29.4 and *Selections* 1.26.3).

Passage 23 Some of the Pythagoreans, according to the reports of Aristotle and Philip of Opus, maintain that [the Moon is eclipsed] by reflection and interposition, sometimes of the Earth, sometimes of the Counter-Earth. Those who say the Earth is not located at the centre say it moves around this in a circle, and not only the Earth but the Counter-Earth as well, as we pointed out earlier. Some think that several such bodies may travel about the centre but they are invisible to us because the Earth blocks our vision of them. That is supposedly why eclipses of the Moon occur more frequently than eclipses of the Sun. For each of these moving bodies blocks [the light to] the Moon, and not just the Earth (Graham, 2015:225-6; see also Diels, 1906:277, 1879:360).

So it appears that Philolaus set both the Earth and the Antichthon in orbit around the Central Fire in an attempt to explain why there are more eclipses of the Moon than of the Sun. Aristotle reports that some people believed there were several Antichthons, and the Christian theologian Hippolytus of Rome (c. 170 – c. 235) identifies one of these people as Anaxagoras of Clazomenae (c. 510 – c. 428 BC), when he ascribes to him the idea that several sublunar objects are responsible for Moon eclipses (in *Refutation of All Heresies* 1.7).

Passage 24 [Anaxagoras said that] the Moon is eclipsed when the Earth blocks it, but sometimes it is also blocked by the objects below the Moon. The Sun is eclipsed during New Moons when the Moon blocks it (Hippolytus, 2016:38; see also Diels, 1906:301).

Now, Anaxagoras was about forty years Philolaus' senior, but this fact alone is not enough to determine whether the idea of the sublunar planets originated with one man or the other. Perhaps both men influenced each other or learned it from someone else. In any case, the very idea has some implications.

Passages 10 and 17 make it clear that the Antichthon is closer to the Central Fire than the Earth is. Hence, it must move faster than the Earth does and complete a revolution in less than one (earthly) day. In other words, the Antichthon and the Earth have different periods of revolution, just like the other planets. Should they not (that is, should the Earth and the Antichthon have the same period of revolution and be always aligned with the Central Fire, whether before or behind it), it would be impossible for the Antichthon to cast its shadow on the Moon in such a way that the resulting eclipse could be seen from the far side of the Earth. Clearly, this is not what Philolaus meant. In any case, the Antichthon always lies below the horizon for those living on the far side of the Earth, just as the Earth always lies below the horizon for those living on the far side of the Moon.

Now, the far side of the Earth would be a very cold, dark place indeed if it was not for the Sun. Philolaus envisioned the latter as a glassy, mirror-like object that reflected the light and heat of the Central Fire (Aetius, *Opinions* 2.20.12, *Selections* 1.25.3d).

> **Passage 25** Philolaus the Pythagorean says that the Sun is like glass, receiving the reflection of the fire in the cosmos, straining the light and heat through to us, so that in a way there turn out to be two suns, both the fiery one in the heaven and that which is from it and fiery in reflection; unless someone will also say that there is a third, the light that is spread from the mirror to us by reflection. For we call this latter the Sun which is, as it were, the image of an image (Huffman, 1993:266; see also Diels, 1906:237, 1879:349).

The hestiocentric model described so far can be summarized as follows.

A summary of Philolaus' model of the universe

- The Centre of the Universe is a fiery place that serves as home to the gods (Passages 10 and 17).
- Starting to grow from the centre, the universe developed into a sphere whose upper and lower halves are roughly symmetrical (Passages 11 and 13).
- As befits the number of completion, which is ten, according to the Pythagoreans (Aristotle, 1933:34), there are ten heavenly bodies (namely, Antichthon, Earth, Moon, Mercury, Venus, Sun, Mars, Jupiter, Saturn, and Sky) moving in circles around the Central Fire (Passages 7, 10 and 18).

- The presence of the Antichthons explains why eclipses of the Moon are more frequent than those of the Sun (Passage 23).
- The Earth and Moon always show the same face to the Central Fire (Passage 17).
- We live on the far side of the Earth, from where neither the Antichthon nor the Central Fire can ever be seen (Passage 17).
- The Moon is inhabited by beautiful plants and animals that are fifteen times as large and strong as those on Earth (Passage 14).
- The Sun is a glassy mirror of the light and heat produced by the Central Fire (Passage 25).
- A great year takes 59 solar years (of 364½ earthly days each) or 729 lunar months (of 29½ earthly days each) (Passage 16).

The Roman orator Marcus Tullius Cicero (106 – 43 BC) provides further detail on a variant of this model when he says that Theophrastus attributed to Hicetas the idea that everything in the universe remains motionless, except the Earth, which spins on its axis, giving the impression that everything moves but us (*Lucullus* 123).

> **Passage 26** The Syracusan Hicetas, as Theophrastus asserts, holds the view that the heaven, Sun, Moon, stars, and in short all of the things on high are stationary, and that nothing in the world is in motion except the Earth, which by revolving and twisting round its axis with extreme velocity produces all the same results as would be produced if the Earth were stationary and the heaven in motion (Cicero and Rackham, 1933:626; see also Diels, 1906:265, 1879:492).

Cicero's statement is clearly wrong, but as Sarton (1952:290) explains, the exaggeration is understandable in the mouth of a man who was not an astronomer and who overemphasized the idea expressed by Hicetas and Theophrastus that 'it is the Earth that turns around its axis every day, not the starry heavens'. Aetius reports (in *Opinions* 3.13.3) that this idea was held by two other men: Heraclides of Pontus (c. 390 – c. 310 BC) and the Pythagorean Ecphantus of Syracuse (4th century BC).

> **Passage 27** Heraclides of Pontus and Ecphantus the Pythagorean make the Earth move, not in the sense of translation, but by way of turning as in an axle, like a wheel, from west to east, about its own centre (Heath, 1913:251; see also Fortenbaugh and Pender, 2009:158; Schütrumpf et al., 2008:141, 143; Diels, 1906:266, 1879:378).

Nicolaus Copernicus (1473 – 1543) quoted Passages 26 and 27 (along with Passage 9) as two (or rather three) of his most inspirational sources (*Praefatio*, 1543, folio 4r; Africa, 1961:405). According to these sources, Hicetas, Heraclides, and Ecphantus declared that the Earth spins on its axis. Whether this idea started with one of these men (who were all contemporaries) or with someone else is something the quoted sources do not tell. However, they are clear on one point: these men kept the Earth spinning on its axis, but not moving through space, or as Aetius puts it (in *Opinions* 3.13.3 and Heath, 1913:282), moving in the sense of rotation, but not in the sense of translation.

Later authors mentioned Heraclides several times, repeating the statement that he kept the Earth spinning on its axis at the centre of the universe. The philosopher Proclus of Athens (AD 412 – 485), for example, said that Heraclides made the Earth go round and round, while Plato kept it still (*Timaeum*, 281E).

> **Passage 28** How can we, when we are told that the Earth is wound round, reasonably make it turn round as well and give this as Plato's view? Let Heraclides of Pontus, who was not a disciple of Plato, hold this opinion and move the Earth round and round; but Plato made it unmoved (Heath, 1913:255; see also Diehl, 1906:138).

A century later, Simplicius mentioned Heraclides four times in his commentaries on Aristotle's *On the Heavens* and *Physics*, saying,

- in *Caelo* 2.7 (Passage 29), that Heraclides and Aristarchus both thought the Earth spins on its axis, rotating once relative to the stars in about the same time as it does relative to the Sun—the word 'about' accounting for the Sun's apparent motion against the stars (Fortenbaugh and Pender, 2009:212; Schütrumpf et al., 2008:147; Heath, 1913:254);

> **Passage 29** [Aristotle] thought it right to take account of the hypothesis that both [the stars and the heaven as a whole] are at rest—although it would appear impossible to account for their apparent change of position on the assumption that both are at rest—because there have been some, like Heraclides of Pontus and Aristarchus, who supposed that the phenomena can be saved if the heaven and the stars are at rest while the Earth moves about the poles of the equinoctial circle from the west [to the east], completing one revolution each day, approximately; the 'approximately' is added because of the daily motion of the Sun to the extent of one degree (Heath, 1913:254; see also Heiberg, 1894:444);

- in *Caelo* 2.13 (Passage 30), that Heraclides supposed that the Earth is at the centre, rotating, while the sky is at rest (Fortenbaugh and Pender, 2009:171, 212; Schütrumpf et al., 2008:145; Heath, 1913:254);

> **Passage 30** But Heraclides of Pontus believed that he was saving [that is, explaining] the phenomena by hypothesizing that the Earth is in the center [of the All] and moves in a circle, while the heavens are at rest (Fortenbaugh and Pender, 2009:171; see also Schütrumpf et al., 2008:145); Heiberg, 1894:519

- in *Caelo* 2.14 (Passage 31), that Heraclides thought the Earth rotates about its centre, while the heavenly bodies are at rest (Schütrumpf et al., 2008:145; Heath, 1913:255);

    Passage 31 If the Earth rotated about its centre while the heavenly bodies were at rest, as Heraclides of Pontus supposed, then, on the hypothesis of rotation towards the east, if it so rotated about the poles of the equinoctial circle [the equator], the Sun and the other planets would not have risen at different points of the horizon, and, if it so rotated about the poles of the zodiac circle, the fixed stars would not always have risen at the same points, as in fact they do; so that, whether it rotated about the poles of the equinoctial circle or about the poles of the zodiac, how could the translation of the planets in the direct order of the signs have been saved on the assumption of the immobility of the heavens? (Heath, 1913:255; see also Heiberg, 1894:541);

- and in *Physica* 2.2 (Passage 32), that Heraclides thought that the apparent motion of the Sun can be explained by assuming it to be still, while the Earth moves (Fortenbaugh and Pender, 2009:162; Schütrumpf et al., 2008:149; Kidd, 1999:80; Eastwood, 1992:235; Gottschalk, 1980:65; Heath, 1913:276);

    Passage 32 A certain Heraclides Ponticus has come forward and says that it is possible to account for the apparent irregularity of the Sun if the Earth moves in a certain way and the Sun stays still in a certain way (Schütrumpf et al., 2008:149; see also Diels, 1882:292).

The latter fragment is part of a longer text which is quoted verbatim by Simplicius from Alexander of Aphrodisias (c. AD 200), who in turn, quotes it verbatim from Geminus of Rhodes (c. AD 50), who in turn quotes it (possibly verbatim) from a book by Posidonius of Apamea (c. 135 – c. 51 BC) called *Meteorology*. Of these texts, only Simplicius' survives.

In order to understand what follows, it may help to summarize the information on the Earth's rotation contained in the above fragments. This is done in Table 1.2, where authors are arranged chronologically and the information they give is categorized, showing (by means of ticks) who said what and when.

It is important to note that none of the fragments in Table 1.2 say that Heraclides developed any form of heliocentrism. On the contrary, one of these fragments (Passage 30) says that he placed the Earth, not the Sun, at the centre of the universe. Another fragment (Passage 29) says that the Sun moves (relative to the fixed stars), while three other fragments (Passages 26, 31 and 32) deny this, saying it is still. However, when these fragments say that 'the Earth moves and the Sun is still', they are not proposing any heliocentric model; rather, they are oversimplifying the original idea that 'the Earth spins, while the sky stands still'.

Table 1.2 Information contained in some early comments on the Earth's motion

author	date	Passage	Hicetas	Heraclides	Ecphantus	Aristarchus	Earth is centre	Earth spins	Earth moves	Sky is still	Sun is still	Sun moves
Posidonius	1c. BC	32*	–	✓	–	–	–	✓	–	–	✓	–
Cicero	1c. BC	26	✓	–	–	–	–	✓	–	✓	✓	–
Plutarch	1c. AD	41	–	–	–	–	–	✓	✓	✓	–	–
Aetius	2c. AD	27	–	✓	✓	✓	–	✓	–	–	–	–
Proclus	5c. AD	28	–	✓	–	–	✓	✓	–	–	–	✓
Simplicius	6c. AD	29	–	✓	–	✓	–	✓	–	✓	–	–
		30	–	✓	–	–	–	✓	–	✓	–	–
		31	–	✓	–	–	–	✓	–	–	✓	–
		32*	–	✓	–	–	–	✓	–	✓	✓	–
Aquinas	13c. AD	33	–	✓	–	–	–	✓	–	✓	–	–
		34	–	✓	–	✓	–	✓	–	✓	–	–

Passage 32, by Posidonius, is preserved only in Simplicius' *Physica* 2.2.

Before Copernicus, people knew exactly what these fragments meant. Saint Thomas Aquinas (1225 – 1274), for example, gets it right from Simplicius when he (1886:162, 205) says that 'Heraclitus Ponticus' (as he spells it) supposed the Earth to be spinning at the centre of the universe (*Caelo* 2.21.5), while the heavens are at rest (*Caelo* 2.11.2).

> **Passage 33** One Heraclitus of Pontus posited the Earth to be in motion in the centre and the heaven to be at rest (Aquinas, 1886:205).

> **Passage 34** Some, positing the stars and the whole heaven to be at rest, posited the Earth on which we live to be moved from west to east around the equinoctial poles [i.e., its axis] once a day. According to this, it is due to our own motion that the stars seem to move in a contrary direction. This is said to have been the opinion of Heraclitus of Pontus and Aristarchus (Aquinas, 1886:162, 1964:396).

Aquinas' fragments are a rewording of Simplicius' *Caelo* 2.7 and *Caelo* 2.13 (Passages 29 and 30), except for the reference to the Sun's motion, which is mentioned by Simplicius alone. Interestingly, both mention Aristarchus along with Heraclides when it comes to 'moving' [that is, rotating] the Earth and stilling the heavens, but significantly enough, only Heraclides is mentioned when it comes to placing the Earth at the centre of the universe.

Four centuries after Copernicus revived heliocentrism and reintroduced the movement of translation of the Earth, the French scholar Thomas Henri Martin (1849:119-23, 419-28) provided a different reading of the fragments in Table 1.2 based on another fragment by the 4th century AD philosopher Calcidius, who in about AD 321, wrote a commented translation into Latin of the first part of Plato's *Timaeus*. In his *Commentary on Plato's Timaeus* 110, Calcidius reported that Heraclides said that Venus is sometimes above, sometimes below the Sun, and that Venus orbits a point lying somewhere along the straight line joining the centres of the Earth and the Sun.

> **Passage 35** Finally, in describing the orbit of Venus and the Sun and in assigning one centre point to the two orbits, Heraclides of Pontus showed that Venus sometimes appears above and sometimes below the Sun. For he says that the location of the Sun, Moon, Venus, and all planets, each and every one, is indicated if a single line is drawn extending out from the centre of the Earth through that of the star (Calcidius and Magee, 2016:303; see also Schütrumpf et al., 2008:147; Neugebauer, 1975:694; Heath, 1913:256; Plato et al., 1876:176).

As it happened, Martin understood that Heraclides said that Venus orbited the Sun. Soon afterwards, the Italian astronomer and historian Giovanni Virginio Schiaparelli (1873:48), inspired by Martin's interpretation, went on to say that Heraclides proposed that all planets orbited the Sun, which in turn orbited the

Earth, just as the Danish astronomer Tycho Brahe (1546 – 1601) proposed at the end of the 16th century.

However, these interpretations read more into the extant sources than is really there. A careful examination of Calcidius' words reveals that the Sun is explicitly said to have an orbit, and Venus is also said to have an orbit whose centre is always aligned with, but not the same as those of the Earth and the Sun. That is, Venus orbits this intervening point, not the Sun. As for the centre of the Sun's orbit, we learn from Simplicius' *Caelo* 2.13 (Passage 30) that Heraclides meant it to be the Earth. Heath (1913:249-283), Pannekoek (1952:373-381), Neugebauer (1975:694), and Eastwood (1992:235) have discussed at length on this issue and pointed out that Mercury is not explicitly mentioned in this whole discussion. Neugebauer has also argued that the words 'sometimes *above*, sometimes *below* the Sun' do not mean '*further* from, or *closer* to the Earth than the Sun is', as would happen in a Tychonic system, but simply that 'Venus is sometimes *right* and sometimes *left* of the Sun'.

Thus, in light of the evidence presented, it seems safe to conclude (with Eastwood, 1992:256) that Heraclides never advanced any form of Tychonic or heliocentric theory, nor did he ever have anything moving around the Sun. Any such attribution is groundless and goes no further back in time than Martin's 19th century interpretation.

There are, however, certain passages by Vitruvius (*Architecture* 9.1.6) and Theon of Smyrna (*Mathematics* 3.37) clearly stating that both Mercury and Venus orbit the Sun, but these passages make no mention of Heraclides and were written two centuries after Aristarchus had introduced heliocentrism.

> **Passage 36** The stars of Mercury and Venus make their retrograde motions and retardations about the rays of the Sun, forming by their courses a wreath or crown about the Sun itself as centre. It is also owing to this circling that they linger at their stationary points in the spaces occupied by the signs [of the zodiac] (Heath, 1913:255; see also Granger, 1931:216; Rowland and Howe, 2001:110; Neugebauer, 1975:694).

> **Passage 37** Hermes and Venus hide the stars which are beyond them when they are similarly placed in a straight line between them and ourselves. They even appear to eclipse each other when one of the two planets is higher than the other or because of their sizes, or the obliquity or position of their circles. The fact is not easy to observe, because the two planets turn around the Sun, and Hermes, in particular, which is only a small star, is the Sun's neighbour and is drowned out by it and is rarely apparent (Lawlor and Lawlor, 1979:125; see also Dupuis, 1892:312).

Four centuries later, Martianus Capella wrote a similar fragment (*On Astronomy* 857), and so did Proclus (*Timaeum* 259A), again without reference to Heraclides.

Passage 38 Venus and Mercury, although they have daily risings and settings, do not travel around the Earth at all; rather they encircle the Sun in wider revolutions. The centre of their orbits is set in the Sun (Stahl and Johnson, 1971:333; see also Capella, 1866:317).

Passage 39 Above the Sun are Venus and Mercury, these planets being solar and fabricating in conjunction with the Sun, and also contributing together with him to the perfection of wholes. Hence their course is equally swift with that of the Sun, and they revolve about him, as communicating with him in the production of things (Taylor, 1820:228; see also Diehl, 1906:65).

Thus, the late association of Heraclides with heliocentrism was just a 19th century mistake. This leaves only one candidate on whom to bestow the high honour of being the first person ever to discover that the Earth orbits the Sun. Namely, Aristarchus of Samos, The Mathematician. 'The ancient testimony is unanimous on the point', says Heath (1921b:2) on the authority of Archimedes (Heiberg, 1881:244, 1972:218; Heath, 1897:221; Gomez, 2013:7).

1.6 The authenticity of *On Sizes*

As mentioned in Section 1.1, the authenticity of the book *On Sizes* was first questioned by Voltaire in the following terms.

Passage 40 As to the pretended Aristarchus of Samos, who, it is asserted, developed the discoveries of the Chaldeans in regard to the motion of the Earth and other planets, he is so obscure, that Wallace has been obliged to play the commentator from one end of him to the other, in order to render him intelligible. Finally, it is very much to be doubted whether the book, attributed to this Aristarchus of Samos, really belongs to him. It has been strongly suspected that the enemies of the new philosophy have constructed this forgery in favor of their bad cause. It is not only in respect to old charters that similar forgeries are resorted to (Voltaire, 1775:410; 1901:40).

Voltaire's conclusion is based on a passage by Plutarch in which Aristarchus is said to have said that Cleanthes, the successor of Zeno as leader of the Stoics in Athens, should be brought to trial for disturbing the hearth of the universe and saying that the Earth spins on its axis as it travels along the ecliptic. This passage can be found in Plutarch's *De facie* (923A), a book whose only sources are two manuscripts dating from the 14th and 15th centuries, each, preserved in the Bibliothèque Nationale. Namely, Grec 1672 and Grec 1675, conventionally called *Parisinus E* and *Parisinus B* (Plutarch, 1957:26).

Both of these manuscripts—the one probably a copy of the other—contain Ἀρίσταρχος in the nominative and Κλεάνθη in the accusative (Grec 1672, folio

810v; Grec 1675, folio 405r), meaning that Aristarchus recommended an action against Cleanthes, and not the other way round. However, most modern translations assume that this is a mistake, and swap the names, as in the following sample (where the square brackets are by the present author).

> **Passage 41** Thereupon Lucius laughed and said, 'Oh, sir, just don't bring suit against us for impiety as [Cleanthes] has thought that the Greeks ought to lay an action for impiety against [Aristarchus] the Samian on the ground that he was disturbing the hearth of the universe because he sought to save the phenomena by assuming that the heaven is at rest while the Earth is revolving along the ecliptic and at the same time is rotating about its own axis' (Plutarch, 1957:55).

Russo and Medaglia (1996) have discussed the possibility that Passage 41 may be correct in the original. However, the presence of the words Κλεάνθη τὸν Σάμιον, meaning 'Cleanthes the Samian', when in fact he was from Assos, and attributing to him the astronomical achievements of the true man from Samos, Aristarchus, significantly increases the chances that the text is indeed in error. Furthermore, according to Laertius (1925*b*:280), one of the books written by Cleanthes bore the title Πρὸς Ἀρίσταρχον, meaning 'Against Aristarchus'. So it seems clear that it was Cleanthes who recommended an 'action' against Aristarchus, and not the other way round, as Voltaire believed after reading the uncorrected form of Passage 41. This explains why he suspected that the author of *On Sizes* was not Aristarchus, since the Paris manuscripts portray the latter man as opposed to heliocentrism, and *On Sizes*, though containing no trace of it, does pave the way to the latter theory by providing strong evidence that the Earth is much smaller than the Sun.

Voltaire's rejection of *On Sizes* as a genuine work is therefore based on both a mistake and no proof other than its language being 'obscure', but this is just a personal opinion by no means universally shared.

The next person to doubt the authenticity of *On Sizes* was Dennis Rawlins (2008:19), who bases his objection on the apparent impossibility that a good astronomer, such as Aristarchus, could have authored mistakes as blatant as the following.

(a) The book *On Sizes* wrongfully says that the angular size of the Moon is 2 degrees, while *The Sand Reckoner* says that Aristarchus found it to be half a degree. According to Rawlins, the former value is the result of someone else's misreading of the latter.

(b) By saying that the Moon is 2 degrees wide, the author of *On Sizes* also says [though not explicitly] that eclipses can last about half a day (rather than about 4 hours in reality).

(c) The Moon in *On Sizes* subtends a daily parallax of about 3 degrees, making it move in retrograde as seen from Earth.
(d) The Sun's daily parallax in *On Sizes* is about 9 arcminutes, which is enough for Venus to show a detectable retrograde motion at near inferior conjunction.

According to Rawlins (2008:19, 21), these 'follies' expose the author of *On Sizes* as a poor astronomer, and so imply that Aristarchus was not its author. It may well be, as Voltaire put it, the result of a 'forgery' at the service of a 'bad cause' (presumably, that of undermining heliocentrism by slandering Aristarchus). Rawlins has thus attempted to explain the discrepancy between some of the data in *On Sizes* and *The Sand Reckoner* that has puzzled humanity for centuries.

However, there are a few weak links in this chain of reasoning. For example, in a personal conversation with Angelo Gioè—the author of a philological thesis on Aristarchus (Gioè, 2007), he revealed to the present author his conclusion that the language in *On Sizes* is that typical of the academic world of third-century BC Greece. So, if *On Sizes* is a forgery, whoever wrote it must have been extremely careful to pass the philological test. But let us suppose that this is so. As we saw in Section 1.2, *On Sizes* can be traced as far back as the fourth century, when Pappus of Alexandria quoted some passages from it, apparently verbatim (though the oldest manuscripts containing *On Sizes* and Pappus' comments themselves date from the ninth or tenth centuries). So the forgery was made before Pappus commented on a book he believed to be genuine, that is, before the fourth century AD. It is roughly estimated that Pappus wrote his comments in about AD 320, at a time when Christianity had just been legalized by Emperor Constantine and the first ecumenical council was yet to take place (in Nicea). It seems therefore unlikely that Christianity, which had been troubled enough with persecutions and was yet to define itself, should have had the foresight to sponsor the mentioned forgery; and as for scientists, it is hard to imagine why they should wish to do so: if a theory is wrong, it shall be exposed; there is no need to change its contents, but to check whether they are true.

But let us go on supposing that *On Sizes* is a forgery. Rawlins explains that the word μέρος (meaning 'part') used in Hypothesis 6 of that book (Heath, 1913:352) meant one thing for Aristarchus and another for the (supposedly fake) author of *On Sizes*. As Neugebauer (1975:652, 671) explains, the ancients divided the Zodiac into either of the following units.

- 12 *parts* (or *signs*) of 30 degrees each,
- 24 *steps* (or *half-signs*) of 15 degrees each,
- 48 *parts* (or *half-steps*) of 7½ degrees each.

According to Rawlins, the supposedly original expression 'one fifteenth of a *half-step*' (that is, 7.5°/15 = 0.5°) was misread by the author of *On Sizes* as 'one fifteenth of a *sign*' (that is, 30°/15 = 2°). However, the whole of the book *On Sizes* assumes that the Moon is 2-degrees wide, and there is no doubt that the author really meant it that way. It is rather odd that an author whom we know was incredibly careful when it came to philological matters should make such a mistake, and all the more so when Aristarchus' supposedly original writings would have repeated the correct use of the word once and again. So the only plausible explanation is not one of carelessness, but one of purposefulness. Thus, the supposed author of the book *On Sizes* was perfectly aware of Aristarchus' true meaning, but deliberately changed it to the wrong value. Perhaps, as Rawlins says, this was part of a malicious plan to slander Aristarchus.

A slight variation of this explanation had been advanced earlier by Manitius (1974:292), when he suggested that the expression πεντεκαιδέκατον μέρος, meaning 'one fifteenth of a part' (or 2 degrees), may have originally been πεντηκοστὸν μέρος, meaning 'one fiftieth of a part' (or 36 arcminutes), which is a value much closer to the final half a degree reported by Archimedes.[9]

Both Manitius and Rawlins consider the apparent mistakes in *On Sizes* as 'nonsense unworthy of Aristarchus', suggesting he did not write that book. But let us now consider the possibility that he did. After all, he was human and therefore very capable of making mistakes no matter how intelligent or skilful. Like any human being, he was not born knowing it all, but learned as he lived, so let us allow that his thoughts, like everyone else's, evolved over time. He is said by Archimedes to have been the first person ever to measure the angular size of the Sun with amazing accuracy, and to have discovered that the Earth revolves around the Sun. These things were unknown before him, and it takes quite an exceptional insight to observe what millions of human beings had overlooked until then and would overlook for centuries afterwards. The 2-degree value may have been taught to him at school, perhaps by his tutor Strato, or he might have learned about it at the Mouseion or the Great Library. In any case, this value and the cosmic centrality of the Earth were state-of-the-art science when he was a child. Is it a sin to have written a book early in one's life using the most advanced knowledge available?

Such is the alternative explanation given by Heath for the discrepancies between the two main sources (*On Sizes* and *Sand Reckoner*) on matters such as the Moon's angular size. According to him, 'the treatise was an early work

---

9 In any case, the author of *On Sizes* clearly defines a fifteenth part of a sign as 1/180th of the whole circle of the zodiac, or 2 degrees (Heath, 1913:367).

written before Aristarchus had made the more accurate observation recorded by Archimedes' (Heath, 1913:312). We have, therefore, access to two snapshots in time in the development of heliocentrism: a book (or perhaps, a draft) written by Aristarchus early in his career, and a book written by Archimedes, describing Aristarchus' achievements at a later (perhaps final) stage. The former reflects important discoveries, but is yet written in the frame of geocentrism and what was then considered to be one of the best estimates of the Moon's angular size; the latter reflects the first ever accurate measurement of the mentioned angle along with the most important scientific breakthrough in the history of astronomy. It all fits. Together, both books reflect a chronologically consistent evolution of thought: poor, yet state-of-the-art values at first, incredibly accurate values later.

Saying that the language in either or both of these books is 'obscure', 'pedantic', 'nonsense', 'unworthy', or 'folly' is not doing justice to the truth. Has the dear reader ever thought how the language of modern mathematics, at the peak of its latest developments, would sound to ancient ears? Would it sound crystal-clear? Well, it goes to Aristarchus' credit that he managed to explain some of the greatest discoveries in science ever in a language that, to a mathematician, such as Heath or the present author, sounds pretty clear.

1.7 The authenticity of *The Sand Reckoner*

As mentioned in Section 1.1, Erhardt and Erhardt-Siebold (1942) wrote an article questioning the authenticity of *The Sand Reckoner*, the book in which Archimedes (if indeed it was him) estimated the number of grains of sand that would fill the largest model of the universe known to him: that of Aristarchus. The Erhardts (1942:587-9) found the following oddities about the book.

(a) *The Sand Reckoner*'s definition of **cosmos** (as the sphere bounded by the Sun's orbit) is quite 'unusual', despite the book presenting it as widely spread.
(b) The Doric dialect of Syracuse is better preserved in *The Sand Reckoner* than in other works of Archimedes.
(c) *The Sand Reckoner* is the only extant astronomical work of Archimedes.
(d) Acquainted as he was with the study of infinitesimals, Archimedes, better than anyone else, should have understood Aristarchus' assertion that the Earth's orbit and the sphere of the fixed stars bear the same ratio as the centre of a sphere bears to its surface.
(e) Instead of using the ratio of the Sun's body to the Earth's orbit (as Aristarchus would), the author of *The Sand Reckoner* uses the ratio of the Earth's body to the Sun's orbit (as any geocentrist would).

(f) Archimedes' writings have been regarded, at all time, as 'a classical monument of simplicity and straight forward thinking', whereas the language in *The Sand Reckoner* sounds 'irrational', 'unreal', 'fanciful', and 'desultory'.

According to the Erhardts, these oddities point towards a possible fraud. However, further considerations indicate that this may not necessarily be the case. To start with, the word κόσμος (meaning 'order') was first used as an astronomical term denoting the (apparent) orderliness of the universe by Anaximander of Miletus (c. 610 – c. 546 BC), for whom the latter was an endless succession of countless worlds, whose source and sink was the ἄπειρον (or 'boundless stuff'). The various objects in one of these worlds (namely, ours), are arranged in order of increasing brightness (that is, stars, Moon, and Sun), in 'wheels' or 'rings' around the Earth that are separated from each other by a distance equal to nine times the Earth's diameter (Diels, 1879:12; Fairbanks, 1898:13).[10]

This original definition of **cosmos** (as the comely order of the universe) was later modified by various authors, such as Anaxagoras, who introduced an intelligent principle, called νοῦς (or 'Mind'), to start and guide the development of the cosmos (Fairbanks, 1898:254), and Empedocles, who introduced four basic elements (earth, water, air, and fire) and two forces (love and strife) to explain Change as combinations of these (Fairbanks, 1898:223).[11] A less abstract definition was given (in Passage 10) by Philolaus, for whom the **cosmos** was the region comprising the five planets, the Sun, and the Moon. For Eudoxus, the **cosmos** was the neat arrangement of the universe into a complex mechanism of crystalline spheres all centred at the Earth, whose combined motions explained those of the planets. Later, Aristotle increased the complexity of this system of spheres and (following his teacher Plato's lead) added a fifth element (aether) to the list of basic elements.

As we can see, the word *cosmos* acquired a variety of meanings depending on who ventured an opinion. To say that the earliest of these cosmologies were 'outmoded' by Archimedes' time and that 'the astronomers of his time believed in a finite universe bounded by the sphere of the fixed stars' (Erhardt and Erhardt-Siebold, 1942:580) may be too simplistic. After all, it is likely that Archimedes, one of the greatest mathematicians ever, knew better than the Erhardts which

---

10 The latter number may be inspired by Hesiod's statement that a brazen anvil takes nine days to fall from Heaven to Earth, and nine days to fall from Earth to Hell (Evelyn-White et al., 1920:130).
11 These elements had already been hallowed by Zarathustra and resemble the sea, earth, sky, and wind gods of the Babylonian *Enuma Elis* cosmology.

definition of *cosmos* was in vogue in his own time, and all the more so knowing that his father, Phidias, as Archimedes himself reports (Heiberg, 1881:248; 1972:220), had estimated the Sun's distance to be 12 times that of the Moon, which by the time, was the best estimate ever reached by anyone, meaning that Archimedes had been exposed to the latest astronomical developments from his earliest childhood.

As for the Doric dialect being more present in this particular book than in other works of Archimedes, Erhardt and Erhardt-Siebold (1942:589) say that this, in itself, is no warranty that the book is genuine, since a clever forger could easily have imitated it. If so, why should anyone make such a forgery, and if not, why should Archimedes use a more homely, less technical language in this particular book? Perhaps the answer lies on the very first line of it, where the book is dedicated to a child called Gelo, king of Syracuse (along with his father, Hiero II). Perhaps it is not too bold to guess that Archimedes was (at least, temporarily) charged with tutoring Gelo, who was about twenty years his junior (just as Strato had been charged with tutoring Ptolemy Philadelphus, who was about twenty-six years his junior). By the time the child was of age (say, about 255 BC), Archimedes would have reached the acme of his life. *The Sand Reckoner* may have been written then as a textbook to introduce Gelo to the wonders of Astronomy and Mathematics in a language that sounds 'fanciful' (among other things) to the Erhardts.

The above dating of *The Sand Reckoner* is consistent with (though slightly earlier than) Knorr's (1978:237, 269) proposal for a chronological arrangement of Archimedes' works, where compelling arguments are given to place this book among the very first he ever wrote. Thus, saying that Archimedes deliberately ignored his own knowledge of infinitesimals in order to expose Aristarchus' (apparent) lack of mathematical rigour supposes that Archimedes had already developed this knowledge. But if *The Sand Reckoner* was an early book, then Archimedes' knowledge of infinitesimals may not yet have been fully developed. In any case, he may have judged that the use of such knowledge (whatever its stage of development) was not appropriate for his intended audience.

As for using the Sun's orbit, rather than the Earth's orbit, it is again obvious that Archimedes is trying to adapt his book to his audience in order to make it more readable. After all, his book is not a monograph on Aristarchus. Neither is it entirely on Mathematics, so Archimedes is forced to abandon the rigorous language of proof of the only science that deals with absolute truths, a science he knew so well. Once you abandon the safe field of Mathematics, you plunge into the shaky grounds of everlasting arguing and the eternal pursuit of ever-fleeting objectivity. That is why Archimedes cannot proof anything astronomical with the same rigour as is found in his mathematical works. In fact, nobody can. But Archimedes achieved his goal: that of stirring the imagination of a child (and of all who read his book) into wondering wide-eyed at the world around us.

So, in the author's view, none of the arguments casting doubts on the authenticity of either *On Sizes* or *The Sand Reckoner* (as presented by the cited authorities) stand to scrutiny. Therefore, the rest of this book will proceed under the assumption that these books are genuine.

2 On Sizes

Abstract: This chapter presents a mathematical analysis of Aristarchus' book *On Sizes*, identifying the kind of problems it deals with and explaining how Aristarchus attempts to solve them. The exact solution to these problems is also found, and a judgement is pronounced on the authenticity of this book.

2.1 Hypotheses, conclusions, and the Euclid connection

The book *On Sizes* begins by laying down six hypotheses, which, according to Aristarchus, lead to the three main conclusions of his treatise. They are all given below (Heath, 1913:352).

Starting hypotheses in *On Sizes*

Hypothesis 1 The Moon receives its light from the Sun.

Hypothesis 2 The Earth is in the relation of a point and centre to the sphere in which the Moon moves.

Hypothesis 3 When the Moon appears to us halved, the great circle which divides the dark and the bright portions of the Moon is in the direction of our eye.

Hypothesis 4 When the Moon appears to us halved, its distance from the Sun is then less than a quadrant by one-thirtieth of a quadrant.

Hypothesis 5 The breadth of the [Earth's] shadow is [that] of two Moons.

Hypothesis 6 The Moon subtends one fifteenth part of a sign of the zodiac.

Main conclusions in *On Sizes*

Conclusion 1 The distance of the Sun from the Earth is greater than eighteen times, but less than twenty times, the distance of the Moon [from the Earth] (as follows from Hypothesis 4).

Conclusion 2 The diameter of the Sun has the same ratio [as aforesaid] to the diameter of the Moon (as follows from Hypothesis 4 and Conclusion 1).

Conclusion 3 The diameter of the Sun has to the diameter of the Earth a ratio greater than that which 19 has to 3, but less than that which 43 has to 6 (as follows from Conclusion 1 and Hypotheses 5 and 6).

In order to show how the given hypotheses lead to the stated conclusions, Aristarchus works through a number of propositions, which he proves one by one. The first three of these are as follows.

> **Proposition 1** Two equal spheres are comprehended by one and the same cylinder, and two unequal spheres by one and the same cone which has its vertex in the direction of the lesser sphere; and the straight line drawn through the centres of the spheres is at right angles to each of the circles in which the surface of the cylinder, or of the cone, touches the spheres (Heath, 1913:355).
>
> **Proposition 2** If a sphere is illuminated by a sphere greater than itself, the illuminated portion of the former sphere will be greater than a hemisphere (Heath, 1913:358).
>
> **Proposition 3** The circle in the Moon which divides the dark and the bright portions is least when the cone comprehending both the Sun and the Moon has its vertex at our eye (Heath, 1913:361).

These propositions bear a close resemblance to Propositions 23 to 27 in Euclid's *Optics* (Burton, 1945:361-3), and awareness of this resemblance can be traced as far back as the fourth century, when Theon of Alexandria made the following remark in his comments on Ptolemy's *Almagest* (Rome, 1943:957).

> **Passage 42** On every side, more than a hemisphere [of the Moon] is illuminated by the rays of the Sun, since the Sun is larger than the Moon; and because of this, the light and dark portions of the Moon are not divided by its full circumference, but by a smaller one. For these things have been demonstrated by Aristarchus and Euclid (Webster, 2014:539).

This remark shows that Euclid and Aristarchus shared a common interest in (at least) Mathematics and Optics, and for some reason, paid special attention to the tiny difference referred to in Passage 42. Euclid is also known to have been active in Alexandria during the reign of Ptolemy I (323 – 283 BC), as we learn from Pappus (*Synagoge* 7.35),

> **Passage 43** [Apollonius] spent a very long time with the pupils of Euclid at Alexandria (Heath, 1921a:356; see also Hultsch, 1877:678),

and Proclus (*On Euclid* 68.19),

> **Passage 44** [Euclid] lived in the time of Ptolemy the First, for Archimedes, who lived after the time of the first Ptolemy, mentions Euclid. It is also reported that Ptolemy once asked Euclid if there was not a shorter road to geometry than through the *Elements*, and Euclid replied that there was no royal road to geometry. He was therefore later than Plato's group but earlier than Eratosthenes and Archimedes, for these two men were contemporaries, as Eratosthenes somewhere says (Morrow et al., 1970:xxii, 56; see also Friedlein, 1873:68).

It is, therefore, likely that Euclid and Aristarchus met (in Alexandria) and shared ideas. Whether the propositions of the former inspired those of the latter, or the other way round, is something we may never know. But one thing is remarkably striking: the proofs in Aristarchus' *On Sizes* are all mathematically correct, while some of those in Euclid's *Optics* are not, as Heath (1913:363) and Webster (2014:540) have pointed out. For example, Euclid's Proposition 25 that 'exactly half a sphere can be seen by two eyes placed as far away from each other as the sphere is wide' is simply wrong, but true only for the great circle of the sphere that is coplanar with the two eyes. A similar argument applies to Proposition 26 (Burton, 1945:362).

Thus, the young Aristarchus has shown himself to be a better mathematician than the author of the book on *Optics* that is attributed to the same person who wrote the most influential book on Mathematics ever (namely, *The Elements*).

2.2 The angular resolution of the human eye

Next in *On Sizes* comes one of the most interesting propositions in the whole treatise. It states the following.

> **Proposition 4** The circle which divides the dark and the bright portions in the Moon is not perceptibly different from a great circle in the Moon (Heath, 1913:365).

An illustration of this proposition is shown in Figure 2.1, where the distance between the observer's eye A and the Moon's centre B has been reduced for clarity. Should it not have been reduced (but arranged in such a way that Hypothesis 6 is satisfied), these points would be nearly 29 times as far away from each other as the Moon is wide, as can easily be found by applying the sine law to the triangle BAD (as will be done in Equations 2.2 and 2.3 later in this chapter). (Aristarchus himself found upper and lower bounds for this value in Proposition 11, as we shall see.)

The proof of this proposition, as given in *On Sizes*, contains a most remarkable statement, which is reproduced below.

> **Passage 45** The angle KAH is less than 1/3960th of a right angle. But a magnitude seen under such an angle is imperceptible to the eye (Heath, 1913:367).

What we have here is nothing less than the first numerical assessment (on extant record) of the limit of human vision. Aristarchus cites no sources for this datum, nor does he say how it was obtained, so we are left wondering whether he found it himself. The reported value is quite acceptable, given that the sharpness of human vision is highly dependent on lighting conditions as well as the

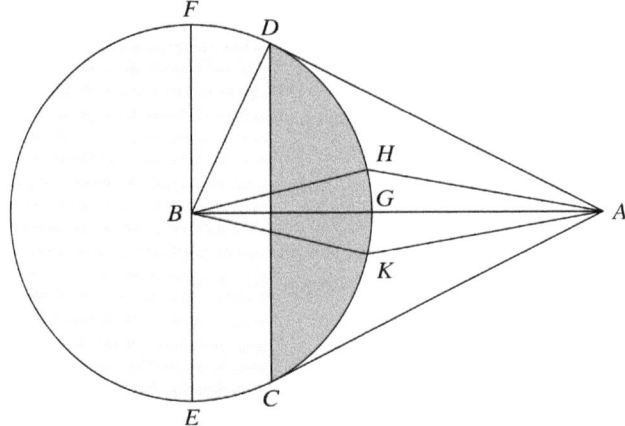

Figure 2.1 Proposition 4

colour of the objects and backgrounds involved. The next time such a measurement was made was nearly two thousand years later, when, in January of 1681, Robert Hooke (1705:97) lectured on the subject, assessing the visual limit of the human eye at about one minute of arc. Later researchers found that the typical human eye can see slightly better than this under 'optimal' conditions, but Hooke's value is widely accepted as a convenient approximation.[12]

Like Hooke's value, Aristarchus' could only have been found empirically, which adds to the evidence gathered in the summary box on page 23 that (as a well-trained student of Strato) Aristarchus was an empiricist. His value, '1/3960th of a right angle' (or $90°/3960 \approx 1'22''$), is only slightly bigger than Hooke's, and is well within the range of possible values that can be obtained depending on the conditions under which the measurement is made.

The fact that the circles CD and EF (in Figure 2.1) are visually indistinguishable from each other plays an important role in the proof of Proposition 4. In this proof, the angle DAF is said to be invisible to the human eye because it is

---

12 Mention of a method for testing eyesight, based on the ability to count mustard seeds at a given distance, was made earlier by the Spanish doctor Benito Daza de Valdes (1623:28), and even earlier by the Persian astronomer Sufi (964:43), who described an old-time test for good eyesight, based on the ability of the eye to distinguish between the stars Mizar and Alcor of the Great Bear. Though these stars are twelve arcminutes apart, the conditions of the test (which is carried out on points of light against a dark background) have been shown to be equivalent to modern tests for normal eyesight (Bohigian, 2008:537).

much smaller than the limit of human vision, which Aristarchus equates to the angle HAK (that is, the angle subtended by the arc HK as seen from A). As it happens, H and K are equidistant from G, the point on the Moon's surface that is closest to A. However (as Passage 45 shows), Aristarchus is aware that the human eye cannot distinguish between either H or K, or any point in between. In fact, these points are defined by the ability of the human eye to tell whether the **lunar terminator** (or line dividing the bright and dark sides of the Moon) lies between them or not. Whenever it does, the Moon looks halved to us, and it does so during the many hours it takes the terminator to move from H to K, or from position (a) to position (d) in Figures 2.2 and 2.3 (which are bird's eye renditions of Figure 2.1).

In these figures, we can see that the lunar terminator never splits the Moon into two equal halves, but one is always bigger than the other, because the Sun is bigger than the Moon (as Aristarchus points out in Proposition 2, Euclid in Proposition 27, and Theon in Passage 42). Because of this, what we call **dichotomy** (that is, the perfect splitting of the Moon's disc) is a physical impossibility. Though the eye can be tricked into believing that such a thing is possible, Aristarchus' sharp mathematical mind knows better: we cannot have both a perfectly straight terminator and a perfectly halved Moon at the same time.

Not even when the terminator sweeps the centre of the Moon's disc (at position (b) in Figure 2.2 or position (c) in Figure 2.3) can we say that the Moon is exactly halved. Only the Moon's equator is, so, in this book, we will call this moment the time of **equatorial dichotomy**. But this is not what Aristarchus is after. What he really wants is just a right-angled triangle (in order to work out the Sun's distance), and there are two such triangles in a lunar month: one at position (d) in Figure 2.2 (when the terminator is on point K), and the other at position (a) in Figure 2.3 (when the terminator is on point H). At these precious moments, the Sun and the Earth are at right-angles to the Moon. Aristarchus chose these very moments (in Proposition 5) because the Moon's phase angle is then exactly 90 degrees. So, here, for lack of a better name, we will call this the time of **lunar orthogony**. Note that this is neither the time of **lunar quadrature** (when the Moon's elongation from the Sun is exactly 90 degrees), nor the (impossible) time of **lunar dichotomy** (or 50% phase). (See Figures 2.4 and 2.5.)

Aristarchus' mathematics are impeccable, there are no flaws in them, and it is actually possible to measure the Moon's elongation from the Sun at the mentioned times using primitive technology (as we shall do in Chapter 3) with results similar to, and even better than that in Hypothesis 4, strongly suggesting

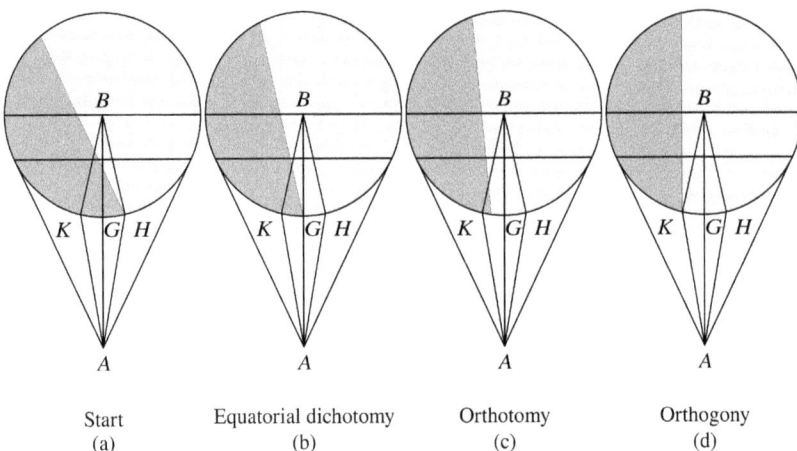

Start	Equatorial dichotomy	Orthotomy	Orthogony
(a)	(b)	(c)	(d)

Figure 2.2 The first half-Moon of the month starts at (a), when the Moon stops looking waxing crescent, and ends at (d), when the Moon starts looking waxing gibbous, as seen from A. (For clarity, sizes and distances are not to scale.)

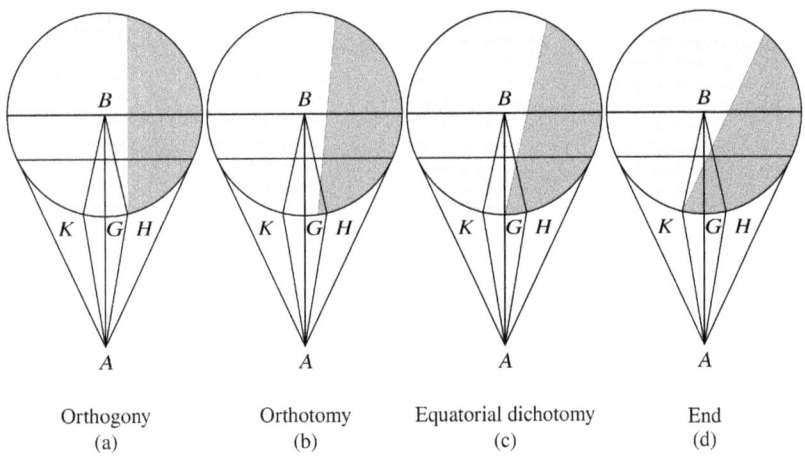

Orthogony	Orthotomy	Equatorial dichotomy	End
(a)	(b)	(c)	(d)

Figure 2.3 The last half-Moon of the month starts at (a), when the Moon stops looking waning gibbous, and ends at (d), when the Moon starts looking waning crescent, as seen from A. (For clarity, sizes and distances are not to scale.)

it was obtained experimentally. However, there are several things that contribute to a poor result: one is the human eye's ability to tell the time of *lunar orthogony*, another is the inaccuracy of Hypothesis 6, and another is the assumption

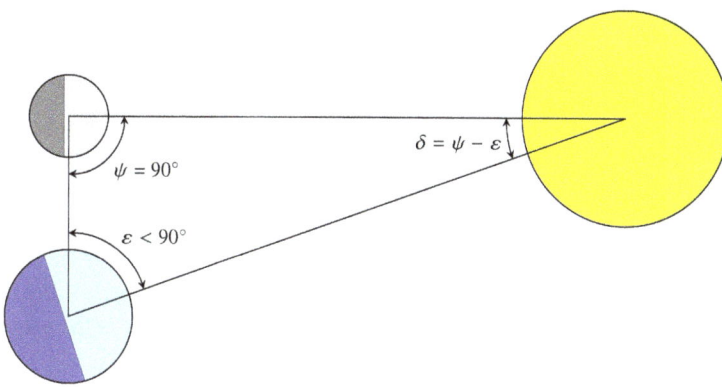

Figure 2.4 At *lunar orthogony*, the Moon's phase angle is $\psi = 90°$ (and the Moon's elongation is $\varepsilon < 90°$).

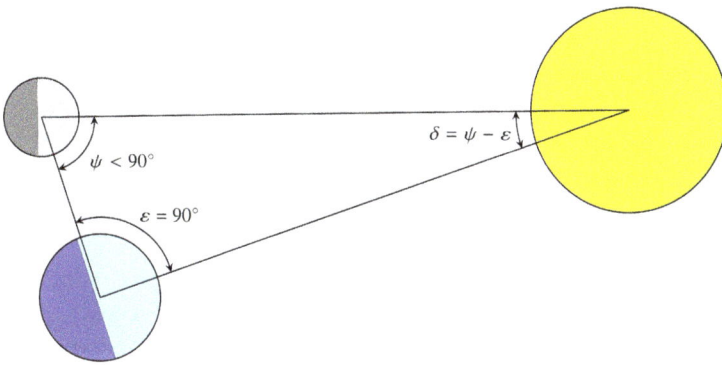

Figure 2.5 At *lunar quadrature*, the Moon's elongation is $\varepsilon = 90°$ (and the Moon's phase angle is $\psi < 90°$).

that the length of the Moon's shadow is equal to the Moon's distance from Earth.

To see how the latter figures, we must first see how it affects Proposition 3, which says that 'the lunar terminator is smallest when the cone containing both the Sun and Moon has its vertex at our eye'. Strictly speaking, this is all true (except at annular eclipses, when it is even smaller). However, the light of the Sun reaches our eye after going through the Earth's atmosphere and is slightly bent by it.

Like a small drop of water suspended in air and scattering the light from the Sun into colourful rainbows, so does the whole planet Earth along with its atmosphere, suspended in space, scatter the Sun's light in such a way that the shadow it casts is slightly longer for blue wavelengths than it is for red (which is why the Moon turns reddish when totally eclipsed). Thus, the Earth's shadow is slightly longer than pure geometry predicts. Not so the Moon's shadow, which is affected by no atmosphere other than the Earth's, and only when it enters it. Then it bends and is slightly shortened (by Snell's law). Thus, the overall effect of the Earth's atmosphere is that of enlarging outgoing shadows (such as that of the Earth on the Moon) and shortening incoming ones (such as that of the Moon on the Earth).

The first of these effects (namely, the enlargement of the Earth's shadow on the Moon) was discovered and measured by the French astronomer Philippe de la Hire (1687:73) and was attributed to the effects of the Earth's atmosphere by the Italian-born French astronomer Giovanni Domenico Cassini (1740:34). Let us now consider the second of the above effects (namely, the shortening of the Moon's shadow on Earth that occurs during an eclipse of the Sun). Aristarchus said that during a total eclipse of the Sun, the tip of the Moon's shadow is at our eye. This is not exactly so, because annular eclipses are slightly more frequent than total ones, but even if we take the man's word as a valid simplification, there is a little problem. If the Earth's atmosphere has any effect on the incoming shadow of the Moon, then the above simplification may not be so valid after all. But what is this effect?

As a first answer, we may think it is completely negligible. (Not worth a thought.) Pictures of the Moon's shadow on Earth taken from space show it to be very blurry. But this is because most of it is penumbra, with the umbra being just a tiny dot right in the middle and completely indistinguishable from its dark, blurry surroundings. Besides, the distance that light travels through the atmosphere is very small compared with the length of the Moon's shadow. So the bending effect of the Earth's atmosphere must be negligible. However, for a person like Aristarchus, who is observing a total eclipse of the Sun, this air bending effect is not so negligible, because the tip of the Moon's shadow is not where geometry says it should. In fact, due to refraction, the breadth of the Moon's shadow on Earth is less than it should be without the atmosphere. So, looking from a point on the Earth's surface, the Moon appears slightly larger.

Nevertheless, Aristarchus' Proposition 3 that the tip of the Moon's shadow is at our eye is correct whenever the Moon exactly overlaps the Sun. Then, the *geometric* (or unrefracted) length of the Moon's shadow is the same as the Moon's distance from the observer's eye. The relation between this distance and the angle

BAD (in Figure 2.1), as derived from the sine law, is given by

$$\frac{M}{m} = \frac{1}{\sin \rho_{\mathbb{C}}}, \tag{2.1}$$

where M is the Moon's distance, m is the Moon's radius, and $\rho_{\mathbb{C}}$ is the angle BAD (or half the angular size of the Moon).

Taking $m = 1$ (as a unit of length) and $\rho_{\mathbb{C}} = 1°$ (as given in Hypothesis 6), the Moon is

$$\frac{1}{\sin 1°} \approx 57 \tag{2.2}$$

Moon radii away from A, or, equivalently,

$$\frac{1}{2 \sin 1°} \approx 29 \tag{2.3}$$

times as far away from us as it is wide (as said earlier). According to Proposition 3, this is also the length of the Moon's shadow (in lunar radii).

Substituting modern values (such as $m \approx 1737/6367 \approx 0.273$ Earth radii and $\rho_{\mathbb{C}} \approx 0.26°$) into Equation 2.1, we obtain

$$\frac{M}{e} \approx \frac{0.273}{\sin 0.26°} \approx 60. \tag{2.4}$$

So the Moon is about 60 Earth radii away from Earth, and, according to Proposition 3, this is also the length L of the Moon's shadow. However, the latter is not always the case. The fact that total eclipses of the Sun are slightly less frequent than annular ones (Espenak and Meeus, 2006a:19, 2009a:17) suggests that, on average, L is slightly shorter than M. From a strictly geometric point of view, the length of the Moon's *geometric* shadow is given by any of the following equations (as derived in Appendix A).

$$L = \frac{m\sqrt{S^2 - M^2}}{s - m} \tag{2.5}$$

$$= \frac{m}{\sin \zeta_{\mathbb{C}}} \tag{2.6}$$

$$= \frac{M \sin \rho_{\mathbb{C}}}{\sin \zeta_{\mathbb{C}}} \tag{2.7}$$

$$= \left(\frac{\sqrt{S^2 - M^2}}{s - m}\right) M \sin \rho_{\mathbb{C}}, \tag{2.8}$$

where L is the length of the Moon's *(geometric)* shadow, S is the Sun's distance from the observer, M is the Moon's distance from the observer, s is the Sun's radius, m is the Moon's radius, $\rho_{\mathbb{C}}$ is the angle *BAD* (or half the Moon's *apparent* size), and $\zeta_{\mathbb{C}}$ is half the angular size of the Moon as seen from the tip of its *(geometric)* shadow.

Substituting modern values (such as $S \approx 23455$ Earth radii, $M \approx 60$ Earth radii, $s \approx 109$ Earth radii, $m \approx 0.273$ Earth radii, $\rho_{\mathbb{C}} \approx 0.26°$, and $\zeta_{\mathbb{C}} \approx 0.266°$) into any of Equations 2.5 to 2.8, we have

$$L \approx \frac{0.273\sqrt{23455^2 - 60^2}}{109 - 0.273} \approx 59. \tag{2.9}$$

So, the length of the Moon's shadow is about 59 Earth radii (as given by Approximation 2.9), which is less than the Moon's distance (as given by Approximation 2.4). As a consequence, the angles *DBF*, *KBH*, and *BAD*, which Aristarchus equates in his proof of Proposition 4 (Heath, 1913:368), take now slightly different meanings. The angle *DBF* (or angle of incidence of the Sun's rays on the Moon) is none other than $\zeta_{\mathbb{C}}$, and the angle *KBH* (which is a rotation of *DBF*) is now seen from a point *A* which is slightly further from *B* than geometry predicts. So the angle *BAD*, which is none other than $\rho_{\mathbb{C}}$, is not exactly equal to, but slightly smaller than the angles *DBF* or *KBH*.

Since Aristarchus equated the angles *BAD* and *KBH*, if follows (from Hypothesis 6) that the angle *KBG* is exactly half a degree. But in reality, both $\rho_{\mathbb{C}}$ and $\zeta_{\mathbb{C}}$ are about half this value, so the point *D* lies closer to *F* than originally thought by about half the distance between *K* and *G*. So, assuming that a half-Moon lasts for about twelve hours, position (c) in Figure 2.2 occurs roughly three hours earlier, and position (a) in Figure 2.3, roughly three hours later than Aristarchus said. Had he taken his measurements at the specified moments, Hypothesis 4 would have been more accurate.

As for the visual limit in Passage 45, it might be interesting to know what exactly did Aristarchus mean by 'less than 1/3960th of a right angle'. Fortunately, this calculation has already been made by Berggren and Sidoli (2007:229), who correctly found the angle *KAH* to be 0.0178 degrees (to four decimal places), without showing their work. Neither will the present author show his, and will rather leave it to the reader as an exercise, but at least he will give a formula that yields the required result. Namely,

$$KAH = 2\tan^{-1}\left(\frac{1 - \cos(\rho_{\mathbb{C}})}{\sec(\rho_{\mathbb{C}}/2) - \sin(\rho_{\mathbb{C}})}\right). \tag{2.10}$$

Taking $\rho_{\mathbb{C}}$ to be equal to the angle BAD in Figure 2.1, that is, taking $\rho_{\mathbb{C}}$ to be one degree (as specified in Hypothesis 6), we have

$$KAH = 2\tan^{-1}\left(\frac{1-\cos(1°)}{\sec(1°/2)-\sin(1°)}\right) = 0.017762...°, \quad (2.11)$$

which (expressed sexagesimally) is about $1'4''$.

We may never know whether Aristarchus ever handled such a figure (which is almost identical to Hooke's empirical value), but we can be certain that he handled the figure he gives in Passage 45 (which may be based on observations of the Moon in broad daylight). Let us now move on to the next proposition, which provides a confirmatory check on the work presented so far. It runs as follows.

> **Proposition 5** When the Moon appears to us halved, the great circle parallel to the circle which divides the dark and the bright portions in the Moon is then in the direction of our eye; that is to say, the great circle parallel to the dividing circle and our eye are in one plane (Heath, 1913:371).

Aristarchus gives us here a clear definition of what he calls 'the moment when the Moon appears to us halved'. Reading carefully, we see that he equates this moment to either position (d) in Figure 2.2 or position (a) in Figure 2.3. (Note that this is not what we understand by *equatorial dichotomy* in the mentioned figures.) However, since the eye cannot distinguish between any of these positions, Aristarchus assumes that the terminator can be taken to be coplanar with the eye at the specified times: namely, at the end of a first half-Moon or start of a last half-Moon, or, to put it more succinctly, at the very birth or death of a gibbous Moon. This is the result condensed in Hypothesis 3 and Proposition 5, which can now be merged as follows.

> **Passage 46** At the very birth or death of a gibbous Moon, the terminator is coplanar with our eye. Right then, it looks perfectly straight, and we call this the time of **orthotomy**. This is not exactly the same as the more desirable time of **orthogony** (or 90 degree phase angle), but the difference is imperceptible to the eye, so we can take the one for the other.

Note that Aristarchus explicitly identifies *orthogony* (in Proposition 5) and *orthotomy* (in Hypothesis 3) as the right times to make the measurements required to estimate the Sun's distance. He knows that, technically, these are different times (the former corresponding to a slightly gibbous Moon whose phase angle is exactly 90 degrees, the latter corresponding to a terminator that looks perfectly straight on a Moon whose phase angle is slightly less than 90 degrees). He knows the difference, but he also knows that the human eye cannot distinguish between these events. So he takes the one to be the same as the other. His next move is to make use of this simplifying assumption.

2.3 Estimating sizes and distances

In Proposition 6, Aristarchus proves that when the Moon is halved, its elongation from the Sun is less than a right angle. The wording of this proposition and its proof is interesting in that it shows that the young author of *On Sizes* was still a geocentrist.

> **Proposition 6** The Moon moves [in an orbit] lower than [that of] the Sun, and, when it is halved, is distant less than a quadrant from the Sun (Heath, 1913:371).

So the Sun and Moon are both said to orbit the Earth. The illustration of this proposition in *On Sizes* and the words 'the sphere on which the centre of the Sun moves' in its proof leave no doubt. As Heath (1913:312) explains, Aristarchus has not yet discovered that the Earth orbits the Sun.

Furthermore, as worded, the second half of Proposition 6 is strictly true. However, it is worth noting that there is a slight difference between the phrase 'when the Moon is halved' (used in this proposition) and the phrase 'when the Moon appears to us halved' (used earlier). The difference is that Aristarchus uses the real thing (in this proposition), rather than what it looks like (as in previous ones). That is, he takes the Moon to be actually 'halved' when it looks halved. But this is not true. In reality, *orthogony* (as defined on page 51), does not occur at the very birth or death of a gibbous Moon, but many hours before the former, or after the latter event, at a time the naked eye is completely unable to tell. In fact, it cannot even tell the time of *quadrature*. It is simply not sharp enough. So, unbeknown to Aristarchus, by the time the human eye says that a half-Moon is on the brink of turning gibbous, or gibbous on the brink of turning halved, the angle between the Sun and the Moon has already been obtuse for many hours!

So, the Moon is less than a quadrant from the Sun during some of the long hours during which it looks halved to us, but more than a quadrant during the others! This little realization will bear significantly in what follows. But first, let us meet Proposition 7, in which Aristarchus explains the main conclusion of his treatise. Namely, Conclusion 1. It states that 'the Sun is eighteen to twenty times as far away as the Moon'.

> **Proposition 7** The distance of the Sun from the Earth is greater than eighteen times, but less than twenty times, the distance of the Moon from the Earth (Heath, 1913:377).

That is,

$$18 < \frac{S}{M} < 20, \tag{2.12}$$

Estimating sizes and distances 59

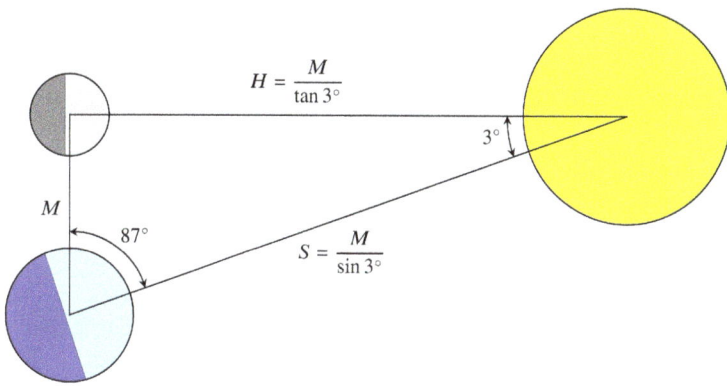

Figure 2.6 Aristarchus' triangle. (Sizes and distances are not to scale.)

where S is the Sun's distance from Earth, and M is that of the Moon.

This result follows directly from the value given in Hypothesis 4, for which Aristarchus gives neither reference nor indication of how it was obtained. So, as with previous values introduced in the same manner, we are left wondering whether he found it himself.

Aristarchus proves this proposition using a geometric approach (involving the use of equalities, inequalities, and approximations) which yields a lower and an upper bound for the Sun's distance, and he states his meaning in an unambiguous language: the Sun is no closer than eighteen times the Moon's distance, and no further than twenty times this distance.

Today, from the vantage point of millennia of accumulated knowledge, we would use trigonometry to obtain the exact solar distance corresponding to the value in Hypothesis 4, according to which, 'When the Moon appears to us halved, its distance from the Sun is then less than a quadrant by one-thirtieth of a quadrant'. That is, when the Moon is at position (d) in Figure 2.2, or position (a) in Figure 2.3, its elongation from the Sun is then less than 90° by 90°/30 = 3°, and is therefore 90° − 3° = 87° (Heath, 1913:353). (See Figure 2.6.)

Thus, the ratio between the distance S (to the Sun) and the distance M (to the Moon) is

$$\frac{S}{M} = \frac{1}{\sin \delta} \tag{2.13}$$

$$= \frac{1}{\sin 3°} = 19.107..., \tag{2.14}$$

where δ is the Moon's elongation from the Earth as seen from the Sun. Alternatively,

$$\frac{S}{M} = \frac{1}{\cos \varepsilon} \qquad (2.15)$$

$$= \frac{1}{\cos 87°} = 19.107..., \qquad (2.16)$$

where ε is the Moon's elongation from the Sun as seen from the Earth. (Note that $\delta = \psi - \varepsilon$, where ψ is the Moon's phase angle, as illustrated in Figures 2.4 and 2.5.)

So far so good. Aristarchus' sound mathematical mind takes for granted that the Moon's elongation is less than a quadrant. He has dedicated Proposition 6 to showing that this is so. The triangle in Figure 2.6 requires it. It wouldn't make sense otherwise, would it? It seems logical. Even a properly calibrated Antikythera-like device suggests that this is so, but in fact, it is not. Something odd has happened. In order to understand what it is, we must first think about elongation itself.

Elongation is the angular distance between two celestial bodies (as seen from Earth). In the case of the two luminaries, it is smaller the closer they are to each other (that is, at New Moon) and bigger the farther away they are in the sky (that is, at Full Moon). The smallest value it can take is 0 degrees (at a perfectly central eclipse of the Sun) and the greatest, 180 degrees (at a perfectly central eclipse of the Moon). So the more illuminated the Moon is, the greater its elongation. But the times chosen by Aristarchus to read the Moon's elongation are precisely the birth and death of a *gibbous* Moon, as determined by the naked eye. So the Moon's elongation at these times is not less, but more than a quadrant by about 3°. So Aristarchus has chosen times when the Moon's elongation is not $90° - 3° = 87°$, but rather $90° + 3° = 93°$ (as the reader can easily check using a modern sky simulator).

Closely connected to this phenomenon is the difference between *planar* and *spherical* trigonometry. For observers on Earth, the sky looks like a spherical dome, producing effects like the *Moon's paradox* (Meeus, 2007:29) or *spherical triangles* (Meeus, 2009:21) like the one in Figure 2.7, which represents an *orthogony* (as defined on page 51): specifically, the death of a first half-Moon (or birth of a gibbous Moon). At times like this, Aristarchus thinks his instrument is reading 87 degrees (because that is the 'logical' thing to assume), when in reality, it is reading 93 degrees. Yet, amazingly enough, his method of measurement

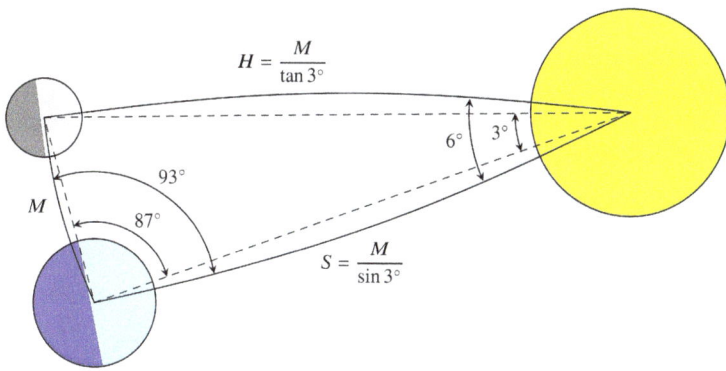

Figure 2.7 Spherical version of Aristarchus' triangle. (Sizes and distances are not to scale.)

(whichever it was) works and (unbeknown to him) correctly yields the solar distance corresponding to a lunar elongation of 87 degrees! (We shall see why in Chapter 3.)

The fact that 87 and 93 degrees are, respectively, the Moon's elongation roughly six hours either side of *equatorial dichotomy* (as defined on page 51) is a very strong indication that Aristarchus did actually measure these angles, then chose the one that made more sense to him, and we are very fortunate that he did so, because his choice is one that reveals what his method of measurement exactly (or rather, most likely) was. But, as promised, this is something we will see in a later chapter. Let us now go on with our analysis of the book *On Sizes*. The next two propositions are dedicated to showing that the Sun's size can also be determined by making the following simplifying assumption.

> **Proposition 8** When the Sun is totally eclipsed, the Sun and the Moon are then comprehended by one and the same cone which has its vertex at our eye (Heath, 1913:383).

Proposition 8 builds on Proposition 1 (which says that 'the same cylinder inscribes two equal spheres; and the same cone, two unequal ones'). Aristarchus takes it for granted that the tip of the cone inscribing the Sun and Moon is at our eye during a total eclipse of the Sun, a fact he says 'is manifest from observation'. In reality, the Moon is sometimes closer, sometimes farther away from us, so that the tip of its shadow is not always at our eye. As Heath (1913:383) points out, 'it is evident from this that [by the time he wrote *On Sizes*] Aristarchus had not observed…an annular eclipse of the Sun'. It may well be so, given the rarity of such an event. However, the wording of Proposition 8 excludes annular eclipses, since, strictly speaking, the Sun is not 'totally eclipsed' during such events. In any case, Aristarchus is correct in making the simplifying assumption that, on average, things are roughly as he says in Proposition 8. (This is a perfectly valid procedure in mathematical modelling.) His next proposition follows beautifully from what he has discovered so far.

Proposition 9 The diameter of the Sun is greater than eighteen times, but less than twenty times, the diameter of the Moon (Heath, 1913:383).

That is,

$$18 < \frac{s}{m} < 20, \tag{2.17}$$

where s is the Sun's radius, and m is that of the Moon.

This follows directly from Equation 2.13. Thus, the Sun's radius s, in terms of the Moon's radius m, is

$$\frac{s}{m} = \frac{1}{\sin 3°} \approx 19. \tag{2.18}$$

He even dares with volumes too. But the exact formula for that of a sphere had not yet been found by the time he writes his book. It was discovered by Archimedes and published in about 225 BC (Heath, 1897:189), about thirty years after *The Sand Reckoner* was written (according to our estimate on page 45). Undaunted, Aristarchus works out the solution that exactly corresponds to the ratio he has found in Propositions 7 and 9 (that is, that the Sun is eighteen to twenty times as far away and wide as the Moon). He does so in his next proposition.

Proposition 10 The Sun has to the Moon a ratio greater than that which 5832 has to 1, but less than that which 8000 has to 1 (Heath, 1913:385).

He is able to find the exact solution because he works with ratios. So the exact formula for the volume of a sphere is not really needed (since all common factors cancel down). Thus, if the Sun and Moon are as described in Proposition 9, then their volumes $V_\odot$ and $V_\mathbb{C}$ compare as follows.

$$\frac{V_\odot}{V_\mathbb{C}} = \frac{4\pi s^3 /3}{4\pi m^3 /3} = \frac{s^3}{m^3}, \tag{2.19}$$

where s is the radius of the Sun and m is the radius of the Moon. That is,

$$\left(\frac{18^3}{1^3} = 5832\right) < \frac{V_\odot}{V_\mathbb{C}} < \left(\frac{20^3}{1^3} = 8000\right). \tag{2.20}$$

So, Proposition 10 is correct (whenever Proposition 9 is correct).

The next proposition finds upper and lower bounds for the lunar distance that satisfies both Hypothesis 6 and Proposition 4. That is, it finds out how many Moons away the Moon must be to subtend two degrees from the observer's location.

> **Proposition 11** The diameter of the Moon is less than 2/45ths, but greater than 1/30th, of the distance of the centre of the Moon from our eye (Heath, 1913:387).

Using his masterful command of geometry (which typically involves juggling equalities, inequalities, and approximations), Aristarchus shows that, to subtend an angular size of two degrees from the observer's position, the Moon must be somewhere between 45/2 and 30 times as far away as it is wide. That is,

$$\frac{45}{2} < \frac{M}{2m} < \frac{30}{1}, \quad \text{or, equivalently,} \quad 45 < \frac{M}{m} < 60, \quad (2.21)$$

where M and m are, respectively, the Moon's distance and radius. The exact answer (as given by Equation 2.3) is

$$\frac{M}{2m} = \frac{1}{2\sin\rho_{\mathbb{C}}} = \frac{1}{2\sin 1°} = 28.6493\ldots \quad (2.22)$$

That is, the Moon is about 29 times as far away as it is wide.

The next proposition finds upper and lower bounds for the tiny difference between the width of the Moon's terminator and body (that is, between the line segments CD and EF in Figure 2.1).

> **Proposition 12** The diameter of the circle which divides the dark and the bright portions in the Moon is less than the diameter of the Moon, but has to it a ratio greater than that which 89 has to 90 (Heath, 1913:389).

Here, again, Aristarchus resorts to geometry to find his upper and lower bounds. Today, with the advantage of trigonometry, any mathematician would spot at once that the required ratio is simply the cosine of any of the angles DBF, CDB, KBH, or BAD, which are all shown to be the same in Proposition 4 (Heath, 1913:369), and which we denoted by $\zeta_{\mathbb{C}}$ in the discussion that followed. If, however, instead of using the length L of the Moon's shadow (as Aristarchus did), we use the Moon's distance M (so that A is now slightly further from B than Aristarchus thought), then the angle BAD is no longer equal to, but slightly smaller than $\zeta_{\mathbb{C}}$, and we denoted this new angle by $\rho_{\mathbb{C}}$ in Equation 2.1. Since Aristarchus equated L to M (in Propositions 4 and 8), he did not distinguish between $\zeta_{\mathbb{C}}$ and $\rho_{\mathbb{C}}$. But, in fact, the tiny difference between these these angles was a minor source of inaccuracy in the calculations that followed. The major source of inaccuracy, however, was the exaggerated angular size Hypothesis 6 assigns to the Moon. Namely, two degrees. According to this value, $\zeta_\mathbb{C} = \rho_\mathbb{C} = 1°$, and $\cos 1° = 0.99984...$, which which lies, as Aristarchus said, within the bounds of Proposition 12. Namely,

$$\frac{89}{90} < \frac{CD}{FG} < 1. \tag{2.23}$$

It is to be noted that, sometime after writing the book we are currently analysing, Aristarchus corrected the exaggerated angle in Hypothesis 6, reducing it to the more accurate half a degree reported in *Sand Reckoner* 1.10 (Heiberg, 1881:248; Heath, 1897:223). This surely led Aristarchus to immediately and thoroughly revise his early book *On Sizes*, but the extraordinary consequences of this revision will be considered in Chapter 4. For now, let us move on to the next proposition, which is a crucial, yet challenging one.

> **Proposition 13** The straight line subtending the portion intercepted within the Earth's shadow of the circumference of the circle in which the extremities of the diameter of the circle dividing the dark and the bright portions in the Moon move is less than double of the diameter of the Moon, but has to it a ratio greater than that which 88 has to 45; and it is less than 1/9th part of the diameter of the Sun, but has to it a ratio greater than that which 22 has to 225. But it has to the straight line drawn from the centre of the Sun at right angles to the axis and meeting the sides of the cone a ratio greater than that which 979 has to 10125 (Heath, 1913:393).

These words have been interpreted differently throughout history by the many copyists and translators who did their best to pass the now lost original down to us. According to Sidoli (2007:527, 547), the difficulty arises, this time, in the impossibility to draw a single, mathematically coherent diagram, since we can have the Moon either completely engulfed in (and tangent to) the Earth's shadow (as in most of the diagrams in the Arabic tradition, such as Thabit's and Tusi's) or drawn such that only its visible face is completely engulfed in (and tangent to) the Earth's shadow (as in virtually all the diagrams in the Greek and Latin tradition, such as *Codex Vaticanus Graecus* 204's or Commandino's 1572:22), but we cannot have it both ways.

However, it is possible to make sense of this proposition even though we can have it only one way. In fact, only one of the above interpretations satisfies all the geometric demands of the proposition. The other does not. Neither does Sidoli's (2007:535) halfway compromise, as he himself owns, nor the one based on a (seemingly) literal reading of the words in Proposition 13, according to which, 'the circle dividing the dark and bright portions in the Moon' is one and the same as the circle separating the near and far sides of the Moon during a central eclipse of the Sun (that is, the circle whose diameter is the line CD in Figure 2.1), as Aristarchus himself had explicitly had it in previous propositions (namely, in Propositions 3, 4 and 12). Since we have a Moon eclipse now (rather than a Sun eclipse), we may naively think that his 'dividing circle' can be found by simply reversing the shading in Figure 2.1, but soon, problems arise with this literal interpretation. First, there is no terminator on a totally eclipsed Moon, and even if we drew it, its foremost and rearmost points would not move in a circle (as the proposition requires), but rather (since their positions depend on those of the Earth and the Sun), in an interweaving path of coils whose complexity need not worry us here.

However, if we take Aristarchus to mean something as simple as 'the edge of the Moon's disc' by what he keeps calling 'the circle which divides the dark [unseen] and bright [seen] portions in the Moon', just as Carman (2014:44) proposes, then things start to make sense. In fact, Aristarchus himself defines it this way in his next proposition (Heath, 1913:400). Furthermore, this is exactly the traditional Greek and Latin interpretation, which is illustrated in Figure 2.8 below (where the author has fancied to add the Moon's terminator, even though, in fact, there is none on a totally eclipsed Moon).

There is no contradiction this time. The Moon (or rather, its disc), as seen from Earth, is both totally engulfed in, and tangent to the Earth's shadow, just as the proposition requires. The requirement that the lines LC and ON (in Figure 2.8 and Heath, 1913:394) are parallel to each other is also satisfied, as is the requirement that the fore and aft of the Moon's disc (that is, the ends of the arc LN in Figure 2.8) both move in a circle around the Earth.[13] Furthermore, Hypothesis 5 also holds, since there are exactly two such arcs (or lunar discs) between the points where the Earth's shadow cuts the circle in which the points L and N move.

According to Proposition 13, the Earth's shadow (where the Moon cuts it) is less than twice as wide as the Moon's sphere (hence the protrusion in Figure 2.8), but bears to it a ratio greater than that which 88 bears to 45. At the same time, to

---

13 This is a valid simplifying assumption. In reality, the Moon's orbit can be described as a nearly circular, wobbling ellipsoid which is influenced by the gravitational pull of all other bodies in the Solar System (Meeus, 1997:14, 2002:11).

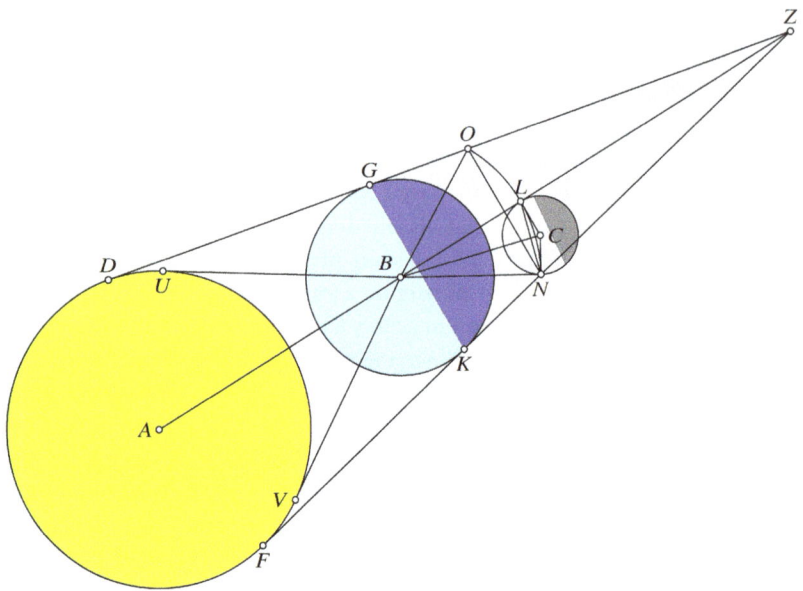

Figure 2.8 Start of a total Moon eclipse. (Sizes and distances are not to scale.)

an observer on Earth, this very same shadow appears to be twice as wide as the Moon's disc (as required by Hypothesis 5). It all fits.

These things said, the results of Proposition 13 can be summarized in the following mathematical expressions.

$$\frac{88}{45} < \frac{ON}{2m} < \frac{2}{1} \quad \text{and} \quad \frac{22}{225} < \frac{ON}{2s} < \frac{1}{9}, \tag{2.24}$$

or, in decimal notation,

$$1.9555... < \frac{ON}{2m} < 2 \quad \text{and} \quad 0.0977... < \frac{ON}{2s} < 0.1111..., \tag{2.25}$$

where m is the Moon's radius and s is the Sun's radius.

Trigonometric calculations (as developed in Chapter B) show that the required ratios (under Aristarchus' assumptions) are exactly given by

$$\frac{ON}{2m} = \frac{\sin 2\rho}{\tan \rho} \quad \text{and} \quad \frac{ON}{2s} = \frac{\sin 2\rho \sin 3°}{\tan \rho}, \tag{2.26}$$

where ρ is half the angular size of either the Sun or the Moon. Putting $\rho = 1°$ (as required by Hypothesis 6), gives the following results,

$$\frac{ON}{2m} = \frac{\sin 2°}{\tan 1°} = 1.9993... \quad \text{and} \quad \frac{ON}{2s} = \frac{\sin 2° \sin 3°}{\tan 1°} = 0.1046..., \quad (2.27)$$

which lie within the bounds in Inequalities 2.24 and 2.25. So, Aristarchus' outstanding geometric work is correct.

It is to be noted that (according to Hypothesis 6 and Proposition 8) the angular sizes of the Sun and Moon (that is, the angles UBV and NBL) are both equal to two degrees, though this is not immediately obvious in the unscaled arrangement of Figure 2.8 (where, for illustrative purposes, $2° < NBL < UBV$). In order to have it as meant (that is, in order to have $2° = NBL = UBV$), the Sun and Moon must both be moved further from the Earth, and in fact, they can be placed in such a way that all the above conditions hold. That is, there is a very special set of sizes and distances that satisfies each and every single one of the above conditions. This is exactly what Aristarchus is after. He dedicates the rest of his book to finding such a set.[14] At this point, we are also in a position to find the exact solution to this fabulous optimization problem. The necessary mathematics are all given in Chapter B. Thus, using Equation B.17 (in the mentioned appendix), the Moon's distance that exactly satisfies Aristarchus' early model of the cosmos is

$$\frac{M}{e} = \frac{1}{\sin(2\rho + \zeta)\cos\rho} \quad (2.28)$$

$$= \frac{1}{\sin(2 \times 1° + 0.850862...°)\cos 1°}$$

$$= 20.109060..., \quad (2.29)$$

where e is the Earth's radius (which is used here as a unit of length), ρ is half the angular size of the Moon, and ζ (whose value is given exactly by Equation B.25) is half the angular size of the Earth (as seen from the tip of its own shadow). Note that the precision displayed in Equation 2.29 is of mathematical interest only. In reality, this value is a very crude approximation which is not even good to one significant figure, but this is not the point. The point is that Aristarchus undertook to solve a very complex optimization problem in his early book and did

---

14 The possibility of finding such an optimal set of variables may be explored by using an online interactive illustration that can be found by googling the words 'Equations related to Proposition 13 of *On Sizes*'.

succeed in finding upper and lower bounds for the actual solution. His estimate of the Moon's distance—crude as it may look to us now—was once humanity's best, and his book—soon as it was surpassed by the very hand who wrote it—was once humanity's finest understanding of the cosmos. It goes to his credit as a mathematician, that he achieved his goal without the powerful tools that modern mathematicians have at their disposal, such as a handy numerical system, trigonometry, and Cartesian geometry (as used in Chapter B), none of which were then available.

The rest of the values that exactly meet his every requirement can be found as follows. Using Equations 2.13 and 2.29, the solar distance is

$$\frac{S}{e} = \frac{M}{e \sin 3°} = \frac{1}{\sin(2\rho + \zeta)\cos\rho \sin 3°} = 384.230302..., \qquad (2.30)$$

using Equations 2.1 and 2.29, the lunar radius is

$$\frac{m}{e} = \frac{M \sin\rho}{e} = \frac{\sin\rho}{\sin(2\rho + \zeta)\cos\rho} = 0.350951..., \qquad (2.31)$$

and using Equations 2.18 and 2.29, the solar radius is

$$\frac{s}{e} = \frac{m}{e \sin 3°} = \frac{\sin\rho}{\sin(2\rho + \zeta)\cos\rho \sin 3°} = 6.705743..., \qquad (2.32)$$

all in terms of the Earth's radius. This completes our solution.

A comparison of the relative sizes of the Earth, Sun, and Moon, as derived from the exact solution to Aristarchus' optimization problem (worked out above for the first time in history) is shown in Figure 2.9 (where everything is to scale except the distance from the Sun to either of the other two bodies).

It is now clear that Aristarchus' eyes were the first ever to see that there was something odd about the Sun's awkwardly prominent role in this game. But he was still a geocentrist by the time he wrote *On Sizes*. Let us now see what his own solution to the above problem was. Proposition 14 was his next step towards it.

> **Proposition 14** The straight line joined from the centre of the Earth to the centre of the Moon has to the straight line cut off from the axis towards the centre of the Moon by the straight line subtending the (circumference) within the Earth's shadow a ratio greater than that which 675 has to 1 (Heath, 1913:399).

This proposition states that the centre of the Moon is more than 675 times farther away from us than it is from the centre of the Earth's shadow at the distance where the Moon cuts its edges during a perfectly central eclipse of the

Figure 2.9 Sizes in *On Sizes*. (Relative sizes and distances are to scale, except for the distance from the Sun to either of the other two bodies.)

Moon. That is, the distance BC is greater than 675 times the distance CS in Figure 2.10 (where sizes and distances are not to scale, and where the author has wished to draw the Moon's terminator, even though there is none to be seen on a totally eclipsed Moon).

The line BC is just the Moon's distance M from the observer, and the line CS is the difference between BC and BS, where BS is given by Equation B.12. That is,

$$CS = M - M \cos \rho \cos 2\rho,$$

where ρ is half the Moon's angular size. So, the ratio referred to by Aristarchus in this proposition is

$$\frac{BC}{CS} = \frac{M}{M - M \cos \rho \cos 2\rho} = \frac{1}{1 - \cos \rho \cos 2\rho}. \tag{2.33}$$

Taking $\rho = 1°$ (as required by Hypothesis 6) gives

$$\frac{BC}{CS} = \frac{1}{1 - \cos 1° \cos 2°} = 1313.3959... \tag{2.34}$$

So, strictly speaking, Aristarchus' statement that this ratio is greater than 675 is correct. He used this knowledge in the geometric work of his next proposition.

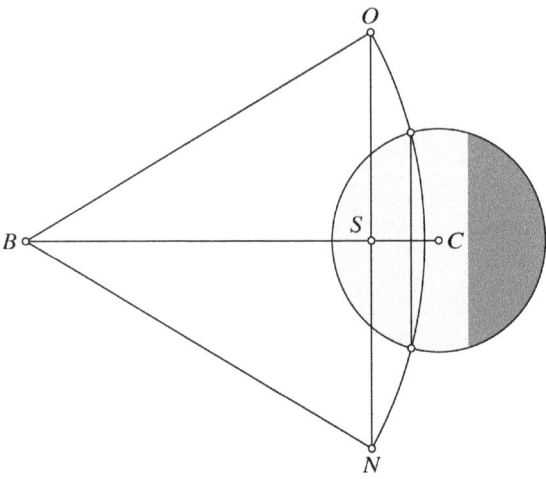

Figure 2.10 Proposition 14

Proposition 15 The diameter of the Sun has to the diameter of the Earth a ratio greater than that which 19 has to 3, but less than that which 43 has to 6 (Heath, 1913:403).

According to this proposition,

$$\frac{19}{3} < \frac{s}{e} < \frac{43}{6}, \tag{2.35}$$

or, equivalently,

$$6.3333... < \frac{s}{e} < 7.1666...,$$

where s is the Sun's radius and e is the Earth's radius. The exact solution (as given by Equation 2.32) lies almost exactly halfway between the latter bounds.

In his next proposition, he compares the volumes of the Sun and Earth (in a similar way as he compared those of the Sun and Moon in Proposition 10). Again, undaunted by the fact that the formula for the volume of a sphere had not yet been discovered, he finds the exact solution to this problem too.

Proposition 16 The Sun has to the Earth a ratio greater than that which 6859 has to 27, but less than that which 79507 has to 216 (Heath, 1913:409).

Thus, if the Sun and Earth are as described in Proposition 15, then their volumes $V_\odot$ and $V_\oplus$ compare as follows.

$$\frac{V_\odot}{V_\oplus} = \frac{4\pi s^3/3}{4\pi e^3/3} = \frac{s^3}{e^3}, \qquad (2.36)$$

where s is the radius of the Sun and e is that of the Earth. That is,

$$\left(\frac{19^3}{3^3} = \frac{6859}{27}\right) < \frac{V_\odot}{V_\oplus} < \left(\frac{43^3}{6^3} = \frac{79507}{216}\right). \qquad (2.37)$$

So, Proposition 16 is correct (whenever Proposition 15 is correct).

In his next proposition, he finds upper and lower bounds for the Moon's size and concludes as follows.

Proposition 17 *The diameter of the Earth is to the diameter of the Moon in a ratio greater than that which 108 has to 43, but less than that which 60 has to 19 (Heath, 1913:409).*

So, according to this proposition,

$$\frac{19}{60} < \frac{m}{e} < \frac{43}{108}, \qquad (2.38)$$

or, equivalently,

$$0.3166... < \frac{m}{e} < 0.3981...,$$

where m is the Moon's radius and e is that of the Earth. The exact solution (as given by Equation 2.31) lies almost exactly halfway between Aristarchus' bounds.

He then compares the volume of the Moon to that of the Earth and reaches the following conclusion.

Proposition 18 *The Earth is to the Moon in a ratio greater than that which 1259712 has to 79507, but less than that which 216000 has to 6859 (Heath, 1913:411).*

Following the simple procedure of comparing ratios (as he did in Propositions 10 and 16), he gets round the problem of having no exact formula for the volume of a sphere and finds the exact answer to this problem. Thus, if the Earth and Moon are as described in Proposition 13, then their volumes $V_\oplus$ and $V_\mathbb{C}$ compare as follows.

$$\frac{V_\oplus}{V_\mathbb{C}} = \frac{4\pi e^3/3}{4\pi m^3/3} = \frac{e^3}{m^3}, \qquad (2.39)$$

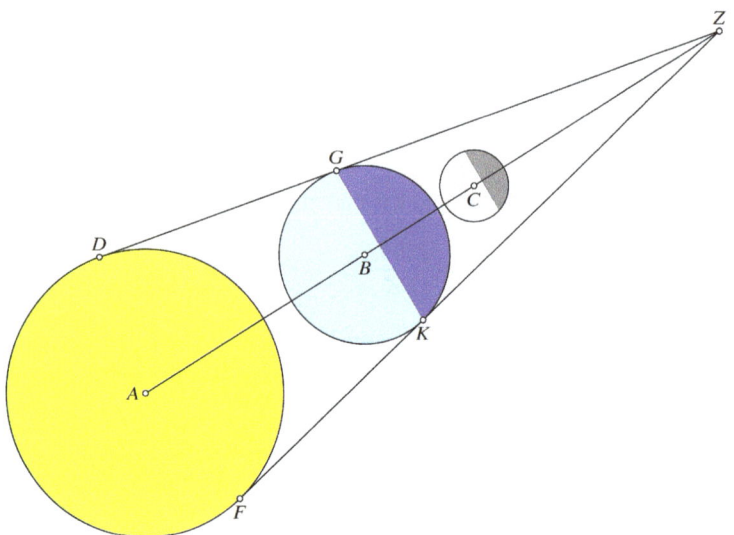

Figure 2.11 Middle of a total Moon eclipse. (Sizes and distances are not to scale.)

where e is the radius of the Earth and m is that of the Moon. That is,

$$\left(\frac{108^3}{43^3} = \frac{1259712}{79507}\right) < \frac{V_\oplus}{V_☾} < \left(\frac{60^3}{19^3} = \frac{216000}{6859}\right). \tag{2.40}$$

So, Proposition 18 is correct (whenever Proposition 13 is correct).

The final proposition of *On Sizes* has not been preserved in the Greek manuscripts, but only in the Arabic translations. It runs as follows.

> **Proposition 19** The ratio of the distance of the vertex of the shadow cone from the centre of the Moon, when the Moon is on the axis of the cone containing the Earth and the Sun, to the distance of the centre of the Moon from the centre of the Earth is greater than the ratio 71 to 37 and less than the ratio 3 to one (Berggren and Sidoli, 2007:250).

According to this final proposition, the ratio of the distance V, from Z to C, to the distance M, from B to C (in Figure 2.11), lies somewhere between 71/37 and 3/1. That is,

$$\frac{71}{37} < \frac{V}{M} < 3, \tag{2.41}$$

or, equivalently,

$$1.9189... < \frac{W-M}{M} < 3,$$

where W is the geometric length of the Earth's shadow and M is the Moon's distance from Earth. Now, W is given either by Equation B.16 or Equation B.20 (in Chapter B), and M is given by Equation 2.29. So, the exact value of the ratio in Proposition 19 is

$$\begin{aligned}
\frac{V}{M} &= \frac{W-M}{M} \qquad\qquad\qquad\qquad\qquad\qquad\qquad\qquad (2.42)\\
&= \frac{W}{M} - 1 \\
&= \frac{e}{\sin\zeta} \times \frac{\sin(2\rho+\zeta)\cos\rho}{e} - 1 \\
&= \frac{\sin(2\rho+\zeta)\cos\rho}{\sin\zeta} - 1 \\
&= 2.348787..., \qquad\qquad\qquad\qquad\qquad\qquad\qquad (2.43)
\end{aligned}$$

where $\rho = 1°$ and ζ is as given by Equation B.25. The above result lies almost halfway between Aristarchus' bounds.

But there is more. As a bonus, the Arabic text contains extra information regarding the length W of the Earth's shadow, which is explicitly compared at several points in the reasoning to the distance M (between Earth and Moon) and to the distance S (between Earth and Sun). Thus, gathering this information, the ratio of W to M is

$$\frac{108}{37} < \frac{W}{M} < \frac{15}{4}, \qquad (2.44)$$

and that of W to S is

$$\frac{6}{37} < \frac{W}{S} < \frac{3}{16}. \qquad (2.45)$$

It is to be noted that the very last line of the Arabic text contains a little oddity. When it says that the ratio of W to M is less than 15/4, and hence that the ratio of V to M (as defined above) is less than 12/4 (or simply, 3), it should have said that this ratio is $15/4 - 1 = 11/4$, which is an even better estimate for the upper bound referred to. Thus, Inequality 2.41 can be improved to

$$\frac{71}{37} < \frac{V}{M} < \frac{11}{4}, \qquad (2.46)$$

which brings the mean of these bounds even closer to the exact solution (in Equation 2.43).

There is a myriad possible explanations for the above oddity, and it is up to the reader to choose thy favourite one. In the author's personal opinion, Aristarchus' original text was as in Inequality 2.46, and that in Inequality 2.41 is the result of a later copyist trying thy best to make out a damaged or unclear original.

Finally, though this is not explicitly stated in *On Sizes*, an expression for the Moon's distance in terms of the Earth's radius can be deduced by reasoning as follows.

According to Inequality 2.21, the Moon's distance (in terms of the Moon's radius) is

$$45m < M < 60m,$$

which (using Inequality 2.38) can be rewritten as

$$45\left(\frac{19e}{60}\right) < M < 60\left(\frac{43e}{108}\right),$$

or equivalently,

$$\frac{57}{4} < \frac{M}{e} < \frac{215}{9}. \tag{2.47}$$

Also, an expression for the Sun's distance in terms of the Earth's radius can be found by reasoning as follows. According to Inequality 2.12, the Sun's distance (in terms of that of the Moon) is

$$18M < S < 20M,$$

which (using Inequality 2.47) can be rewritten as

$$18\left(\frac{57e}{4}\right) < S < 20\left(\frac{215e}{9}\right),$$

or equivalently,

$$\frac{513}{2} < \frac{S}{e} < \frac{4300}{9}. \tag{2.48}$$

Likewise, rearranging Inequality 2.44, we have

$$\frac{108}{37}M < W < \frac{15}{4}M.$$

Substituting for M (from Inequality 2.47) into this, we have

$$\frac{108}{37}\left(\frac{57e}{4}\right) < W < \frac{15}{4}\left(\frac{215e}{9}\right).$$

Hence, the length of the Earth's shadow in terms of the Earth's radius is

$$\frac{1539}{37} < \frac{W}{e} < \frac{1075}{12}. \tag{2.49}$$

In summary, the sizes and distances in Aristarchus' first book *On Sizes* are as follows.

A summary of Aristarchus' first book *On Sizes*

- By Proposition 7, the ratio of the Sun's distance to that of the Moon is

$$18 < \frac{S}{M} < 20. \tag{2.12}$$

- By Proposition 9, the ratio of the Sun's radius to that of the Moon is

$$18 < \frac{s}{m} < 20. \tag{2.17}$$

- By Proposition 11, the ratio between the Moon's distance and radius is

$$45 < \frac{M}{m} < 60. \tag{2.21}$$

- By Proposition 15, the ratio of the Sun's radius to that of the Earth is

$$\frac{19}{3} < \frac{s}{e} < \frac{43}{6}. \tag{2.35}$$

- By Proposition 17, the ratio of the Moon's radius to that of the Earth is

$$\frac{19}{60} < \frac{m}{e} < \frac{43}{108}. \tag{2.38}$$

- By Proposition 19,
  (a) the ratio of the length of the Earth's shadow to the Moon's distance is

$$\frac{108}{37} < \frac{W}{M} < \frac{15}{4}, \tag{2.44}$$

  (b) the ratio of the length of the Earth's shadow to the Sun's distance is

$$\frac{6}{37} < \frac{W}{S} < \frac{3}{16}, \tag{2.45}$$

(c) the ratio of the length of the Earth's shadow minus the Moon's distance to the latter distance is

$$\frac{71}{37} < \frac{V}{M} < \frac{11}{4}. \tag{2.46}$$

- Finally (though this is not explicitly stated in *On Sizes*),
  (a) the ratio of the Moon's distance to the Earth's radius is

$$\frac{57}{4} < \frac{M}{e} < \frac{215}{9}, \tag{2.47}$$

  (b) the ratio of the Sun's distance to the Earth's radius is

$$\frac{513}{2} < \frac{S}{e} < \frac{4300}{9}, \tag{2.48}$$

  (c) the ratio of the length of the Earth's shadow to the Earth's radius is

$$\frac{1539}{37} < \frac{W}{e} < \frac{1075}{12}. \tag{2.49}$$

So (according to Inequality 2.47) the distance to the Moon is $14\frac{1}{4}$ to $23\frac{8}{9}$ Earth radii, and (according to Inequality 2.48) the distance to the Sun is $256\frac{1}{2}$ to $477\frac{7}{9}$ Earth radii.[15]

Now, we may ask the question, Why is the latter proposition not included in *On Sizes*? Was it removed at some point in history, or was it never included in the first place? Again, a myriad possibilities arise and it is up to the reader to choose thy own. The author's belief (as of this writing) is that Aristarchus decided not to include this 'phantom' proposition simply because he did not like it. The bounds of all other propositions are reasonably (almost magically) close to the true solutions, whereas those of this one are too (embarrassingly) far apart from each other. He sensed there was room for improvement and on his way to achieve it, he eventually stumbled upon the greatest astronomical discovery ever. Namely, heliocentrism. But this occurred when his first book *On Sizes*, the only one by him we have access to, had already been written.

---

15 The reader may check an online interactive illustration of the above summary by googling up the words 'Sizes in *On Sizes* 1'.

2.4 Conclusion

The present analysis of the book *On Sizes* is now concluded. It has shown that this book, rather than being too bad to be the work of a genius, is too good to be the work of a forger. In fact, it is a mathematical masterpiece, hardly intended to slander its author, Aristarchus. Instead, he is shown to be a man capable of dealing with a problem whose exact solution has only been found here, in the preceding pages, twenty-three centuries after it was posed! His upper and lower bounds for the actual solution are truly the work of a genius who was born at a time when the mathematical tools necessary to solve such a problem exactly were not yet available; a time when the most advanced knowledge on the subject was that of Phidias, Archimedes' father, who reckoned the Sun to be twelve times as far away as the Moon; a time when the finest estimate of the Sun's and Moon's angular size was about four times its actual value. Saying that Aristarchus should have known better at this early stage of his life is unfair. No one could. No one did.

Next, we are going to investigate how he might have measured the Moon's elongation from the Sun when the former is at *orthogony* (as defined on page 51), and hence gauged the Sun's distance.

3 Measuring the Sun's distance

Abstract: In this chapter, we address the question of whether the Sun's distance can be measured experimentally with technology such as we imagine would have been available to Aristarchus. We shall also study the methods and results of all those who have tried in the past.

3.1 Introduction

In 2011, the author released a working paper on Aristarchus which was never published. It did, however, elicit feedback from a wide range of respondents, including teachers who wanted to engage their pupils in the study of astronomy, writers who were writing about him, philologists who had spent much of their lives studying him in the original and kindly shared their expert knowledge of the ancient Greek and Babylonian languages (both of which figured in the paper), engineers who were curious about his methods and devices, and mathematicians who scrutinized every detail and provided new food for thought. Among these letters, there was one that started what proved to be a most fruitful line of investigation. This chapter is dedicated to it.

It started when the American psychiatrist and historian of astronomy William Patrick Sheehan sent his first email to the author three days before the end of the year. He posed the question of how Aristarchus could have measured the Sun's distance (if at all), and eventually, invited the author to join him and his friend, the American geography professor John Edward Westfall, in the quest for an answer. This was the beginning of a frantic exchange of letters that lasted a good many years of unquenched enthusiasm.

We had not been the first to try. Attempts to estimate the Sun's distance had been made well before Aristarchus was a lustful twinkle in his father's eye, but it was *his* estimate that set the reference for the two thousand years that followed. A record of those who tried was made by Riccioli (1651:110) and more recently by Hughes (2001b:2). These records show that before the invention of the telescope astronomers (other than Posidonius) had not ventured far from the estimates of the Sun's distance that result from updating those of the Moon in Aristarchus' Proposition 7 (or Inequality 2.12, as tagged in this book). But before seeing how we answered the above question, let us see how others coped before us.

3.2 Thomas Harriot

The invention of the 'instrument for seeing far', as it was first described (Helden, 1977:36), kindled new hopes. Could now the Sun's distance be estimated with greater accuracy? Could now the time of *orthotomy* be finally pinned down? The first to try was the English mathematician and astronomer Thomas Harriot. On January 21, 1611 (or January 11, 1610, in the Old Style), he turned his 'instrument' on the Moon several times from his garret at Syon House, Isleworth, and estimated the time of dichotomy, or rather, **orthotomy** (that is, the time when the lunar terminator looked as straight as possible) to have occurred nearly eleven hours after local noon, as measured with a 'clock or watch', as stated in folios 8 and 11 of his *Moon Papers* (namely, the ones containing a sketch of Aristarchus' triangle and a pair of sketches of the half-Moon he saw that night).[16] His comments on these papers (as selected and pieced together by the present author) are given in Table 3.1.

In reality (that is, by modern reckoning), *orthotomy* occurred twenty minutes after midnight, when the Sun was well below the horizon and the Moon was about to set (as Harriot's drawings show). Harriot's comments also show that it is much easier to estimate the time of *orthotomy* using a spyglass than using the naked eye, since, according to his report, the terminator started to look straight to the naked eye well over half a day after the true event!

By a method involving the measurement of the angles the Sun and Moon made with the edges of Aquarius and Taurus (such as shown on a planisphere or sky chart), he was able to estimate the Moon's elongation from the Sun at near *orthotomy* to be

$$\varepsilon_1 = 90° - (1°39'29'' - 59') = 89°19'31'', \tag{3.1}$$

which he rounded up to 89°20′ (as stated in line 13 of folio 8). With this angle, he estimated the Sun's distance from Earth to be 85.9453 times greater than that of the Moon (as stated in line 14 of the same paper). The presence of the symbol ψ, which Harriot uses to denote the secant of an angle (Schemmel, 2008:49, 393), indicates how he found the latter distance. The same value (but for a slight discrepancy due to rounding) can also be found using Equation 2.15 as

$$\frac{S}{M} = \frac{1}{\cos \varepsilon_1} = \frac{1}{\cos(89°20')} = 85.9456..., \tag{3.2}$$

where S and M are as usual, and ε_1 is the Moon's elongation from the Sun as reported by Harriot on this momentous occasion.

---

16 Harriot's papers were never published in his lifetime. After his death, they were scattered and vanished altogether only to reappear nearly two centuries later among the stable accounts of the former property of his friend and patron Henry Percy, Earl of Northumberland, in Petworth, Sussex.

Table 3.1: Harriot's half-Moon observation of January 21, 1611.

clock time	true time	field notes	estimated UT of take*	event†
	$4^h 22'$	Tempus apparens quo ☽ distabat a ☉ 87°.	[18:07]	[17:37]
$8^h 40'$	$8^h 55'$	Not yet a right line, but almost.	[22:40]	
		☉ 1°39′29″ ♒ [Aquarius].		
		☽ 59′ ♉ [Taurus] verius.		
		Differentia longitudinis Solis et ☽, 89°19′1/2.		[23:18]
		Distantia ☽ a ☉ (in circulo magno), 89°20′.		[23:19]
		ψ [secant of said angle], 859453/10000.		
$9^h 5'$	$9^h 20'$	Not, but imadged right.	[23:05]	[23:45]
		Distantia ☽ a ☉, 89°32′48″.		
$9^h 10'$	$9^h 25'$	Right.	[23:10]	[23:50]
		Distantia ☽ a ☉, 89°35′21″.		
	$10^h 13'$	Tempus apparens quo ☽ distabat a ☉ 90°.	[23:58]	[00:37]
$10^h 40'$	$10^h 55'$	Right. It did sensibly lack of a right line by playne sight, without instrument.	[00:40]	

* The time of note-taking, based on Harriot's drawings, is UT ≈ clock time + 14 hours.
† The time of actual astronomical event, as given by the JPL Horizons Online Ephemeris System, is UT ≈ true time + 14 hours + 24 minutes.

Also, in the lower half of his paper, he went on to compare the latter elongation to that in Aristarchus' first book *On Sizes* (namely, 87 degrees) and used it to calculate both the true time at which lunar **quadrature** had occurred (that is, the time at which the Moon was 90 degrees from the Sun), giving it as 10 hours and 13 minutes, and the true time at which the Moon must have been 87 degrees from the Sun, giving it (in the very last line of this particularly interesting paper of his) as 4 hours and 22 minutes. Unbeknown to anyone else but himself, his lucky assistants (Sir Nicholas Sanders and Christopher Took), and whoever they cared to tell, this was the first time in (recorded) history that someone (other than Posidonius) had departed significantly (on sound scientific grounds) from the range of values that follows from Aristarchus' Proposition 7 (or Inequality 2.12).

But this was not the only elongation Harriot recorded that night. There are two more on the upper right corner of folio 11, corresponding respectively to his takes at $9^h 20'$ (when he judged that the lunar terminator was about to look straight) and $9^h\,25'$ (when it did start to look straight to him). Using them, we obtain the ratios

$$\frac{S}{M} = \frac{1}{\cos \varepsilon_2} = \frac{1}{\cos(89°32'48'')} = 126.3890\ldots \qquad (3.3)$$

and

$$\frac{S}{M} = \frac{1}{\cos \varepsilon_3} = \frac{1}{\cos(89°35'21'')} = 139.4635\ldots, \qquad (3.4)$$

which are more accurate than the one in Equation 3.2.

Unfortunately, neither of these ratios nor the calculations leading to them can be found in the mentioned paper, but it is likely that these calculations were actually made and the paper on which they were made is either lost or waiting to be found among his many others. Promisingly enough, the author has already examined one containing something very similar. Namely, folio 127 of his *Spherical Trigonometry Papers*. Lines 15 and 16 of this paper contain the numbers $89°20'$ and 85.9453 all over again, along with the secant version of Equation 3.2 explicitly connecting them together as well as the new numbers $89°29'$ and 112.7946, which (by contextual similarity) may represent a new lunar elongation and solar distance (though it is not explicitly stated that they are so). The last two numbers, however, do not exactly correspond to each other, since the said angle is not the arcsecant of the said value, but rather the truncation of this arcsecant to the nearest minute. Thus,

$$\varepsilon_4 = \lfloor \sec^{-1}(112.7946) \rfloor = \lfloor 89°29'31''17'''59''''11'''''\ldots \rfloor = 89°29', \qquad (3.5)$$

or, working backwards,

$$\frac{S}{M} = \frac{1}{\cos \varepsilon_4} = \frac{1}{\cos(89°29'31''18''')} = 112.7946\ldots, \qquad (3.6)$$

where ε_4 is an angle that might have been obtained during a yet unidentified observation, or averaged out of previous elongations. In fact, to the nearest minute,

$$\varepsilon_4 = \frac{\varepsilon_1 + \varepsilon_2 + \varepsilon_3}{3} \approx 89°29'. \qquad (3.7)$$

Significantly enough, a further new angle appears in the second line from the end of this paper, namely, $89°34'$, which is equal to the average of ε_2 and ε_3 rounded down to the nearest minute (though it is not stated that this is so either).

The presence of all these data in the same paper strongly suggests that Harriot considered solar distances in the range of Equations 3.3, 3.4 and 3.6, but (as with everything historical) we cannot be sure of this unless we find a paper explicitly stating it as unambiguously as the papers containing the 'distantia Lunae a Sole' of Equation 3.2 do. As for the unidentified observation mentioned above, there is hope in what follows.

On April 19, 1611 (or April 9, in the Old Style), he turned his 'instrument' on the Moon again and estimated the time of *orthotomy* to occur at the strike of midnight, or twelve hours from noon, as measured with a 'clock or watch' (Rigaud, 1833:20; Bean, 1978:585). In reality (that is, by modern reckoning), it had occurred only 26 minutes earlier! This was remarkably accurate, especially considering that his watch may have been running fast, as the last of his field notes for the day suggests! In full, these notes (as recorded in folio 3 of his *Moon Papers*) are given in Table 3.2.

As mentioned above, the last of these notes suggests that Harriot's clock had been running slightly fast. So his timings h can be corrected by an approximation such as

$$h \approx \frac{287}{292} t + \frac{355}{1752}, \qquad (3.8)$$

which is valid for any time t between $6^h\, 30'$ and $20^h\, 30'$ and assumes his clock was running at a steady pace. (The times thus corrected are included in square brackets in the above transcription of his notes.)

There is more information to be found in his other papers for the day which is most worthy of note. For example, folio 7 captures in exquisite detail the moment of *orthotomy* as he saw it through his spyglass at **mean midnight** (or midnight o'clock), when (mark this) '*all* said it was a straight line'. Unfortunately, despite all the care that went into this particular observation, something went wrong. Hopefully, the notes in Table 3.3, which (have been picked by the present author from those that) can be found in folio 2, may reveal what it is.

These notes and the calculations accompanying them may sound unnecessarily repetitive, but they were not idle. They were simply trying to deal with a difficulty that did not arise in the January observation. Namely, that on this occasion, no matter what sort of time we choose (whether mean or true), quadrature preceded orthotomy (rather than the other way round, as should be expected of a First Quarter Moon)![17] This was no small issue. In fact, it was one that

---

17 As it happens, the time between quadratures and Full Moons is an average of 10 minutes shorter than the time between quadratures and New Moons (Meeus, 2007:12). Also, though

Table 3.2: Harriot's half-Moon observation of April 19, 1611

clock time	field notes	estimated UT
$6^h 30'$	I observed the moone by an instrument of $\frac{10}{1}$ and the line of division of light and darkness was far from right, and more without instrument, by playne sight.	[18:35]
$6^h 58'$	The sonne set. Then I set my minute watch.	[19:03]
$7^h 30'$	Chr[istopher] observed. Sensibly crooked.	[19:34]
$8^h 0'$	I observed. Sensibly crooked.	[20:04]
$9^h 46'$	Per stellas, $9^h 43'$ [or $9^h 43' 12''$, as calculated in folio 5].	[21:48]
	Altitude [of] Canis minoris, $19°56'$ by my catholic astrolabe; not yet a right line but almost.	[21:52]
$10^h 5'$	Not yet a right line.	[22:07]
$10^h 20'$	Very nere, but yet not perfect.	[22:22]
$10^h 45'$	Yet doubtful to be perfect.	[22:46]
$10^h 52'$	As I judge, now unsensibly a right line, an accidentall rag or two at and nere the lower corner being abstracted. But others say not yet perfect.	[22:53]
$11^h 15'$	Yet a right line and not contrary, but if not right, rather wanting by the lower corner.	[23:16]
$11^h 30'$	Yet continuing; others say wanting.	[23:30]
$11^h 47'$	Per stellas, $11^h 39'$ [or $11^h 38' 52''$, as calculated in folio 5].	[23:47]
	Altitude of Cap. II australianis $20°0'$, as before.	
$12^h 0'$	All say it is in a right line; at $12^h 0'$, I described the appearance as soone as I could by my instrument of $\frac{32}{1}$.	[00:00]
$12^h 30'$	Unsensibly different. And then we departed to bed. Sir Nicholas Sanders and Christopher [Took] were with me and also observed in my garret. Instruments: two $\frac{10}{1}$, one $\frac{15}{1}$, one $\frac{32}{1}$, one $\frac{11}{1}$.	[00:29]
$20^h 30'$	The next morning, about $8^{h}\frac{1}{2}$, my watch was forward from the sonne by a $\frac{1}{4}^h$ and somewhat more.*	[8:21]

*By Approximation 3.8, this happened at about 8:21:06 UT. Subtracting $11'37''$ gives the local mean solar time as 8:09:29. Adding the equation of time, $1'14''$, gives the local apparent solar time as 8:10:43. So Harriot's clock was ahead of the Sun by $8:30 - 8:10:43 = 19'17''$, which is somewhat more than a quarter of an hour, as he said.

---

it is always true that *geocentric* orthotomies (that is, as seen from the centre of the Earth) precede First Quarter Moons and follow Last Quarter Moons by an average of 18 minutes (Meeus, 2009:22), it sometimes happens that observers on the Earth's surface, especially

Table 3.3: Harriot's half-Moon timings of April 19, 1611.

recorded time	field notes	estimated UT	exact UT
$6^h\,17'$	Tempus medium quo distant 87°	[18:23]	[17:03]
$11^h\,50'$	Tempus medium quo distant 90°	[23:50]	[23:50]
$6^h\,25'36''$	Tempus apparens quo distant 87°	[18:31]	
$11^h\,58'36''$	Tempus apparens quo distant 90°	[23:58]	
$5^h\,15'36''$	Tempus apparens quo distant 87° pro Syon	[17:22]	
$10^h\,48'36''$	Tempus apparens quo distant 90° pro Syon	[22:50]	

Note that Harriot takes (true time) = (mean time) + (8 minutes + 36 seconds) and (true time for Syon) = (true time) − (1 hour + 10 minutes).

compromised the whole experiment to the point that all efforts went into trying to sort it out, and so, no lunar elongation was ultimately recorded in his papers (other than Aristarchus' 87 degrees) and neither was any solar distance (at least, none that the author has found).

However, the problem that troubled Harriot this time was one of human making, since, in reality (according to modern estimates for his particular time and location), orthotomy (23:34 UT) did precede quadrature (23:50 UT), as expected. So everything was fine and everything should have been fine had he followed *his* own judgement (rather than his assistants'), adopting the time when the terminator first started to look 'unsensibly a right line' to *him* (rather than to them). At this time (that is, at about 22:53), the Moon's elongation was about the same as that of Equations 3.5 and 3.6. In fact, the realization that the only solution to the above problem was to assume that orthotomy had occurred earlier than 'all' said may have led to these equations (which agree with Equation 3.7, which is the average of his January findings). So he seems to have saved the day, after all.

But, as fate would have it, his findings were to remain hidden from the world for centuries. So, when others tried again, such as Kepler and Wendelin, they knew nothing of what had happened at Syon House those clear winter nights. But at least, they knew how to call the 'instrument for seeing far', since five days before the last of those nights, the word 'telescope' had been pronounced for the first time ever at a banquet held to make Galileo a member of the prestigious

---

when the half-Moon is just above the horizon (as is the case here), may observe *topocentric* orthotomies occurring slightly sooner or later than expected, but in no case is the order reversed.

Lincean Academy by its very founder, Prince Federico Cesi, who had first heard it from the Greek mathematician John Demisiani (Rosen, 1947:31).

3.3 Johannes Kepler

The second person ever to think of using telescopes to shed new light on the solar distance conundrum was the German astronomer and mathematician Johannes Kepler, who, in the preface to his *Ephemerides* for the year 1619, urged scientists to devote themselves to this worthy cause (Kepler, 1617:125, 1868:516; Helden, 1985:80). But, after trying himself, he was soon disappointed and complained that a major difficulty thwarted his efforts. Namely, the unevenness of the Moon's surface made it impossible (for him) to determine the exact time of *orthotomy* (Kepler, 1630:165, 1868:528; Helden, 1985:80). So, he eventually despaired and abandoned the project.

However, something good came out of it all, since he learned from this experience that the Sun was farther away than had traditionally been thought. The question now was not 'whether it was', but 'by how much'. In fact, by the time he made his appeal to the scientific community, he himself had come up with three different answers, to which he added a fourth one three years later.

- The first one, published (even before the invention of the telescope) in *De Stella Nova*, was that the Sun is 1432 Earth radii away from Earth (Kepler, 1606:86, 1859:674; Helden, 1985:78).
- The second, published in his *Ephemerides* for 1617, was that the Sun is 1800 Earth radii away, or about 30 times as far away as the Moon, if the Moon is placed 60 Earth radii away from Earth (Kepler, 1617:2, 1868:483; Helden, 1985:79).
- The third, published in the same *Ephemerides*, was that the Sun is about 23 times as far away as the Moon (Kepler, 1617:6, 1868:486; Helden, 1985:79).
- The fourth, published in his *Epitome Astronomiae Copernicanae*, was that the Sun is 3469 Earth radii away from Earth (Kepler, 1620:483, 1866:326; Helden, 1985:84; Swerdlow, 2002:26).[18]

It is beyond the scope of the present book to go into how each of these solar distances was found or for what purpose, nor there seems to be a need, since this

---

18 This was the result of his belief that the ratio of the Sun's distance S to the Earth's radius e could be expressed as a function of the Sun's apparent size $2\rho_\odot$ alone. Namely,

$$\frac{S}{e} = \left\lfloor \sin^{-\frac{3}{2}} \rho_\odot \right\rfloor = 3469.$$

is already covered in the cited literature. Suffice it to say that none of Kepler's answers were based on actual observations of the Moon at orthotomy, but rather on his personal quest for harmonic ratios between the motions of the different bodies in the solar system. But, as it happened, no matter how intriguing these ratios are even today, when modern science regards them as mere coincidences (Gingras, 2003:265), the world was starting to rely less on theoretical speculations and more on empiricism, and even Kepler, the man who discovered the secrets of the universe, would have benefited from Harriot's secret lunar observations had he known of them.

3.4 Godfrey Wendelin

Prominent among those who heeded Kepler's call to repeat Aristarchus' experiment with the aid of telescopes and did not flinch at the unevenness of the Moon was the Flemish astronomer Godfrey Wendelin. On his first attempt, he (1626:11) got 3460 Earth radii, a figure nearly equal to Kepler's ultimate solar distance. On subsequent attempts, he obtained solar distances of 3600, 7000, 13720, and 14656 Earth radii, in that order, as he reported in a letter dated November 9, 1647, to the Italian astronomer Giovanni Battista Riccioli (1651:109; Helden, 1985:113), and a solar distance of 14720 Earth radii, as he reported in an even earlier letter dated May 1, 1635, to the French astronomer and mathematician Pierre Gassendi (1658:428). Nine years later, he (1644:29) published his final figure as 14656 Earth radii. This was the first time that a radical departure from all previous estimates appeared in print, apart from the huge solar distance of nearly a myriad Earth radii that was first mentioned in Archimedes' *Sand Reckoner* 2.1 (Heiberg, 1881:262; 1972:232), and to which we shall return in Chapter 4.

Four years after Wendelin wrote his letter of May 1635 to Gassendi, two young English astronomers, Jeremiah Horrocks and William Crabtree, measured the distance to the Sun by a totally new method involving timings of the first transit of Venus ever observed with a telescope. Namely, that of December 4, 1639 (Sheehan and Westfall, 2010:75-91; Westfall and Sheehan, 2014:278-84). Horrocks obtained a solar distance of nearly 15000 Earth radii, as reported in his *Venus in Sole visa* (Hevelius and Horrocks, 1662:142; Whatton, 1859:212). Unfortunately, due to the untimely death of both Englishmen, this work remained in the dark for over twenty years and came to light only when the Polish astronomer Johannes Hevelius published it in 1662. So, it seems that despite Wendelin's and Horrocks' solar distances being almost identical, neither men had any influence on the other.

As we learn from the above-mentioned report by Riccioli (a little fragment of which is reproduced below), Wendelin was most careful in his measurements, comparing both the morning and evening dichotomies in order to make up for the 'unevenness' of the Moon. He even allowed for the possibility that the light of the Sun entering the 'lunar air' may advance the evening dichotomies and retard the morning ones. Despite the fact that this 'lunar air' is now known to be virtually non-existing, Wendelin's careful considerations allowed him to obtain (by Aristarchus' half-Moon method) as accurate a Solar distance as Horrocks' (by the Venus transit method).

> **Passage 47** Having compared the dichotomies with each other, the morning as well as the evening ones, to consider how much was conceded by the unevenness of the Moon, and having considered also that the light of the Sun entering into the lunar air advances the evening dichotomy and retards the morning one, with all the circumstances considered, I could not decide other than that the dichotomy occurs [refringere] at a distance of 89°45′, or even more (Helden, 1985:113; see also Riccioli, 1651:109).

Unfortunately, the details of how Wendelin and Harriot made their measurements will not be further explored in this book, interesting though they are. The task is left to future researchers. For now, it is brought to the reader's attention that, as Harriot showed, the Moon's elongation can be measured even when the Sun is below the horizon. But now we are going to make a seemingly reasonable assumption about Aristarchus: namely, that he measured the mentioned angle when both luminaries (Sun and Moon) were above the horizon. Such was the explicit assumption of the next character in our story.

3.5 Arthur Allen Hoag

Unconvinced by a text in which Neugebauer (1975:642) says that Aristarchus' method is 'totally impracticable' and his 87-degree angle is 'purely fictitious' (because the actual angle is too close to 90 degrees and 'must therefore elude direct determination by methods available to ancient observers'), the American astronomer Arthur Allen Hoag felt compelled to try measuring himself what Aristarchus might have measured. He 'guessed that direct measures of the angle between the Sun and Moon, around times of quarter Moon, and at times when the Sun and Moon were at equal altitudes in the daylight sky, would be best' (Hoag, 1989:12). He also guessed that there are many ways to measure such an angle using everyday objects. For example, using an eight-foot vertical pole as a gnomon, the antenna of a radio receiver, a steel tape, and a compact mirror (for viewing the Moon over the top of the pole), he measured the Moon's elongation

Table 3.4: Data collected by Hoag during the first half-Moon of August 1988, as observed from Tucson (Arizona). Here, x is the time (in days from August 1), ε is the Moon's elongation, and ϕ is the Moon's illuminated fraction. (The first two columns are by the present author.)

date	time	x	ε	ϕ
1988-08-03	15:48:00	3.6583	104.3	–
	15:54:00	3.6625	104.2	–
	16:00:00	3.6667	–	0.62
	16:06:00	3.6708	104.1	–
	16:16:00	3.6778	104.2	–
1988-08-05	17:03:00	5.7104	78.8	0.42
	17:05:00	5.7118	78.8	–
	17:12:00	5.7167	78.8	0.40
	17:21:00	5.7229	78.6	–
	17:35:30	5.7330	78.6	–
1988-08-06	16:42:00	6.6958	66.6	0.32
	16:49:00	6.7007	66.6	–
	17:09:00	6.7146	66.4	0.34
	17:34:00	6.7319	66.4	–
	17:42:00	6.7375	66.2	–

from the Sun during some of the days surrounding the first half-Moon in August 1988. His readings are shown in Table 3.4 below (where the first two columns have been derived and added by the present author).

Hoag (1989:16) took five measurements in quick succession every day for three days around his chosen half-Moon. Then, for each day, he averaged three things: namely, the times x of observation (which he counted in days from August 1, 1988), the Moon's elongation ε (which he measured in degrees), and the Moon's illuminated fraction ϕ (which he simply 'guesstimated'). Then, he used the sums of these averages to obtain the typical (least squares) regression lines for these statistical data: one for ϕ and one for ε. (These are reworked in Equations 3.11 and 3.12 below to slightly better accuracy.) The averages and sums in his paper are slightly improved in Tables 3.5 and 3.6 below. In fact, columns 'date' and 'time' (in Table 3.4), which have been deduced from Hoag's column x, show his timings were taken to an accuracy of half a minute, which in turn, allows us to reconstruct column x (and the whole batch of his calculations) to any degree

90 Measuring the Sun's distance

Table 3.5: Statistical data derived from Hoag's half-Moon observation of August 1988, where x is the day of the month and y is the Moon's illuminated fraction.

	x	y	xy	x^2
Aug 3	3.66722	0.62	2.2736764	13.4485025284
Aug 5	5.71896	0.41	2.3447736	32.7065034816
Aug 6	6.71610	0.33	2.2163130	45.10599921
Total	16.10228	1.36	6.834763	91.26100522

Table 3.6: Statistical data derived from Hoag's half-Moon observation of August 1988, where x is the day of the month and y is the Moon's elongation (in degrees).

	x	y	xy	x^2
Aug 3	3.66722	104.20	382.1243240	13.44850253
Aug 5	5.71896	78.72	450.1965312	32.70650348
Aug 6	6.71610	66.42	446.0833620	45.10599921
Total	16.10228	249.34	1278.4042172	91.26100522

of precision. But there is no need to do so, because even rounding column x to whole numbers makes no significant difference. What really makes a difference, however, is the fact that (unlike those in Tables 3.5 and 3.6 below) the sums in Hoag's paper do not exactly match his reported data, leading him to a solar distance that is significantly different from the one that strictly follows from his own data, as we shall see.

According to Hoag (1989:12), the use of regression lines is justified 'because rates of change in elongation and in fractional illumination of the Moon are quite uniform near the quarter phases'.

Now, using the standard (least squares) regression line equation,

$$y = a + bx, \tag{3.9}$$

where

$$a = \frac{(\sum y)(\sum x^2) - (\sum x)(\sum xy)}{n(\sum x^2) - (\sum x)^2} \quad \text{and} \quad b = \frac{n(\sum xy) - (\sum x)(\sum y)}{n(\sum x^2) - (\sum x)^2}, \tag{3.10}$$

we have, for the data in Table 3.5 (where $n = 3$ is the number of days involved),

$$a = \frac{1.36 \times 91.26100522 - 16.10228 \times 6.834763}{3 \times 91.26100522 - 16.10228^2} = 0.969661\ldots$$

and

$$b = \frac{3 \times 6.834763 - 16.10228 \times 1.36}{3 \times 91.26100522 - 16.10228^2} = -0.096196...,$$

and for the data in Table 3.6 (where n is as before),

$$a = \frac{249.34 \times 91.26100522 - 16.10228 \times 1278.4042172}{3 \times 91.26100522 - 16.10228^2} = 149.645315...$$

and

$$b = \frac{3 \times 1278.4042172 - 16.10228 \times 249.34}{3 \times 91.26100522 - 16.10228^2} = -12.395508...$$

Putting these values into Equation 3.9, we find that the illuminated fraction of the Moon (on this particular occasion) is approximated by

$$\phi \approx 0.96966 - 0.096196\, t, \tag{3.11}$$

and the Moon's elongation is approximated (in degrees) by

$$\varepsilon \approx 149.645 - 12.3955\, t, \tag{3.12}$$

where t is the time (in days, as before). Now, from Equation 3.11, we deduce that the time of **dichotomy** (that is, when $\phi = 0.5$) is

$$t \approx \frac{0.96966 - 0.5}{0.096196} = 4.8823... \tag{3.13}$$

That is, *orthotomy* is estimated to have occurred on August 4, at about 21:11 (which is ahead of the truth by just about 48 minutes). Putting this value into Equation 3.12, we have

$$\varepsilon \approx 149.645 - 12.3955 \times 4.8823 = 89.126... \tag{3.14}$$

That is, the Moon's elongation at the estimated orthotomy time is about 89.126 degrees. Substituting this lunar elongation into Equation 2.15 puts the Sun

$$\sec(89.126°) \approx 66 \tag{3.15}$$

times as far away as the Moon. Hoag, however, obtains a lunar elongation of 89.4 degrees, which puts the Sun

$$\sec(89.4°) \approx 95 \tag{3.16}$$

times as far away as the Moon. The data he reports, however, do not strictly yield the solar distance he obtains in Equation 3.16, but rather that in Equation 3.15.

Hoag did not observe just one orthotomy. In all, he reports having observed two First Quarter and three Last Quarter Moons, obtaining an overall lunar elongation of 89.9° ± 0.6° (which puts the Sun no less than 82 times as far away as the Moon) and a mean lunar elongation of 89.852° (which puts the Sun a mean of 387 times as far away as the Moon). However, Equations 3.15 and 3.16 suggest his report should be taken with a pinch of salt. Thus, assuming his other observations are overdone by as much as the one in his paper, a simple rule of three shows that the actual solar distance that follows from his five observations is about $\sec(89.126°) \times \sec(89.852°) / \sec(89.4°) \approx 266$ times the lunar distance. This is indeed much better than the distance ratio of the Sun and Moon in Aristarchus' first book *On Sizes* (Conclusion 1) and slightly better than Wendelin's best estimate.

Successful though Hoag's experiment was, however, it does not prove Neugebauer wrong in his assumptions about Aristarchus, since, after all, Hoag's auxiliary apparatus was confessedly 'non-Aristarchan'. The statistical techniques used above, for example, were developed during the 19th century and are therefore unlikely to have been available to Aristarchus. Neither do Harriot's nor Wendelin's experiments prove Neugebauer wrong, since they used telescopes, which were not available to Aristarchus either. However, a surprising pattern is starting to emerge. Namely, that anyone who tries gets an answer closer to the truth than Aristarchus got in his first book *On Sizes*. As we shall see, this is no accident. What we now need is a method that does not use telescopes or anachronistic mathematical tools or techniques—a method Aristarchus might have truly used, and this is what the next characters in our story set to achieve.

3.6 A lucky conjunction

The present author would have achieved very little, indeed, had he not been approached (by the end of 2011) by the American psychiatrist and historian of astronomy William Patrick Sheehan, through whom he was exposed to a whole new world of astronomical discovery. It is through him, for example, that the author met the American geography professor John Edward Westfall, then director of the Association of Lunar and Planetary Observers (ALPO), and a host of other astronomy enthusiasts who brainstormed possible solutions to the puzzle Sheehan worded as follows (in a mail dated January 14, 2012), 'How might Aristarchus have gone about determining his angles? How did he come up with the 87-degree figure?'

One of the first solutions proposed echoed Evans' (1998:72) theory that Aristarchus measured nothing, but simply 'made up' the value in Hypothesis 4 of *On Sizes*, basing his choice on the following two assumptions: that the month lasts 30 days and that the span of time from last to first dichotomy is 14½ days. In this way, each dichotomy occurs 7¼ days from New Moon (or 7¾ days from Full Moon), and each quadrature occurs exactly 7½ days from either syzygy (New Moon or Full Moon). So, by a rule of three, 7½ days are to 90 degrees what 7¼ days are to 90 × 7.25/7.5 = 87 degrees.

However, this explanation makes too many unjustified assumptions about Aristarchus. First of all, the historical evidence so far (as summarized in the box at the end of Section 1.4) portrays Aristarchus as a committed empiricist, a keen observer of nature (like his tutor Strato), rather than someone who (as Evans proposes) was an indoor theorist. Second, Aristarchus did pay attention to detail, as we know from our mathematical analysis (in Chapter 2) of his book *On Sizes*, so he is unlikely to have rounded up the length of the synodic month (which was known to be about 29½ days since at least the time of Philolaus), knowing the mathematical consequences that such a move would have. Finally, the span of time from last to first dichotomies seems arbitrarily chosen to yield the desired result. No explanation for the choice of this fraction is given other than it does lead to 87 degrees. But this is a circular argument, so it explains nothing at all in the end.[19] So, Evans' explanation was dropped: we were not convinced.

The road was now paved for alternative explanations, and, as it happened, Westfall started to become ever more interested in the Aristarchus problem, as it was somewhat reminiscent of another problem which had long caught the attention of the Venus Section at ALPO: that of estimating the time of dichotomy of Venus. Soon, Westfall saw that the Aristarchus problem combined several of his life-long passions: the study of eclipses, the calculation of distances in the solar system, and the study of the lunar terminator (of which he had already published an Atlas in 2000). The sheer amount of correspondence that we exchanged over the following years was just amazing. We discussed every aspect of the problem we could think of to the best of our abilities: Sheehan's expertise in

---

19 According to Meeus (2007:12), the time between successive phases of the Moon (such as a New Moon and the First Quarter that follows it) can vary by as much as 1⅗ days, and the time between a quadrature and the New Moon closest to it is longer than the time between this quadrature and the Full Moon closest to it by a mean of 10 minutes. Also, dichotomy precedes First Quarter and follows Last Quarter by a mean of about 18 minutes (Meeus, 2009:22). Hence, dichotomy occurs a mean of about 18 − 10 = 8 minutes (rather than Evans' 6 hours) closer to New Moon than to Full Moon.

the psychology of perception proved crucial to explain one of the main objections that had kept humanity from accepting heliocentrism for centuries (as we shall see in Section 5.3); Westfall's immense creativity, enthusiasm, and wisdom kept luring our little team to ever-more fascinating horizons; and the author, being good at nothing but perhaps seeing what others do not—a skill that hopefully has nothing to do with any mental disorder, was only too happy to ride on the shoulders of his giant friends.

Among the many things we discussed was whether Aristarchus knew how to average his presumed measurements. We were aware that some authors (like Geus, 2014:147, and more recently, Raper, 2017:14) had argued that averaging could not have been used so early in history, but others (like Bakker, 2003:3) presented a number of compelling cases in which averages were actually used well before Aristarchus, such as the famous quote in Aristotle's *Nicomachean Ethics* (2.6.6),

> **Passage 48** Let 10 be many and 2 few; then one takes the mean with respect to the thing if one takes 6 (Aristotle, 1956:91).

So averaging was allowed into the equation. Even so (that is, even assuming he had such a mathematical tool at his disposal), Aristarchus might have taken just a single measurement, in which case there would be nothing to average.

Westfall started his observations of lunar dichotomies almost at once, the first one on March 31, 2012. In time, we thought up methods and tools Aristarchus might have used, and both Westfall and the author came up with two different, yet complementary ways to proceed, a comparison of which is as follows.

- Westfall observed many dichotomies (which in the course of the years added up to a total of fourteen morning dichotomies and seventeen evening ones); while the author observed just a few of them (two morning dichotomies and three evening ones in all).
- On each day of observation, Westfall measured the Moon's elongation many times (every half an hour or so), then guessed the time of dichotomy and interpolated the elongation corresponding to it; the author also took many measurements (every half an hour or so), but made no attempt to merge them into a single result.
- Westfall measured (in degrees) the Moon's elongation first, and then deduced the solar distance from it (via Equation 2.15); while the author did not measure any angle at all, but two lengths, the ratio of which mimicked that between the Sun's and Moon's distances from Earth.
- Westfall (like Hoag) discarded all observations greater than 90 degrees, because 'solar distances based on them are mathematically meaningless' (as

he argued in a mail dated July 17, 2018); however, the author never came across this problem, because he used lengths, which are never negative (regardless of the elongation).
- Westfall aligned his instrument Sunwise and then measured the Moon's deviation from the normal; while the author aligned his instrument Moonwise and then measured the Sun's deviation from the normal.
- Westfall's aim was to get the best measurement he could out of the half-Moon method; while the author's aim was to get into the mind of Aristarchus and glimpse at the kind of problems he might have faced and solutions he might have tried.

Do not worry, dear reader, if at first these explanations seem a bit cryptic. They will become clear the moment the instruments we devised are described and their use explained. The fact that we came up with so apparently different, yet complementary methods was due to our agreement that we should develop them more or less independently of each other (that is, in secret), so as to avoid bias.

Curiously, our instruments were completed and ready for use almost simultaneously, the author's preceding Westfall's by just a few days: the former being first used on March 15, 2016 (from Santander, Spain, the author's location at the time); the latter, on March 31, 2016 (from Antioch, California). Previous to using his instrument, Westfall had computed (rather than measured) the elongations he observed.

As for the type of instrument we are about to describe, the author proposed (in a mail dated January 18, 2016) to call it a *dichotometer*. Westfall liked this name and started using it in its shortened form *dichometer*. But after using it for the first time, the author realized this name is inappropriate, because the instrument gives no clue as to when *dichotomy* occurs. If anything, it gives a rough approximation as to when *quadrature* occurs, so he immediately discouraged the use of the latter name. However, Westfall continued to use it (despite the author's warnings). This is a consequence of the differences between our two approaches: Westfall believed his instrument could yield the time of dichotomy (and hence, the Sun's distance) to some degree of confidence, while the author wanted to get into the mind of someone who had always been told that the Sun was at most twelve times as far away as the Moon and just wanted to check whether this is so. Thus, the author's proposal that the instrument be renamed a *quad gauge* (or *quad box*, or even *minicosmos*) met with little success.

3.6.1 The Aristarchus Procedure Series

Our instruments were not made in one go, but developed over time until they reached their final form. At first (by January 21, 2012), the author designed a large triangle with three rods in it forming a right-angled triangle. It was never constructed, but its use was intended to be similar to Tycho's *Semicirculus amplus pro maioribus distantiis coelitus denotandis* ('Great semicircle for the determination of major angular distances in the sky'), or its smaller version, the *Sextans astronomicus trigonicus pro distantiis rimandis* ('triangular astronomical sextant for the determination of distances'), both described in his *Astronomiæ Instauratæ Mechanica* (Raeder et al., 1946:67, 89). With time, this instrument came to be called *Aristarchus Procedure* 1 (or P1, for short). (See Figure 3.1 below.)

This is an interesting instrument that can be used in many ways, illustrating how both the author's and Westfall's methods (or any combination of them) work. For example, it can be arranged such that rods B and C are aligned with the Moon (that is, it can be Moon-aligned), so that all there is to measure is the Sun's deviation from orthogonality (as given by the gap between the shadows cast by rods A and B); or it can be arranged so that rods A and B are aligned with the Sun (that is, it can be Sun-aligned), so that all there is to measure is the Moon's deviation from orthogonality (as given by the gap between the Moon and rod C). These deviations can be measured either (in degrees or whatever) using a protractor, and then, the distance ratio of the Sun and Moon can be found using trigonometry (via Equation 2.15), or the said ratio can be found directly by dividing the distance AB (between rods A and B) by the width of the gap between the shadows cast by these rods.

At quadrature, or rather at **topocentric quadrature** (that is, when the Sun and Moon are at right angles as seen from the observer's location on Earth), the

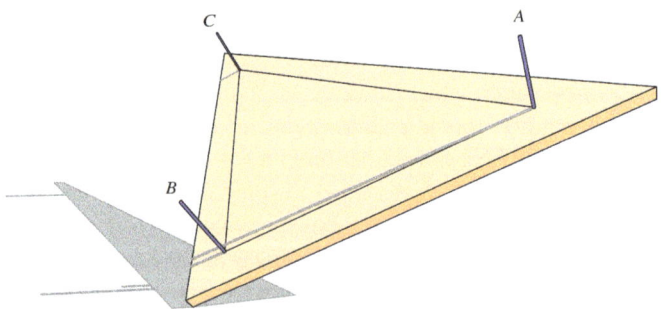

Figure 3.1 Aristarchus Procedure 1

instrument in Figure 3.1 is both Moon-aligned and Sun-aligned at the same time. That is, the shadow cast by rod *A* will exactly coincide with that of rod *B*, and rods *B* and *C* will both be in a straight line with the centre of the Moon. This is why the author said above that our instruments could give a rough approximation to the time of *topocentric quadrature*, but there is no way Aristarchus (or any naked-eye observer using this instrument alone) could know when *dichotomy* occurs other than saying that it must occur some time between quadrature and the time when the Moon no longer looks halved to us.

An important principle guiding our work was safety: we were all aware that looking straight into the Sun can cause permanent damage, so we decided that none of our instruments should involve looking directly into the Sun (and assumed Aristarchus would have risked neither his own eyes nor those of his followers). This is why the rods in Figure 3.1 can be used to look straight into the Moon alone, while their shadows indicate the Sun's position. These rods can be replaced by pinnules, or have holes bored in them to let the Sun shine through both of them for better accuracy.

The P1 design is important, because all other instruments are somehow contained in or derived from it. Thus, the next development came on March 21, 2013, when the author developed his *Aristarchus Procedure* 2 (or P2), which was basically a box with a hole in each of its four opposing sides. Letting the Sun shine through two of these holes ensured the box was Sun-aligned. If at the same time, the Moon could be seen through the other two holes, then the instrument would indicate the time of *quadrature*. At *orthotomy*, when the Moon was judged to be perfectly halved, there was a gap between the little hole at the bottom of the box and the sunbeam cast through the other hole. Measuring this gap gives an approximation to the Sun's distance (as we have already explained). (See Figure 3.2.)

This procedure, however, was never tested. It remained a project and was temporarily abandoned, because by then, the author's attention had turned to the possibility of including mirrors in the instrument. (Always ready to try everything he could think of, the author was under the impression that a mirror would increase the gap to be measured.) The result was *Aristarchus Procedure* 3 (or P3), which was basically the same as P2, but with a mirror on the lower side of the box, so as to reflect the sunbeam back to the top side. The gap between the hole it went in and the spot where it was reflected back was expected to be twice the required gap. This instrument was actually constructed, but took the shape of a cross, rather than a box. (See Figure 3.3.)

P3 was tried on the morning dichotomy of August 17, 2014, with disappointing results. In fact, the instrument yielded poorer lunar elongations than

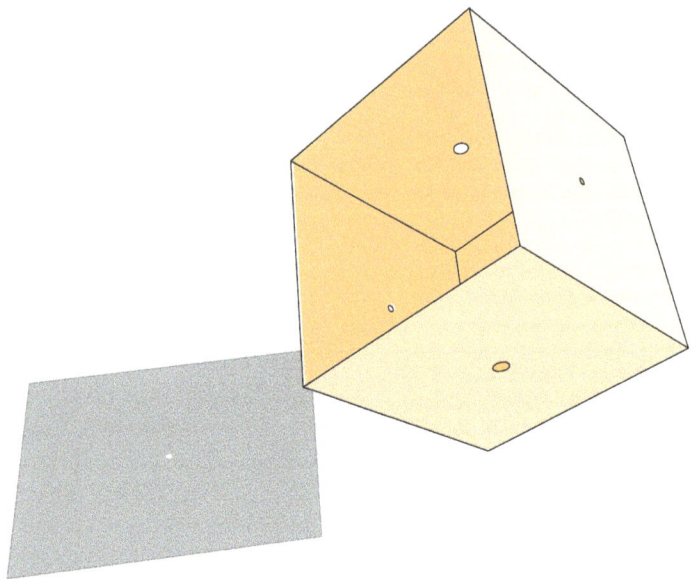

Figure 3.2 Aristarchus Procedure 2

Aristarchus' 87 degrees despite the measurements having been taken at the exact time of dichotomy (which had been precomputed), but there were lessons to be learned from this failure.

At the time, the author was lucky to be corresponding with the German physicist Kurt Guckelsberger over issues concerning the measurement of mountain heights in antiquity and the possible use of mirrors for this purpose. Though mirrors existed in Aristarchus' time—for instance, the use of one is described in Euclid's *Optics* (Burton, 1945:360) and a whole book on *Catoptrics* is attributed to the latter man, Guckelsberger warned the author against them. There were issues concerning their *placement*, *flatness*, and *reflectivity*. For example, the mirror must be perfectly parallel to the tube (since any deviation can send the reflected beam arbitrarily off the mark), perfectly flat (since imperfections can easily distort our readings), and (to be historically accurate) made of either *speculum metal* or polished bronze. These metals are known to reflect at most two thirds of the light hitting them. *Speculum* also has the undesirable property of rusting in the open air.

After such comments, the author attributed the apparent failure of his 2014 observation to the use of the mirror. The next move was to dispose of it. Why

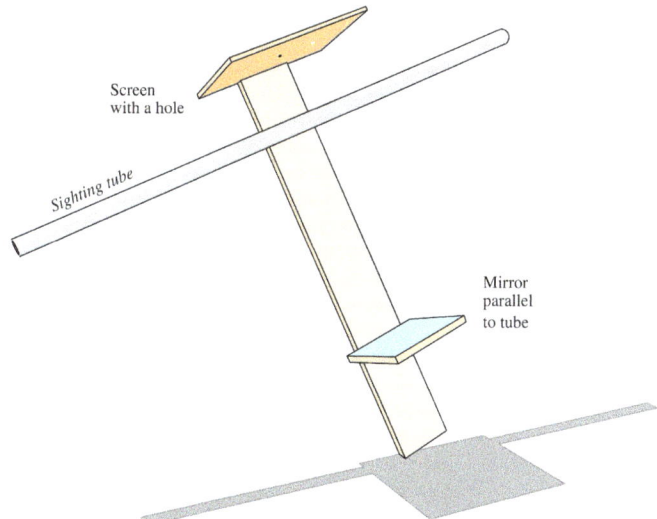

Figure 3.3 Aristarchus Procedure 3

should one be used in the first place? Can the gap not be measured directly? This prompted a return to the P2 idea, or a slight modification of the P3 one: namely, replacing the mirror by a flat surface on which the spot the sunbeam hits at quadrature must be marked, and then also the spot the sunbeam hits at the estimated time of orthotomy. The gap between these points is what we need to measure. Thus (on January 18, 2016) was born the *Aristarchus Procedure 4* (or P4), which includes the mentioned changes to the P3 idea (Figure 3.4), and also (on January 31, 2016), the *Aristarchus Procedure 5* (or P5), which (for the time being) was the final step in the *Aristarchus Procedure Series*.[20]

On trying the P4, the author noticed that too many things had to be aligned for the instrument to work, and realized that things could be greatly simplified if the point he had drawn on the bottom screen was replaced by a line drawn along this screen and normal to the sighting tube. He called this line the **quadrature line** (or **quad line**). (See Figure 3.5.)

The use of this instrument is as follows. One person aligns the box with the Moon by sighting it through the holes at *B* and *C*. Then, the Sun shines through

---

20 Interactive online 3D models of these designs are available at **3D Warehouse > Aristarchus Procedure > models**.

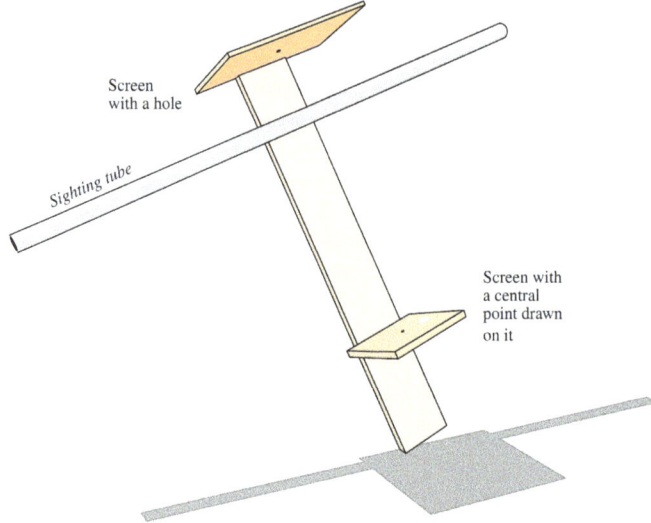

Figure 3.4 Aristarchus Procedure 4

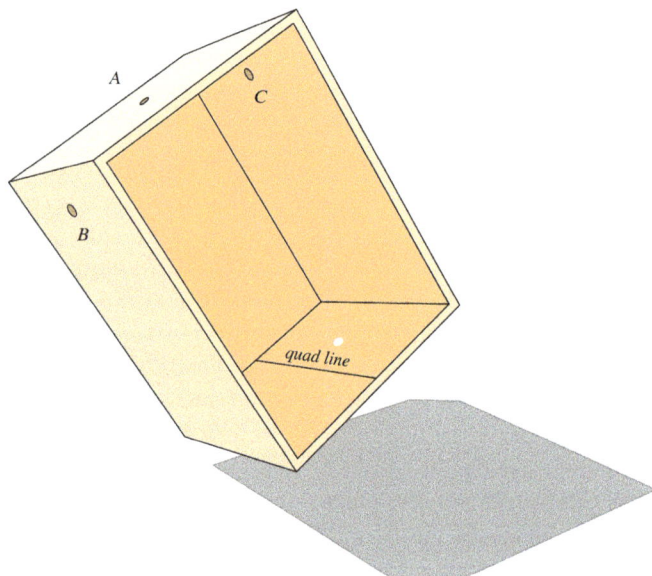

Figure 3.5 Aristarchus Procedure 5

the hole at A and hits the bottom of the box. At the same time, a second observer makes a mark at the spot where the sunbeam meets the bottom of the box. The (shortest) distance from this mark to the *quad line* must then be measured. (We can take either a single measurement or as many as we wish.) Then, we divide the height of the box by the distance (or average of the distances) we have measured. This gives an approximation to the Sun's distance in terms of that of the Moon.

Strictly speaking, these distances (namely, the height of the box and the orthotomy-to-quadrature gap), which we denote here by the letters h and m, respectively, are the sides of a right-angled triangle whose hypotenuse represents the distance between the Earth and the Sun. Hence, by Pythagoras' Theorem, the distance ratio of the Sun and Moon is given by

$$\frac{S}{M} = \frac{\sqrt{h^2 + m^2}}{m}. \tag{3.17}$$

But, in practice, m is so small compared to h that the approximation

$$\frac{S}{M} \approx \frac{h}{m} \tag{3.18}$$

can be used as a quick way to obtain the required ratio. This involves just a simple division, which is child's play.

The P5 method has the disadvantage that, sometimes, it can be difficult to sight the Moon through both holes, but, apart from this, it has the advantage that it can be used as a teaching aid to clearly visualize Aristarchus' triangle (that is, the Sun-Earth-Moon system at the time of orthogony). In fact, a *quad box* is a tiny 'minicosmos' that allows us to see the vast distances out there at a smaller, more manageable scale. Within this box, you can actually see the Sun, Earth, and Moon in real time, if you know where to look. (See Figure 3.6.)

Another disadvantage (if it can be called so) of using a *quad box* is that users often do not pay attention to what side of the *quad line* the sunbeam entering the box lands on. All they need is the gap between this line and the bright spot where the sunbeam meets the bottom of the box. Once they measure this gap, they see how many times it goes into the height of the box to find out how many times the Sun is farther away than the Moon, and that is it. So, the question of whether the Moon is past quadrature or not does not even arise. But, as we have seen, most astronomers, such as Hoag (1989:10) and Westfall (2019:46), are often fussy about this issue and dismiss measurements taken when the Moon's elongation is greater than 90 degrees. According to them, it makes no geometric sense.

However, users of the *minicosmos* method are completely unconcerned about, and even oblivious of this issue. They do not work with angles, but with gaps, and

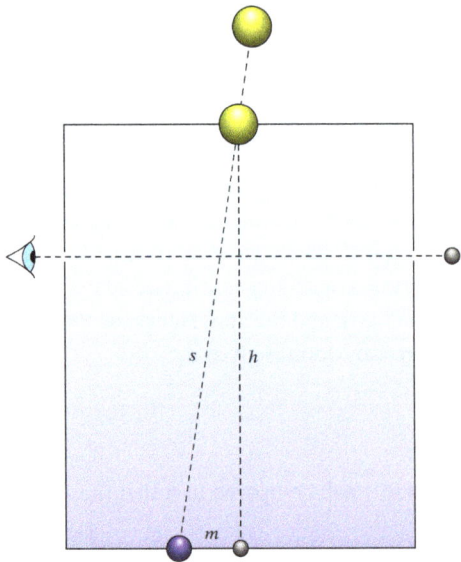

Figure 3.6 A *quad box* (or *cosmometer*) is a 'minicosmos' where (at lunar orthogony) the distances in the box map those in outer space.

as said above, gaps are never negative, so dividing positive numbers by positive numbers always gives positive numbers, no matter what. On the other hand, the distance between the Earth and the Moon does not change much over the course of a half-Moon's life, so measuring it a little before or a little after quadrature should both be possible and yield similar results.

The *minicosmos* method shows that it is perfectly possible to measure (however roughly) the Sun's distance when the Moon's elongation is greater than 90 degrees, and this is just one way around it. Another way, perhaps more appealing to those fond of trigonometry, is to take the modulus of Equation 2.15, which gives positive answers even when obtuse angles are fed into it. (This modulus is both the distance to the Sun's centre and the distance to the antisolar point, since both points are equidistant from the Earth.) There may be more ways around this problem than those proposed by the present author, and it is up to the reader to find them out. Mathematicians are used to find ways around apparently impossible problems and find solutions in the most unbelievable and unorthodox ways. They stop at nothing and are actually trained to think out of the box.

As for Aristarchus, we may now ask, Did he think out of the box? Certainly, he did. But, what about this particular issue? Would he have discarded obtuse angles? We have always been told that he would, but a close reading of his book reveals the answer is not so straightforward. In fact, the moment he prescribes (in Hypothesis 3 and Proposition 5), the one we called *orthogony* (on page 51), turns out to be one when the Moon's elongation is 90° + 3° = 93°, rather than 90° − 3° = 87° (as discussed in Section 2.3). It is possible that the young Aristarchus overlooked this little detail in his first book *On Sizes* (as in fact, those using the *minicosmos* method, often unwittingly do).

The expression in Hypothesis 4 of *On Sizes*, 'When the Moon appears to us halved, its distance from the Sun is less than a quadrant by one thirtieth of a quadrant', is unambiguous. Aristarchus' explanations in Proposition 7 (Heath, 1913:377) confirm that he really meant 87 (not 93) degrees. However, his explanations in Propositions 5 and 6 (Heath, 1913:370) unambiguously use 93 (not 87) degrees. So it seems that the young Aristarchus has made some kind of mistake here—one so subtle that has gone unnoticed to this very day.

Yet, we are fortunate that he made this 'mistake', because it points to his having measured lengths, rather than angles, possibly using an instrument similar (if not identical) to the above-mentioned *minicosmos*—an instrument somewhat reminiscent of the diagram in Aristarchus' Proposition 7 (Heath, 1913:378). However, as with everything historical, we simply cannot be sure (until we learn to travel in time).

3.6.2 The author's observations

The morning half-Moon of August 17, 2014, as seen from Nestares, Spain

As mentioned above, the author's morning half-Moon observation (that of August 17, 2014) yielded results that were slightly worse than those in Aristarchus' first book *On Sizes*. The instrument used was a P3 design (as illustrated in Figure 3.3), with a height h (from screen to mirror) of 1 m. The reflected sunbeam landed back on the screen a distance x of about 12 ± 1 cm from the hole it had come in. With these readings, the Moon's elongation was found to be

$$\varepsilon = \tan^{-1}\left(\frac{2h}{x}\right) \qquad (3.19)$$
$$= \tan^{-1}\left(\frac{2 \times (100 \text{ cm})}{(12 \pm 1 \text{ cm})}\right) \approx 1.5109 \pm 0.0005 \text{ radians} \approx 86°34' \pm 17',$$

and the Sun-to-Moon distance ratio was found (by Equation 2.15) to be

$$\frac{S}{M} = \sec \varepsilon = \sec\left(\tan^{-1}\left(\frac{2 \times (100 \text{ cm})}{(12 \pm 1 \text{ cm})}\right)\right) \approx 16.7 \pm 1.3. \qquad (3.20)$$

Note that, at this stage, the author (like Harriot, Wendelin, Hoag, and Westfall) was still measuring the Moon's elongation first, and then the Sun's distance (in that order). He had not yet realized that the same result can be found more simply either by Equation 3.17 or Approximation 3.18. Thus, using the latter, we have

$$\frac{S}{M} \approx \frac{2h}{x} = \frac{2 \times (100 \text{ cm})}{(12 \pm 1 \text{ cm})} \approx 16.7 \pm 1.3. \tag{3.21}$$

The author was dismayed at these results: they were worse than Aristarchus' even though taken at the exact time of orthogony, when they were expected to be optimal. If taken at another time, they might have been even worse! Something had gone wrong.

After being warned against the use of mirrors by the German professor of Physics and Astrophysics Kurt Guckelsberger, the author decided to drop the P3 method and use mirrors no more. This proved to be a good decision, as we shall see.[21]

The evening half-Moon of March 15, 2016, as seen from Santander, Spain

This observation proved to be spectacularly successful in comparison with the previous one, differing from it in many ways. First of all, the author had carefully avoided learning the time of either *orthogony* or *quadrature* before the event: Aristarchus would have known none of this and neither should the author. The method employed in this observation was P5 (as illustrated in Figure 3.5), and the box employed was a makeshift, cardboard one of height $h = 71$ cm.

Starting at about 15:30 hours UT (when he came back from work) and ending two hours later (nearly an hour before the Sun went down), the author, assisted by his father, took measurements every half an hour or so. Each of these consisted of four alignments of the box and readings (in millimetres) of the gap between orthogony and quadrature (as seen on the bottom of the box). These readings x_i (for $i = 1, 2, 3, 4$) are recorded in Table 3.7 below, along with their mean $\bar{x}$ and median m (which, according to statistical rules, are given to one more figure than the original data). (A cross indicates a failed sighting.)[22]

---

21 An interactive illustration of this observation is available online at 'Measuring the Sun's distance by the P3 method'.
22 Note that the five o'clock readings are all missing. This is because the author's dad's favourite TV show was just then on, and the author decided to spare him the trouble of assisting him just this once.

Table 3.7: Data collected (by the author and his father) during the half-Moon observed from Santander, Spain, on March 15, 2016, using a P5 box of height $h = 71$ cm. The mean $\bar{x}$ and median m of the readings x_i (in millimetres) allow the calculation of the ratios S/M and S/e (of the Sun's distance S from Earth to either that of the Moon M or the Earth's radius e) and the lunar elongation ε at the given times. (The last column was added at Westfall's request.)

UT	x_1	x_2	x_3	x_4	$\bar{x}$	m	S/M	S/e	ε	true ε
15:30	7	8	9	11	8.8	8.5	84	5012	89°19′	89°39′
16:00	5	5	6	8	6.0	5.5	129	7746	89°33′	89°51′
16:30	2	4	6	9	5.3	5.0	142	8520	89°36′	90°02′
17:00	×	×	×	×	×	×	×	×	×	90°13′
17:30	0	1	2	2	1.3	1.5	473	28400	89°53′	90°23′

In Table 3.7, the Moon's elongation ε was calculated using the equation

$$\varepsilon = \tan^{-1}\left(\frac{h}{m}\right), \qquad (3.22)$$

where h is the height of the box and m is the median of the readings taken at the specified time. Thus, for example, for the reading at 15:30 hours, we have

$$\varepsilon = \tan^{-1}\left(\frac{h}{m}\right) = \tan^{-1}\left(\frac{710\,\text{mm}}{8.5\,\text{mm}}\right) \approx 1.5588 \text{ radians} \approx 89°19'. \qquad (3.23)$$

Also, the ratio of the Sun's distance from Earth S to that of the Moon M was calculated using Equation 3.17 (or Approximation 3.18), where h and m are as before.

$$\frac{S}{M} = \frac{\sqrt{h^2 + m^2}}{m} = \frac{\sqrt{(710\,\text{mm})^2 + (8.5\,\text{mm})^2}}{(8.5\,\text{mm})} \approx 84. \qquad (3.24)$$

Finally, the ninth column in Table 3.7 gives the Sun's distance (in Earth radii). This was calculated assuming $M = 60$ Earth radii. Thus, for the reading at 15:30 hours, we have

$$S = \frac{M\sqrt{h^2 + m^2}}{m} = \frac{(60\,\text{e.r.})\sqrt{(710\,\text{mm})^2 + (8.5\,\text{mm})^2}}{(8.5\,\text{mm})} \approx 5012 \text{ e.r.}. \qquad (3.25)$$

This time, the experiment was deemed a success (by the author). It bettered Aristarchus' 87 degrees beyond expectations, and the solar distances obtained were now in the range of Archimedes' reported 'myriad' (or 10,000) Earth radii, which immediately provided food for thought. Looking closely, we see that the observed elongations are always less than the true (computed) ones by a mean (and also median) of about 23′ ± 6′, suggesting that the box may not have been properly calibrated, as it seemed to systematically reduce the data by the said amount.

The important thing to note is that a poorly calibrated, makeshift, cardboard box is capable of estimating the Sun's distance to an accuracy far surpassing that in Aristarchus' first book *On Sizes*, and even reaching numbers that read in terms of thousands of Earth radii.

Curiously, and this was the first time that the issue was brought to the author's attention, Westfall complained (in a mail dated August 23, 2016) that this measurement was invalid, because most of the readings had been taken when the Moon's (true) elongation was over 90 degrees. In his words, 'When computed values are used, Al's 2016-03-15 sighting gives a negative deficiency [from quadrature], so we really can't determine the ratio of the Sun's distance from Earth to that of the Moon'.

However, this was not the way ancient astronomers would have seen things. To start with, they would not have had access to the last column in Table 3.7 (that is, no way of knowing how their readings compared with reality). All they could say was that (according to their instrument) the time of quadrature was slowly approaching (as the elongation grew closer and closer to 90 degrees), and they would even have believed that quadrature occurred some time after the last reading. We now know (thanks to computers) that *topocentric quadrature* occurred at 16:25 hours and *topocentric orthogony* even earlier, at 16:02 hours (in agreement with Footnote 19), but there is no way an ancient observer could have known this. So the instrument's readings would have been regarded as perfectly valid.

Furthermore, and notice the difference between the author's and Westfall's approaches, ancient observers would have had no way of telling which of the solar distances in Table 3.7 was more correct. For all they knew, there was no reason to choose one over the others. While a modern astronomer might be tempted (or biased) to pick the one that is closer to the truth among the 'valid' elongations in the table, or perhaps average the only 'valid' ones there, ancient astronomers would have seen no reason to do so. In any case, they might have been tempted to pick the most conservative one (that is, the shortest distance). But what about Aristarchus? What would *he* have said? As we shall see, he would have given two different answers depending on when he was asked. One is as follows.

The young Aristarchus who wrote the first book *On Sizes* would have discarded all the solar distances in Table 3.7 because none of them were taken at the time of *orthogony* (as defined on page 51). That is, none of the measurements in Table 3.7 were taken at the time when the plane on which the lunar terminator rests is parallel to the plane exactly halving the Moon and going straight through the observer's eye (as explained in Hypothesis 3 and Proposition 5). Since the Sun is bigger than the Moon, the terminator never splits the Moon into two perfectly equal halves, but one is always slightly bigger than the other. So, at *orthotomy*, when the terminator points in the eye's direction, the Moon is at the turning point where a half-Moon becomes gibbous (as in the case here), or the other way round (a fortnight later).

Aristarchus would have taken his own reading(s) when he deemed appropriate (that is, at the time of *orthotomy*), many hours after the last takes in Table 3.7. By then, the Sun would long have gone below the horizon. So this particular half-Moon would not have been ideal in Aristarchus' view. Furthermore, by the time of *orthotomy*, the Moon's elongation would have become greater than 90 degrees, but his instrument would still produce a measurable gap, which he would then measure unaware that the sunbeam was now landing on the 'wrong' side of the *quad line*. In other words, his instrument and method would always map obtuse angles to acute ones, but he would not notice, because he would not use angles, just lengths.

The second answer he would have given is based on what we know of his second book *On Sizes* (which has not survived but was described by Archimedes in *The Sand Reckoner*). This new book by Aristarchus contained many new things: a new angular size for the Sun and Moon, a new solar distance, a new lunar distance, and a brand-new model of the universe. (We shall see more on these in Chapter 4.) This time, rather than discarding the solar distances in Table 3.7, Aristarchus would have taken a more modern stand and try to merge them into one, for example, by noting that the Pythagorean means of these distances (when measured in Earth radii) all round either up or down to a **myriad**, a conveniently ambiguous term whose meaning is both precise ('exactly 10,000') and imprecise ('a very large number').

To summarize, in his first book *On Sizes* (and contrary to what people might believe), Aristarchus was playing conservative and choosing a solar distance that was as small as it was mathematically tenable, justifying his choice on the ability of the human eye to determine the time of *orthotomy*. By then, the Moon's elongation is truly in the range of 93 degrees, but unwittingly, he interpreted this as 87 degrees because his theory demanded acute (rather than obtuse) elongations and also because his instrument facilitated this slip by measuring lengths (rather than angles).

In his second book *On Sizes*, however, Aristarchus must have realized that his previous solar distance was too small, possibly after pondering on the parallactic effects arising from the possibility that the Earth moved around the Sun: if the

Table 3.8: Data collected (by the author and his father) during the half-Moon observed from Santander, Spain, on May 13, 2016, using a P5 box of height $h = 131$ cm. The mean $\bar{x}$ and median m of the readings x_i (in millimetres) allow the calculation of the ratios S/M and S/e (of the Sun's distance S from Earth to either that of the Moon M or the Earth's radius e) and the lunar elongation ε at the given times. (The last column was added at Westfall's request.)

UT	x_1	x_2	x_3	x_4	x_5	x_6	$\bar{x}$	m	S/M	S/e	ε	true ε
15:15	−1.0	2.5	3.5	5.0	5.5	6.5	3.67	4.25	308	18494	89°49′	89°47′

planets are to show no detectable diurnal parallax, they must all (along with the Sun) be sent exponentially farther away than originally thought. Such was the theory—first developed by Rawlins (2008:24)—behind the 'myriad' word in *The Sand Reckoner*. Now it seemed to the author that his first P5 experiment shed a more empirical light on the origin of this enigmatic value. But further experiments were needed to check on such hypotheses. So he afforded himself a new box: this time, bigger, more carefully made, and hopefully better calibrated. All that remained was to wait for the right day and hope for good weather.

The evening half-Moon of May 13, 2016, as seen from Santander, Spain

The author knew that May 13, 2016, was one of those days when a half-Moon can be seen in broad daylight (that is, when the Sun and Moon are both above the horizon), but he had been careful not to learn the exact time of either orthogony or quadrature before the event so as not to introduce bias into the experiment. The day was mostly cloudy, but fortunately, the skies cleared at 15:15 hours UT for just long enough to give the author a brief chance. Thus, assisted by his father, he aligned his new box of height $h = 131$ cm and made a single observation consisting of six readings. These readings x_i (for $i = 1, 2, ..., 6$) are given (in millimetres) in Table 3.8 below, where all other variables ($\bar{x}, m, S/M, S, M, \varepsilon$) are as in Table 3.7.

The Moon's elongation in Table 3.8 was found (using Equation 3.22) to be

$$\varepsilon = \tan^{-1}\left(\frac{h}{m}\right) = \tan^{-1}\left(\frac{1310\,\text{mm}}{4.25\,\text{mm}}\right) \approx 1.5676 \text{ radians} \approx 89°49'. \tag{3.26}$$

The Sun's and Moon's distance ratio was found (using either Equation 3.17 or Approximation 3.18) to be

$$\frac{S}{M} = \frac{\sqrt{h^2 + m^2}}{m} \approx \frac{h}{m} = \frac{1310\,\text{mm}}{4.25\,\text{mm}} \approx 308. \tag{3.27}$$

The Sun's distance (in Earth radii) was found by putting $M = 60$ Earth radii into the previous equation. Thus,

$$S = \frac{M\sqrt{h^2 + m^2}}{m} \approx \frac{Mh}{m} = \frac{(60\,\text{e.r.}) \times (1310\,\text{mm})}{(4.25\,\text{mm})} \approx 18494\,\text{e.r.}. \qquad (3.28)$$

Not until the experiment was over did the author learn that *topocentric orthogony* had occurred at 15:25 hours and *topocentric quadrature* at 15:50 hours. Only then did he realize how lucky he had been that the skies cleared at just about the right time! The results were awesome! They would have astonished any ancient observer, though they were still somewhat short of the truth. The Sun was deemed to be about 308 times as far away as the Moon (versus 389, in reality). Had the mean $\bar{x}$ been used instead of the median m, the results would have been even better, but somehow, the author prefers the latter value for this kind of measurements, as it is often more reliable. He hopes the reader will forgive him this little anachronism, since the idea of the median was first proposed in 1599.[23] The Pythagorean means, on the other hand, were well known to the ancients, as reported by the first to second century AD mathematician and music theorist Nicomachus of Gerasa (1866:122).

The reader may note that one of the readings in Table 3.8 (namely, x_1) is negative (meaning that the sunbeam fell on the 'wrong' side of the *quad line*). Whether we choose to discard it, take its absolute value, or leave it as it is, the above statistics will not change significantly. Whatever an ancient astronomer would have done, the result of this experiment seemed to show again that the Sun was thousands of Earth radii away. In fact, this time, the result was even better than that found by Horrocks. Whether this was due to sheer good luck or the use of a bigger, more carefully made box still awaited confirmation by further experiments.

The evening half-Moon of August 10, 2016, as seen from Mave, Spain

The year 2016 still offered the author another opportunity to test his theories on August 10. Unfortunately, the skies were overcast in his hometown on this occasion, so he travelled south a number of miles in search of clearer skies. At the chosen location, Mave, he deployed his box and (aided by his sister) took six measurements, each consisting of seven sightings, resulting in a total of $6 \times 7 = 42$

---

23 The word 'median' was used for the first time in a book published in 1599 (namely, *Certaine Errors in Navigation*) by the English mathematician and cartographer Edward Wright. In it, he argued that this value is likely to be the more correct in a series of observations.

Table 3.9: Data collected (by the author and his sister) during the half-Moon observed from Mave, Spain, on August 10, 2016, using a P5 box of height $h = 131$ cm. The mean $\bar{x}$ and median m of the readings x_i (in millimetres) allow the calculation of the ratios S/M and S/e (of the Sun's distance S from Earth to either that of the Moon M or the Earth's radius e) and the lunar elongation ε at the given times. (The last column was added at Westfall's request.)

UT	x_1	x_2	x_3	x_4	x_5	x_6	x_7	$\bar{x}$	m	S/M	S/e	ε	true ε
14:40	27.0	29.5	30.0	30.5	32.0	33.0	33.5	30.79	30.50	43	2578	88°40′	89°01′
15:10	20.0	23.0	23.0	27.0	27.0	28.0	33.5	25.93	27.00	49	2912	88°49′	89°12′
15:40	24.5	25.0	26.5	27.0	27.5	29.0	32.0	27.36	27.00	49	2912	88°49′	89°22′
16:10	9.0	12.0	17.0	19.5	19.5	29.5	34.0	20.07	19.50	67	4031	89°09′	89°31′
16:40	13.5	14.5	16.0	18.5	20.0	21.5	22.0	18.00	18.50	71	4249	89°11′	89°40′
17:10	×	16.0	17.0	17.0	18.0	20.5	×	17.70	17.00	77	4624	89°15′	89°49′

readings in all (two of which were marred by clouds). These were all measured (in millimetres) using the same box and procedure as on the previous occasion. The results of this experiment are given in Table 3.9 below.

After the experiment, the author learned that *topocentric orthogony* occurred at 17:15 UT and *topocentric quadrature* at 17:46 UT. So the box correctly predicted that the latter event had not yet occurred (since the sunbeam kept approaching, but never crossed the *quad line*). Unexpectedly, this observation yielded poorer results than those of March, with solar distances measuring in thousands of Earth radii, but not close enough to the 'myriad' Earth radii reported in *The Sand Reckoner*. The box, though presumably better calibrated than that of March, had systematically underrated the true elongations by about $22' \pm 10'$, which is about the same as in March. So maybe the March box was not so poorly calibrated after all: maybe these discrepancies were caused by some other factor (or factors), such as atmospheric refraction (as Westfall suggested in a letter dated August 23, 2016). Indeed, the effect of the Earth's atmosphere on each of the author's three observations is as follows.

- On March 15, 2016, at 16:00 UT (that is, at the time of the measurement closest to topocentric orthogony), the Sun's true (unrefracted) altitude and azimuth (as seen from Santander) were, respectively, $\lambda_\odot = 23°52'50.2''$ and $\theta_\odot = 242°10'56.8''$, and those of the Moon were $\lambda_\mathcal{C} = 50°36'44.2''$ and $\theta_\mathcal{C} = 119°52'14.1''$. On feeding these data into the typical elongation formula (Meeus, 1998:115),

$$\varepsilon = 2\sin^{-1}\sqrt{\frac{1-\cos(\lambda_{\mathbb{C}}-\lambda_{\odot})}{2} + \frac{1-\cos(\theta_{\mathbb{C}}-\theta_{\odot})}{2}\cos\lambda_{\odot}\cos\lambda_{\mathbb{C}}}, \quad (3.29)$$

we find that the true elongation between the two luminaries was $\varepsilon_{\text{true}} \approx 89°50'37''$.

The apparent (refracted) altitude and azimuth of the Sun (for the same time and location) were, respectively, $\lambda_{\odot} = 23°55'04.3''$ and $\theta_{\odot} = 242°10'56.8''$, and those of the Moon were $\lambda_{\mathbb{C}} = 50°37'33.5''$ and $\theta_{\mathbb{C}} = 119°52'14.1''$. On feeding these data into Equation 3.29, we find that the apparent elongation between the Sun and Moon was $\varepsilon_{\text{app}} \approx 89°48'4''$.

So, on this occasion, the Earth's atmosphere brought the Sun and Moon closer to each other by about two and a half minutes of arc (since $\varepsilon_{\text{true}} - \varepsilon_{\text{app}} \approx 2'33''$).

- On May 13, 2016, at 15:15 UT (that is, at the time of the only measurement taken, which happened to occur close to topocentric orthogony), the Sun's true altitude and azimuth (as seen from Santander) were, respectively, $\lambda_{\odot} = 44°17'44.0''$ and $\theta_{\odot} = 251°52'19.8''$, and those of the Moon were $\lambda_{\mathbb{C}} = 37°48'58.4''$ and $\theta_{\mathbb{C}} = 113°12'48.9''$. On feeding these data into Equation 3.29, we find that the true elongation between the two luminaries was $\varepsilon_{\text{true}} \approx 89°47'23''$.

  The apparent altitude and azimuth of the Sun (for the same time and location) were, respectively, $\lambda_{\odot} = 23°33'38.4''$ and $\theta_{\odot} = 269°23'31.1''$, and those of the Moon were $\lambda_{\mathbb{C}} = 32°28'09.3''$ and $\theta_{\mathbb{C}} = 163°33'50.6''$. On feeding these data into Equation 3.29, we find that $\varepsilon_{\text{app}} \approx 89°45'22''$.

  So, on this occasion, the Earth's atmosphere brought the Sun and Moon closer to each other by about two minutes of arc (since $\varepsilon_{\text{true}} - \varepsilon_{\text{app}} \approx 2'1''$).

- On August 10, 2016, at 17:10 UT (that is, at the time of the measurement that was closest to topocentric orthogony), the Sun's true altitude and azimuth (as seen from Mave) were, respectively, $\lambda_{\odot} = 23°31'22.0''$ and $\theta_{\odot} = 269°23'31.1''$, and those of the Moon were $\lambda_{\mathbb{C}} = 32°26'35.5''$ and $\theta_{\mathbb{C}} = 163°33'50.6''$. On feeding these data into Equation 3.29, we find that the true elongation between the two luminaries was $\varepsilon_{\text{true}} \approx 89°49'29''$.

  The apparent altitude and azimuth of the Sun (for the same time and location) were, respectively, $\lambda_{\odot} = 44°18'45.5''$ and $\theta_{\odot} = 251°57'19.8''$, and those of the Moon were $\lambda_{\mathrm{C}} = 37°50'15.6''$ and $\theta_{\mathrm{C}} = 113°12'48.9''$. On feeding these data into Equation 3.29, we find that $\varepsilon_{\mathrm{app}} \approx 89°47'25''$.

So, on this occasion, the Earth's atmosphere brought the Sun and Moon closer to each other by about two arcminutes (since $\varepsilon_{\mathrm{true}} - \varepsilon_{\mathrm{app}} \approx 2'4''$).

But the Earth's atmosphere does more than just this. According to Westfall, who made an exhaustive analysis of the author's observations (and shared it on January 10, 2017), the *apparent quadrature* corresponding to the present observation (which he iterated using spherical trigonometry) must have happened at 17:55:55 UT, about 10 minutes later than the unrefracted one. This may explain the apparently disappointing results of the author's August observation, the last take of which was still very far from the time of *apparent topocentric quadrature*.

In general, as pointed out by Westfall (in many personal letters), 'the atmosphere has the effect of significantly reducing the observed elongations, and it is unlikely that Aristarchus would have adjusted, or been able to adjust for refraction'. It is, however, possible that he knew something about it, since 'refraction in transparent media, such as glass or water, was well appreciated in the third century BC' (Lehn and van der Werf, 2005:5625). In any case, the author's experiments seem to show that the best measurements are those taken when both luminaries are about as high as each other at about the time of topocentric orthogony. Any deviation from this will noticeably send any estimates of the solar distance down, but not so much that they cannot still be measured in terms of thousands of Earth radii. Under these circumstances, those seeking a balanced solution may look upon the word 'myriad' as a graceful compromise between the extreme values found.

In his next observation, the author had something completely different to test.

The morning half-Moon of August 25, 2016, as seen from Nestares, Spain

On August 25, 2016, topocentric orthogony occurred at 3:13 hours UT, when both luminaries were *below* the horizon, as seen from Nestares, Spain. On this occasion, the author had been careful to learn the exact time of the event ahead of time, because he wanted to test something different: namely, he wanted to know whether his instrument would yield something close to Aristarchus' 87 degrees if used six hours after orthogony. Assisted by his father, he took two measurements, each consisting of seven alignments of the same box that had been used on the previous two occasions. (The measurement corresponding to the take at 7:10 was made for sport and does not really count.) The recorded readings (in millimetres) can be found in Table 3.10 (where all variables are as in previous tables).

Table 3.10: Data collected (by the author and his father) during the half-Moon observed from Nestares, Spain, on August 25, 2016, using a P5 box of height $h = 131$ cm. The mean $\bar{x}$ and median m of the readings x_i (in millimetres) allow the calculation of the ratios S/M and S/e (of the Sun's distance S from Earth to either that of the Moon M or the Earth's radius e) and the lunar elongation ε at the given times. (The last column was added at Westfall's request.)

UT	x_1	x_2	x_3	x_4	x_5	x_6	x_7	$\bar{x}$	m	S/M	S/e	ε	true ε
07:10	28.5	31.0	32.5	33.0	35.0	37.0	38.5	33.64	33.00	40	2383	88°33′	89°21′
09:10	42.0	44.0	44.5	46.5	46.5	46.5	47.5	45.36	46.50	28	1691	87°58′	87°35′

The motivation for this experiment was the author's suspicion that Aristarchus' 87 degrees were somehow related to measurements taken a quarter of a day either side of local orthogony, but he did not yet know exactly why Aristarchus should have made such a choice. The reason for this choice was deduced much later from a close reading of Hypothesis 3 and Proposition 5 in *On Sizes*. This reading revealed Aristarchus' preference for a very special moment which the author called the time of *orthogony* (on page 51). The result of the present experiment suggests that the Moon's elongation six hours either side of topocentric orthogony is not far from 87 degrees, even though, on this occasion, it rounded up to 88 degrees. So, it seems that even here the author's box got slightly better results than that in the first book *On Sizes*. In fact, one has to do really very badly to obtain worse results (as the author did when he used a mirror in 2014).

As for the estimate of six hours mentioned above, it is related to what the author calls a **half-Moon's life**, which is the period of time that begins when the Moon starts to look halved to the naked eye and ends when it ceases to do so. It is not easy to assess how long this period lasts, as it varies with each observer and observation, but the author's estimate is based on a statistical study of the most comprehensive set of experiments carried out so far on the half-Moon method. Namely, Westfall's.

3.6.3 Westfall's *dichotometer*

So far, the author has tried to put things in chronological order, but in this particular case, it has not been easy. Strictly speaking, the author's instruments preceded Westfall's, which is why they appear earlier in this book, but the fact is that we worked together as a team, continually influencing each other. So, in order to better understand how events unfolded and how inseparably linked they are, it may pay to go a little back in time to the beginning of our story.

If the author was to put a date to the start of the investigation that our little team so joyfully undertook, he would probably say that it all started with Hoag's experiments back in 1988. Back then, Hoag and Sheehan were already close friends. Not only had the former mentored the latter, but also sponsored his research at the Lowell Observatory (of which he was director) and fostered his passion for the study of all things astronomical. One of these things was, of course, Aristarchus' puzzle. Sheehan remembers exquisite details from Hoag's approach to this puzzle, such as his mentioning (in one of his letters) that the 8-foot vertical pole he used (as a gnomon) was an outdoor clothesline in the backyard of his home in Tucson.

As far as Sheehan remembers (in letters to the author dated July 12 and September 17, 2018), Hoag was not trying to be a purist and restrict himself to instruments and methods Aristarchus would have had at hand. Instead, he used simple instruments and ready-to-hand equipment to measure what Aristarchus might have measured (if anything). But (as he confided to Sheehan) on finding how easy it is to do better than the 87 degree figure, Hoag doubted (at least for some time) whether Aristarchus had measured anything at all (though this did not transpire in his paper).

Doubting is typical of anyone doing genuine research (specially of the historical type). In fact, it is the only way forward as we grope our way through the maze of extant and missing data, of theories and counter-theories, of thoughts and afterthoughts. So, Hoag's intimate doubt did not prove Neugebauer right. It only proved that the puzzle was still unsolved. Thus, feeling as if some stones could still be upturned, Sheehan auspiciously put Westfall and the author in contact, prompting us, as a team, to revisit Aristarchus once more. Our mission: to investigate how he might have done it.

Westfall simply fell in love with this project the moment he heard of it. It combined several of his lifelong passions: measuring distances in outer space, using eclipses, observing the Moon's terminator, and trying to determine the time of dichotomy of a celestial object. He started making observations of the Moon's dichotomy almost immediately and carried them on enthusiastically for the rest of his life. He noted that there are not many opportunities in a year to watch the dichotomy of the Moon in broad daylight (that is, with both luminaries above the horizon), perhaps only two or three per year (weather permitting), but he observed as many as he could.

On December 23, 2015, Westfall announced that he was working on a design for an apparatus for measuring deviations of a few degrees from a right angle which could have been constructed in Aristarchus' time. Things started to gain momentum on January 9, 2016, when the author prepared a special calendar on which favourable lunar dichotomy dates had been highlighted. It was intended to aid our observations, and, inspired by this idea, Westfall prepared alternative calendars for different locations in America and Europe, one for each of the observers involved.

On February 22, 2016, Westfall confided some details on how his design was doing. He said his 'dichometer' (as he called it on this occasion) would have slits, rather than pinholes (to make it easier to focus), would be reversible (to allow for viewing at both First and Last Quarters), and have numerical scales (to measure the elongation deficiency from a right angle). He also mentioned that it would be difficult to use handheld (unless aided by an assistant) and that, probably, a mounting with a triaxial head or camera tripod would be needed (to hold it steady). On March 20, 2016, Westfall announced that he had completed his instrument and sent a picture of it to us (an artistic rendering of which can be seen in Figure 3.7).

The device in Figure 3.7 is shaped like a cross. At each of the ends of one of its beams there is a screen with a hole in it. The instrument is Sun-aligned whenever the Sun shines through both holes. At each of the ends of the other beam there is a sight with a notch and a scale. This scale allows viewers to gauge the Moon's deviation from the aligned notches. The roles of this pair of ends can be switched for morning and evening observations. The size of the actual instrument used by Westfall was 608 mm between sights and one degree (or $608 \times \tan 1° \approx 10.6$ mm) between the marks on the scale.

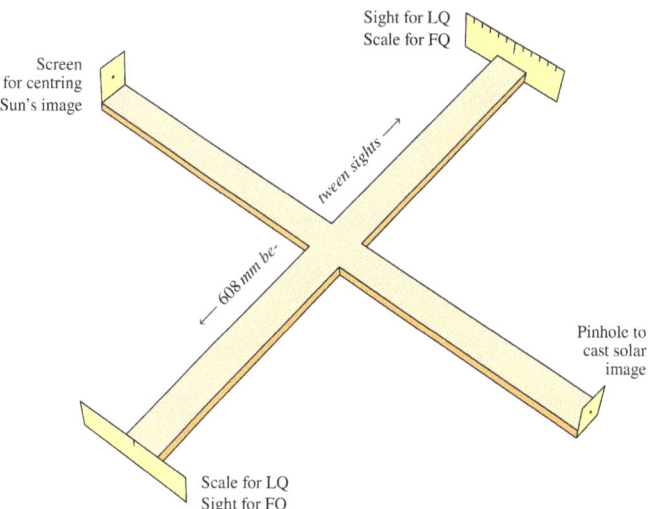

Figure 3.7 *dichotometer* (or *dichometer*)

It is to be noted that, before the invention of the *dichotometer*, Westfall's observations consisted of eyeballed estimates of the time of dichotomy (as can be deduced from his words in a mail dated January 1, 2016), from which the Moon's elongation was then computed, and from it, the Sun's distance was finally estimated using trigonometry (via Equation 2.15).

The morning half-Moon of March 31, 2016, as seen from Antioch, California

Westfall used his *dichotometer* for the first time on March 31, 2016, and later sent us a photograph he said he took 'rather casually' while setting up the instrument for his visual elongation measurement. (An illustration based on this photograph is shown in Figure 3.8.) As he explained (in a letter dated July 15, 2018), this photograph does not show an actual measurement (since such a thing would involve superimposing the front and rear sights on the lunar terminator and centring the Sun's image on the viewing screen on the right). Instead, he said 'the purpose of this photo was simply to show the *dichotometer* set up on a tripod'. The original photographs (on which Figures 3.7 and 3.8 are based) can be seen in Westfall's (2019:47) posthumously published paper.

The picture in Figure 3.8 was taken 'casually' (as Westfall said), but even so it can be used to show how the instrument is used. Thus, tracing a straight line from the front notch to the Moon's centre, we see that this line meets the rear sight at about the point where the scale reads half a degree, so the Moon behind the scale exceeds quadrature by about half a degree. Putting this into the modulus of Equation 2.13, we obtain a Sun-to-Moon distance ratio of

$$\frac{S}{M} = \left| \frac{1}{\sin(-30')} \right| \approx 115. \tag{3.30}$$

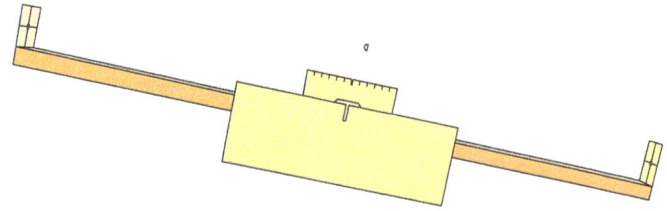

Figure 3.8 A *dichotometer* mounted on a triaxial tripod (not shown), so as to lie steadily in the same plane as the Sun, Earth, and Moon. The Moon's deviation from quadrature can be read off the scale on the rear sight, where each mark represents a degree.

Alternatively, the Moon's elongation is $90° + 30' = 90°30'$. Putting this into the modulus of Equation 2.15, we obtain a Sun-to-Moon distance ratio of

$$\frac{S}{M} = \left|\frac{1}{\cos(90°30')}\right| \approx 115. \tag{3.31}$$

Either way, this is a much better ratio than that in *On Sizes*, dramatically agreeing with Hoag's (unwritten) conclusion that it is very easy to do better than Aristarchus' early ratio.

This also illustrates a few more points. For example, Westfall would never have used this particular measurement (should it have been considered one in the first place) because here the Moon exceeds (rather than is short of) quadrature. Thus, the Moon's elongation is greater than 90 degrees, or, in other words, the Moon's deviation from quadrature is a negative number, which put into Equation 2.13, gives a negative ratio, which apparently is impossible. We saw at the end of Section 3.6.1 that the author sees no problem using negative elongations, but not so Westfall.

The way the author would have proceeded is slightly different. Instead of using angles, he would have used lengths: namely, the distance (of 608 mm) between sights and the distance (of about $10.6/2 = 5.3$ mm) between the rear notch and the point on the numerical scale just in front of the Moon's centre as seen from the front notch. Thus, using Approximation 3.18, we obtain

$$\frac{S}{M} \approx \frac{608\,\text{mm}}{5.3\,\text{mm}} \approx 115, \tag{3.32}$$

which agrees (to the nearest whole number) with the result in Equations 3.30 and 3.31.

Aristarchus himself, would have taken his measurement about five hours earlier, when the Moon was much further right (than in Figure 3.8), about 3 degrees in excess of quadrature, just when it ceased to look gibbous and started to look halved (to his eye). Then, the Moon's elongation was about 93 degrees, as seen from a place where the Sun and Moon were both above the horizon (which, in this particular case, would have been somewhere in the middle of the Atlantic). But whatever he used, whether lengths or angles, he, like the author, never met negative numbers (which had not yet been invented, anyway).

But let us now turn to the way Westfall actually performed his experiment. He reported (in mails dated April 3, 4, and 8, 2016) having used his 'dichotometer' (as he called it this time, when all he knew beforehand was that 'dichotomy' would occur sometime on the morning of March 31) to make 14 measurements of the Moon's elongation over a period of four and a half hours, describing the terminator as

'slightly convex'	at 13:57 and 14:26,
'slightly convex?'	at 14:59,
'straight?'	at 15:27,
'straight'	at 16:00 and 16:26,
'slightly concave?'	at 17:00 and 17:36,
'slightly concave'	at 17:59 and 18:33.

He took the mean of his two 'straight' observations as an estimate for the time of dichotomy (namely, 16:13 UT). At this time, the Sun and Moon were both at about the same height above the horizon, as seen from Antioch, California.

Then, he consulted the JPL Horizons Online Ephemeris System to find the exact times at which *true topocentric quadrature* and *dichotomy* occurred. The system gave these times, respectively, as 16:05:54 and 16:32:59, and also gave the Moon's elongation corresponding to the latter time as $\varepsilon_H = 89°51'07''$ (implying that the value of the Sun-to-Moon distance ratio at this time was $S/M = \sec \varepsilon_H \approx 387$). He also consulted Horizons for the lunar elongation corresponding to his estimated time of dichotomy (16:13 UT), which the system gave as $\varepsilon_W = 89°57'42''$ (implying $S/M \approx \sec \varepsilon_W \approx 1495$).

Then, he compared each of his readings with their computed *apparent* analogues, and found the mean difference to be $2'13'' \pm 2'10''$, indicating no significant alignment error in the instrument. (The individual sightings, though, showed a standard deviation of $8'2''$.) Then (by means of Equations 3.9 and 3.10), he found the line of best fit for his observed elongations and times. Unfortunately, he never sent us his actual readings nor the equation of his trend line (nor did the author ever had the foresight to ask him for them), but he said that using them, he was able to interpolate the *true topocentric elongation* corresponding to his estimated dichotomy time, finding it to be $89°57'14''$, which, as he pointed out, is almost the same as the value given by the Horizons website.

Westfall's *apparent* elongation for this and other observations can be found in the table summarizing his six eventful years investigating the Aristarchus puzzle (Westfall, 2019:53). This table has been split into two and simplified in Tables 3.11 and 3.12 below.

We might argue that the (apparent topocentric) elongations measured directly on the dichotometer should have been easier and more straightforward to use, but Westfall preferred to work with **true** (or unrefracted) values (that is, corrected for atmospheric refraction), rather than **apparent** (or refracted) ones (that is,

Table 3.11: Sun-to-Moon distance ratio estimates based on the *apparent* and *true* elongations of the 17 waxing half-Moons observed by Westfall. (His original table does not include the negative ratios, corresponding to obtuse elongations, which have been added here by the author.)

☽ date and UT	offset	ε (°) app.	ε (°) true	S/M app.	S/M true
2012-03-31 01:04	+7:05	92.80	92.86	−20	−20
2012-04-29 02:29	−8:03	86.14	86.31	15	16
2012-06-26 23:43	−3:25	88.59	88.61	41	41
2012-06-27 01:27	−1:41	89.21	89.25	73	76
2012-08-24 01:48	−11:49	83.45	83.50	9	9
2012-12-19 23:39	−7:09	87.33	87.39	21	22
2013-02-18 01:41	+6:43	92.00	92.38	−29	−24
2013-07-16 01:47	−1:03	89.46	89.50	106	115
2013-07-16 03:22	+0:32	89.82	90.03	318	−1910
2013-08-13 19:08	+7:47	85.37	85.43	12	13
2013-10-12 01:26	+3:47	91.14	91.34	−50	−43
2014-01-08 00:20	−4:16	88.31	88.41	34	36
2014-07-05 02:29	−9:26	85.55	85.61	13	13
2014-09-02 02:26	−9:11	85.28	85.47	12	13
2014-12-29 00:41	+7:02	93.13	93.37	−18	−17
2016-07-11 23:14	+0:15	89.90	89.93	573	819
2017-06-30 21:57	−1:07	89.37	89.42	91	99
Mean		88.638	88.754	42	46
Median		89.210	89.250	73	76

uncorrected for atmospheric refraction). He gave no reason for doing this other than the circular argument that it is necessary 'to remove the bias caused by refraction' in order to obtain 'a best, unbiased mean solar distance' from the several in his table, by which he meant the one obtained by equally weighing the [unrefracted] means of the 14 morning observations and the 17 evening observations (that is, the mean of the means). On doing this, he (2019:49) obtained his 'best, unbiased' Sun-to-Moon distance ratio as

$$\frac{S}{M} \approx \sec\left(\frac{88.754 + 90.593}{2}\right) \approx 175, \tag{3.33}$$

Table 3.12: Sun-to-Moon distance ratio estimates based on the *apparent* and *true* elongations of the 14 waning half-Moons observed by Westfall. (His original table does not include the negative ratios, corresponding to obtuse elongations, which have been added here by the author.)

☽ date and UT	offset	ε (°) app.	ε (°) true	S/M app.	S/M true
2012-06-11 14:19	+4:02	88.46	88.52	37	39
2012-08-09 15:19	−5:45	91.96	92.01	−29	−29
2012-11-06 18:35	−6:21	93.37	93.43	−17	−17
2013-03-04 16:59	−6:06	93.09	93.13	−19	−18
2013-08-28 14:00	+5:25	87.84	88.02	27	29
2013-11-25 15:04	−5:43	91.69	92.08	−34	−28
2014-05-21 15:07	+1:44	89.20	89.24	72	75
2014-06-19 14:05	−6:35	92.47	92.65	−23	−22
2015-03-13 16:36	−2:53	90.99	91.02	−58	−56
2016-03-31 16:13	−0:20	89.93	89.96	819	1432
2017-04-19 15:22	+6:20	87.74	87.78	25	26
2017-07-16 15:48	−5:35	92.41	92.42	−24	−24
2017-10-12 16:02	+4:14	88.30	88.35	34	35
2018-02-07 17:18	+0:30	89.66	89.69	169	185
Mean		90.508	90.593	−112	−97
Median		90.460	90.490	−125	−117

which is less than half the true value (namely, 389), but renders S as about a myriad Earth radii (when M is taken to be 60 Earth radii), providing food for thought.

It is now worth asking ourselves the following question, Is it true that the atmosphere introduces bias into this experiment? Is it really necessary to remove it? The answer lies on what we understand by 'this experiment'. Westfall understood the experiment was about trying to get the best possible solar distance out of the half-Moon method, while the author thought it was about trying to understand what Aristarchus did. If you think like Westfall, then you need to remove the effects of atmospheric refraction, but if you think like the author, then you do not.

Aristarchus himself, though he might have known something about refraction (in solids like glass and liquids like water), most likely had no means to correct for it. So we will assume that he did not. He most likely had no choice but to use only *apparent* elongations (which is why the author never worried about anything else). But, as Westfall said (and we saw in Section 3.6.2), the atmosphere does have an effect. Namely, that of slightly reducing the elongation between two heavenly bodies (by slightly rising them above the horizon, or bringing them closer to the zenith). Let us bear this in mind. We shall come back to it shortly, but first, let us just have a closer look at Westfall's tables. We can learn a lot from them. Westfall (2019:48) himself derives the following conclusions.

1. Naked-eye estimates of the time of dichotomy result in almost as many obtuse elongations as acute ones.
2. The Earth's atmosphere slightly (but significantly) shrinks both the Moon's elongation from the Sun and the Sun-to-Moon distance ratio.
3. There is a marked difference between morning (☽) and evening (☾) observations, with overall larger elongations and Sun-to-Moon distance ratios in the former case, resulting in obtuse elongations being more common in the former than in the latter case.

As Westfall (2019:49) explains, the latter phenomenon may be due to the fact that 'at First Quarter, the sunlit lunar surface near the terminator is mainly fairly bright highlands, while at Last Quarter, the sunlit terrain near the terminator is largely darker maria'. So we tend to advance (and hence reduce) orthotomy readings more at First Quarter than we tend to retard (and hence reduce) them at Last Quarter. This phenomenon was first noticed (but not properly explained) by Wendelin (in Passage 47). A proper explanation of it lies in the *irradiation illusion*, first discovered by Galileo in 1610, 37 years earlier than Wendelin's report to Riccioli was written (Galilei, 1610:16, 2016:57; 1632:328, 1967:336; Sheehan, 2018:19), and to which we shall return in Section 5.3.

But there is more that we can learn from these tables. For example, the 'offset' columns (which have been added by the author) indicate the difference between Westfall's naked-eye estimates and the actual time of *orthogony* (as computed using the Horizons system). Taking the absolute value of these time offsets, we see that their rounded mode and median are both 6 hours, from which the author concluded that the typical length of a half-Moon's life is about twice this number (since we took absolute values). That is, about half a day.

We may also note that, as expected, the 'apparent' columns contain values that are never greater than those in the 'true' columns, unless we remove the minus signs, in which case, we will find that the acute elongations (that is, the ones favoured by Westfall) follow the above rule, but not so the obtuse elongations (that is, those discarded by Westfall), which follow the opposite. That is, taking the modulus of all the ratios in Tables 3.11 and 3.12, we see that the role of acute and obtuse elongations is reversed when their true and apparent values are compared. Of course, we cannot see this if we discard obtuse elongations. But this raises another question: Did Aristarchus discard obtuse elongations?

To start with, he had no JPL's Horizons or the like to tell him when quadrature occurred. At most, the *quad gauge* he was presumably using could give him a rough approximation to the time of this event. But this is not what he wanted. What he really wanted was the time of *orthogony*, and he correctly assumed that, at this precious time, the Moon's elongation from the Sun is always an acute angle. So he had no need to worry. He would therefore discard none of the elongations he read at all, because he assumed all of them were acute (as in fact, they should).

Curiously, if we discard no elongation at all, then, in reality, some of them will be acute and some obtuse. From Westfall's tables, we see that statistically speaking, there seems to be about the same number of each type. So, assuming we read as many acute elongations as we read obtuse ones, the 'shrinking' effect of the former will exactly cancel the 'expanding' effect of the latter. So, serendipitously (and probably also unwittingly), Aristarchus' method automatically corrects for refraction!

However, there is a way in which Aristarchus' early readings were biased in this respect, an explanation of which is as follows. The young Aristarchus (the one who wrote the first book *On Sizes*) favoured the time of *orthogony* (as defined on page 51), which, in reality, we know always happens when the Moon's elongation is acute. If we were able to tell this time with accuracy, our measurements would be shrunk by the atmosphere. But, as it happens, the human eye cannot reliably tell when this event occurs. So by the time Aristarchus judged it did (that is, at the time prescribed by Hypothesis 3 and Proposition 5), the Moon's elongation is always obtuse (though he did not know it), so his readings were systematically enlarged by the atmosphere, albeit slightly.

Did Aristarchus ever notice? We shall address this question in Chapter 5, but for now, Westfall's impressive achievement, which he condensed in Equation 3.33, provides, as we said there, food for thought.

3.6.4 The *cosmometer*

The instruments described so far are but a springboard for the readers' creativity to make thy own. A possible hybrid of the author's and Westfall's creations, which the author intends to make and use hopefully soon is as in Figure 3.9.

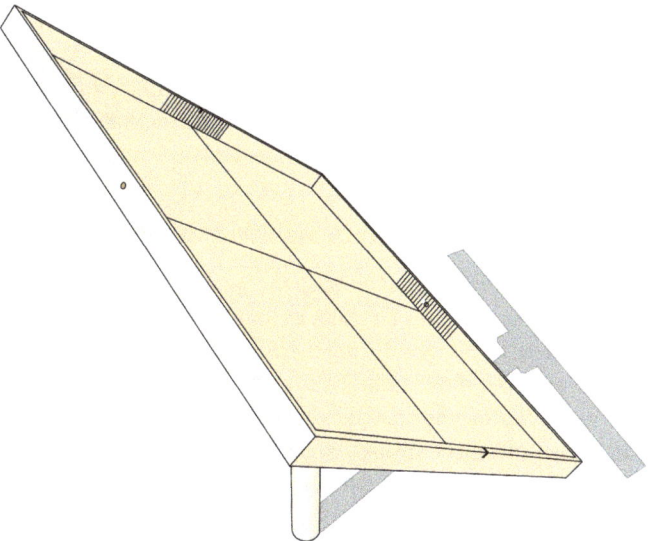

Figure 3.9 A *cosmometer*

This instrument is called a **cosmometer** (or **cosmeter**) because it is intended to measure the **cosmos** (or sphere whose radius is the distance between the Earth and Sun), as defined by Archimedes in *The Sand Reckoner* (Heiberg, 1881:244). That is, this instrument measures (however roughly) the distance to the Sun by Aristarchus' half-Moon method.

It can be used exactly the same as Westfall's *dichotometer* (that is, Sun-aligned and using degrees) or exactly the same as the author's *quad box* (that is, Moon-aligned and using lengths). If the distance between sights is set to $1/\tan(1°) \approx 57.3$ cm, then the marks in the scales will be one centimetre as well as one degree of arc apart. Twice this distance, and two centimetres will be the same as one degree of arc. It all depends on how big we want to make it. (It may be interesting to know that on windy days, the cosmometer in Figure 3.9 would benefit from having large chunks of the mother board cut out to let the air through.)

The cosmometer in Figure 3.9 has been Moon-aligned. That is, the observer has aligned the front and rear notches with the centre of the Moon. At the same time, it has been set such that the Sun shines though the topmost hole and lands close to the hole at the bottom. The observer would then measure the distance between the centre of the bright spot where the sunbeam hits the scale at the bottom and the line through the centre of the hole nearby (that is, the *quad line*). The length of the distance between sights in this particular illustration is 1000 mm, and the distance between bright spot and hole is about 12.5 mm. So, the distance to the Sun (in this hypothetical illustration) can be estimated (by Approximation 3.18) to be about (1000 mm)/(12.5 mm) = 80 times that of the Moon. It is as simple as that.

3.7 An overall assessment of the half-Moon method

It was not the author's original intention to merge the results of his observations into a single one. In fact, he sees no point in doing so (because the Moon always looked equally halved to his eyes during these experiments, and never attempted to pick a 'best' time). But if he was asked to do this, he would say that one way to proceed is to do as Westfall did and take the mean of the means. That is, the arithmetic mean of the arithmetic means, which is the only one of the Pythagorean means that does not give problems when some of the readings fall exactly on the *quad line*.[24]

Thus, discarding medians (of which the author is so fond, but whose use is not attested in antiquity) and using only arithmetic means, then the Sun-to-Moon distance ratios in Tables 3.7 to 3.9 are, respectively, $\bar{x}_1 \approx 225.7$, $\bar{x}_2 \approx 357.27$, and $\bar{x}_3 \approx 58.84$, and the arithmetic mean of these arithmetic means is

$$\bar{x} = \frac{\bar{x}_1 + \bar{x}_2 + \bar{x}_3}{3} \approx \frac{225.7 + 357.27 + 58.84}{3} \approx 214, \tag{3.34}$$

which can be considered as the author's estimate for the ratio of the Sun's distance from Earth to that of the Moon, as derived from his three observations.

But this is not the only way to proceed. If we first average each reading in the author's tables (obtaining, respectively, $\bar{r}_1 \approx 5.3$ mm, $\bar{r}_2 \approx 3.67$ mm, and $\bar{r}_3 \approx 23.30$ mm), and then we derive the Sun-to-Moon distance ratio corresponding to each of these separately (obtaining, respectively, $\bar{x}_1 \approx 134$, $\bar{x}_2 \approx 357$, and $\bar{x}_3 \approx 56$), then the average of these averaged ratios is

$$\bar{x} = \frac{\bar{x}_1 + \bar{x}_2 + \bar{x}_3}{3} \approx \frac{134 + 357 + 56}{3} \approx 182, \tag{3.35}$$

which is a somewhat smaller ratio than the previous one and perhaps more appealing to ancient conservative minds, to whom both would be mind-boggling

---

24 The Neopythagorean Nicomachus of Gerasa (1866:122) reports that these means were well known to ancient mathematicians.

An overall assessment of the half-Moon method

Table 3.13: Some post-telescopic measurements of the Sun's distance obtained mainly by the half-Moon method. S/M and S/e ratios (where S is the Sun's distance, M is that of the Moon, and e is the Earth's radius) are given to the nearest whole number. Grey values are reconstructed by the present author, rather than found in the original sources.

year	observer	ε	S/M	S/e	solar parallax
1606	Johannes Kepler (1606:85)[1]	87°35′55″	24	1432	2′24″
1611	Thomas Harriot (*Moon Paper* 8)	89°19′31″	85	5095	41″
1617	Johannes Kepler (1617:2)[1]	88°05′23″	30	1800	1′55″
1617	Johannes Kepler (1617:6)[1]	87°30′29″	23	1380	2′29″
1620	Johannes Kepler (1620:483)[1]	89°00′32″	58	3469	59″
1626	Godfrey Wendelin (1626:11)	89°00′25″	58	3460	1′
1631	Philip Lansberge (1631:59)	87°42′19″	25	1499	2′18″
1635	Godfrey Wendelin (Gassendi, 1658:428)	89.76646°	245	14720	14″
1640	Jeremiah Horrocks (1662:142)[1]	89°46′00″	246	14733	14″
1644	Godfrey Wendelin (1644:29)	89.76543°	244	14656	14″
1646	Giovanni Riccioli (1651:108)[2]	89°28′26″	109	7260	29″
1648	Giovanni Riccioli (1651:108)[2]	89.47867°	110	7327	28″
⋮	⋮	⋮	⋮	⋮	⋮
1988	Arthur Hoag (1989:4)[3]	89.785°	266	15960	13″
2016	Alberto Gomez	89.73217°	214	12836	16″
2018	John Westfall (2019:49)	89.6735°	175	10529	20″
1976	IAU[4]	89.8527°	389	23455	9″

[1] Not based on the half-Moon method.
[2] Based on a lunar distance of 66.6 Earth radii.
[3] Reconstructed by the present author.
[4] Based on a lunar distance of 60.3 Earth radii.

enough. Anyway, let us just gather all we have learned so far in this chapter in just one table, where solar parallaxes have been calculated using the equation

$$P = \tan^{-1}\left(\frac{e}{S}\right), \tag{3.36}$$

where S and e are as usual.

Table 3.13 puts together the most relevant attempts at repeating Aristarchus' experiment since the invention of the telescope (with the addition of Lansberge's, which is not explicitly said to have been obtained empirically, and those by Kepler and Horrocks, who used entirely different methods and are included in this table only for comparison). It shows some patterns. For example, no one there has done worse than Aristarchus' youthful estimate; no one there has taken thy readings at the time Aristarchus prescribed (in Hypothesis 3 and Proposition 5), but rather, at a time thou judged to be sufficiently past the beginning of a half-Moon's life but not too close to its end (hence the improvement). Most importantly, taking the arithmetic mean (or even the median) of all the results in this table that are purely based on the half-Moon method, we see that (assuming a lunar distance of $M = 60$ Earth radii) the average solar distance is

$$\frac{S}{e} \approx \left(\frac{58 + 85 + 109 + 110 + 175 + 214 + 244 + 245 + 266}{9}\right) \times 60$$
$$= 10040, \tag{3.37}$$

which is about a 'myriad' Earth radii—just the number Archimedes and Posidonius speak about, the former in his *Sand Reckoner*, and the latter in his astronomical works as rescued by Cleomedes (Ziegler, 1891:144; Bowen and Todd, 2004:114).

3.8 Conclusion

The origin of Aristarchus' 87 degrees has been a puzzle for over two thousand years. Now, at last, we have a working theory that explains it while fitting all extant data. It runs like this. Aristarchus realized that when the Earth, Moon, and Sun form a right-angled triangle, we can very easily see how many times the Moon's distance goes into that of the Sun. This can be done by means of an instrument we may call a *cosmometer* (as illustrated in Figure 3.9). All we have to do is wait for the Moon to be at right angles to the Earth and Sun, then measure two distances and divide them. This very special time was called *orthogony* by the author (on page 51). Aristarchus was aware that the terminator is slightly gibbous at this particular time. However, indistinguishably close to it is the time when the terminator is perfectly straight, and we gave this new time the name of *orthotomy* (on page 57). Both of these events (along with the whole cycle of a half-Moon's life) are illustrated in Figures 2.2 and 2.3 (on page 52). Aristarchus used the latter to assess the former.

Then, Aristarchus correctly guessed that both of these times occur in the Sun-bound quadrants of the Moon's orbit (which are those closest to the Sun). Then, he made his measurements and concluded that the Moon was less than any of these quadrants by 'one thirtieth of a quadrant' (that is, 87 degrees). In fact, anyone measuring things the way he did, and at the time he said one must, obtains values that round up or down to the same value he obtained. The reason why this is so, however, is not so simple.

Aristarchus may have made a mistake of a physical (rather than mathematical) nature. Namely, he assumed the eye was sharp enough to tell the time of *orthogony*, so he waited until his eye said that the terminator was either straight but on the brink of turning gibbous, or gibbous but on the brink of turning straight. He thought this happened in either of the Sun-bound quadrants. But it did not; the eye is not that sharp; it cannot reliably tell the time of either *orthotomy*, *orthogony*, or *quadrature*. So, by the time he judged appropriate (for making measurements), the Moon was in one of the Sun-shy quadrants (that is, those far from, rather than close to the Sun). He was completely unaware of this because his instrument measured two distances which could always be divided to give the desired distance ratio between the Sun and Moon, and his instrument never told him anything was wrong. (See Figures 3.10 and 3.11.)

To be the work of a young man, the first book *On Sizes* is truly impressive. There are no mathematical mistakes in it (that the author has spotted), only one of a physical nature (as said above). It was this mistake that set the standards for astronomical distances for the two thousand years that followed. Fortunately, this was not Aristarchus' last word on the subject: the data in *The Sand Reckoner*

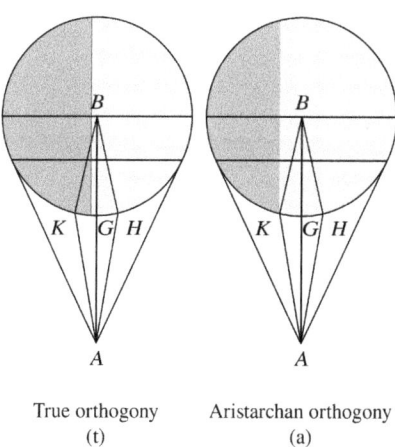

True orthogony Aristarchan orthogony
(t)   (a)

Figure 3.10 Aristarchus thought that the first orthogony of the month happened at (a), when the Moon starts looking waxing gibbous (as seen from *A*). In reality, it happens at (t), many hours before the human eye can tell. (For clarity, sizes and distances are not to scale.)

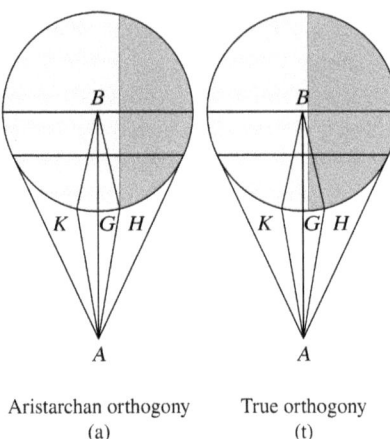

Aristarchan orthogony True orthogony
(a) (t)

Figure 3.11 Aristarchus thought that the last orthogony of the month happened at (a), when the Moon stops looking waning gibbous (as seen from *A*). In reality, it happens at (t), many hours after the human eye can tell. (For clarity, sizes and distances are not to scale.)

points to his having developed a much better theory of the cosmos later in his life. But this is something we shall explore in the next chapter.

4 The Sand Reckoner

Abstract: This chapter analyses *The Sand Reckoner*, one of Archimedes' earliest books, where he describes Aristarchus' model of the universe and finds how much sand is needed to fill up an exaggerated version of it. We shall see whether Archimedes' description is correct and whether Aristarchus' ultimate estimates for the size and distance of the Sun and the universe can be deduced from it. In the process, the exact mathematical solution to the ancient *eclipse method* will be found, all while addressing questions about who developed it.

4.1 Aristarchus' model of the universe

In general, numbers are harder to write the larger they are, and if they are very large, writing them can be next to impossible unless an efficient writing system is used. Archimedes wrote two books dealing with this problem. The first one (now lost) was a formal treatise called Ἀρχαί (or *Principles*) that he sent to a man called Zeuxippus, of whom nothing is known apart from his being mentioned in the second of the books Archimedes wrote on this subject (Heiberg, 1881:242-6, 1972:216-20; Heath, 1897:221-2). This second book was an adaptation intended for King Gelo II, who died shortly before his father, King Hiero II, after having ruled with him in Syracuse for over half a century, as we know from the writings of Polybius (1966:418) and Livy (1876:349). This second book is called Ψαμμίτης (or *Sand Reckoner*), and though small, it is a treasure trove of information that can be found nowhere else.

In it, Archimedes catches everyone's imagination by speaking of a number so large as to equal the number of grains of sand that would fill a sphere as large as the Earth covered to its topmost heights, and not only this, but also one as large as the whole universe, by which he means a sphere as large as that of the fixed stars! In his book, Archimedes uses one word (namely, κόσμος) for two different concepts, which, in the present book, we translate as **firmament**, when he means the sphere of the fixed stars, and **cosmos**, when he means the sphere centred at the Earth and reaching out to the Sun. He claims the latter to be a definition then current among astronomers.

> Passage 49 [Cosmos] is the name given by most astronomers to the sphere whose centre is the centre of the Earth and whose radius is equal to the straight line between the centre of the Sun and the centre of the Earth (Heath, 1897:221; see also Heiberg, 1881:244, 1972:218).

This definition originated with Anaximander (c. 610 – c. 546 BC), for whom nothing was further from Earth than the Sun, not even the stars (Hippolytus,

2016:30), and therefore, its distance from Earth was the radius of the largest possible sphere. Namely, the *cosmos*. Later theories, such as Philolaus' (Figure 1.1), put something else further from Earth than the Sun. Namely, the stars. But, somehow, the old notion of **cosmos** (as the sphere bounded by the Sun's orbit around the Earth) was still alive in Archimedes' time. There is no reason to doubt his word, since he was a rigorous mathematician who frequently corresponded with the leading mathematicians in Alexandria (which is why the above-mentioned Zeuxippus is often counted among them), and so was well acquainted with the latest developments there.

Immediately after defining the **cosmos** as the sphere in which the Sun moves (around the Earth), Archimedes quotes a book in which Aristarchus swaps the roles of the Sun and the Earth, presenting the largest model of the universe ever conceived until that moment. Archimedes describes it in the following terms.

> **Passage 50** Aristarchus of Samos brought out a book consisting of some hypotheses, in which the premises lead to the result that the universe is many times greater than that now so called. His hypotheses are that the fixed stars and the Sun remain unmoved, that the Earth revolves about the Sun in the circumference of a circle, the Sun lying in the middle of the orbit, and that the sphere of the fixed stars, situated about the same centre as the Sun, is so great that the circle in which he supposes the Earth to revolve bears such a proportion to the distance of the fixed stars as the centre of the sphere bears to its surface (Heath, 1897:222; see also Heiberg, 1881:244, 1972:218).

Clearly, these are not the hypotheses in the first book *On Sizes*, but those in a second book by Aristarchus which has not survived and which we will refer to in the present book as Aristarchus' second book *On Sizes*, or *On Sizes* 2. From the above passage and others (in *The Sand Reckoner*, Aetius' *Selections*, and Simplicius' *On the Heavens*), it is possible to reconstruct some of these hypotheses. But before attempting to do so, let us pay attention to one that baffled even Archimedes. Namely, the last of the sentences in the above passage, which is repeated (and rendered into a slightly more direct style) here for convenience.

> **Passage 51** The circle in which the Earth revolves bears such a proportion to the distance of the fixed stars as the centre of the sphere bears to its surface.

Archimedes argues that this hypothesis is impossible, because the centre of a sphere has no magnitude and can therefore bear no ratio whatever to the surface of any sphere. (That is, something that is *sizeless* can bear no ratio to something that is *sized*.) But rather than seizing on this to attack Aristarchus or discredit heliocentrism, he tries to legitimize this theory by offering an interpretation of what he thinks Aristarchus means. In doing so, Archimedes makes it appear as if he is correcting an obvious mistake in Aristarchus' model, when in fact he is adapting this model to serve the purposes of *The Sand Reckoner* (Netz, 2003:255, 266). Archimedes is clearly not against the theory, but wishes to present himself as an impartial referee on an important scientific matter whose originator he admires, as we know from the fact, pointed out by Christianidis et al. (2002:156), that Archimedes never explicitly criticizes other people's work, but is always ready to name those he admires, and the person he happens to name most in the whole body of his extant works is Aristarchus—ten times in *The Sand Reckoner*. The interpretation Archimedes offers is unambiguously expressed in the following passage.

> **Passage 52** We must ...take Aristarchus to mean this: since we conceive the Earth to be ...the centre of the universe, the ratio which the Earth bears to what we describe as the [cosmos] is the same as the ratio which the sphere containing the circle in which he supposes the Earth to revolve bears to the sphere of the fixed stars. For he adapts the proofs of his results to a hypothesis of this kind, and in particular he appears to suppose the magnitude of the sphere in which he represents the Earth as moving to be equal to what we call the [cosmos] (Heath, 1897:222; see also Heiberg, 1881:244, 1972:218).

So, Archimedes takes Aristarchus to mean that the Earth is to the circle in which it moves as this circle is to the sphere of the fixed stars. Or, more succinctly (using the definitions on page 129), the Earth is to the *cosmos* as the *cosmos* is to the *firmament*. (See Figure 4.1.)

This interpretation can be rendered mathematically by saying that the radius of the Earth e is to that of its orbit S as the latter is to the radius of the firmament F. Thus,

$$\frac{S}{e} = \frac{F}{S}. \tag{4.1}$$

So, Archimedes understands that Aristarchus does not take the centre of the mentioned sphere to be a *sizeless* point, in the Euclidean sense of the word (Heiberg, 1888:2), but rather something *sized*, and, in his opinion, this 'something' is the Earth (which, note this, is the centre of the *geocentric* universe). But this is not the only possible interpretation, since there are two other spheres that may be taken as 'sized centres', namely the Sun and the Moon. But the Moon is never mentioned in this discussion (as nothing really moves around it), so it can be discarded, leaving just one other candidate, namely the Sun (which, note this, is the centre of the *heliocentric* universe).

Now, the Sun is explicitly said (in Passage 50) to be the centre of the Earth's orbit and is also said to lie about the same centre as the sphere of the fixed

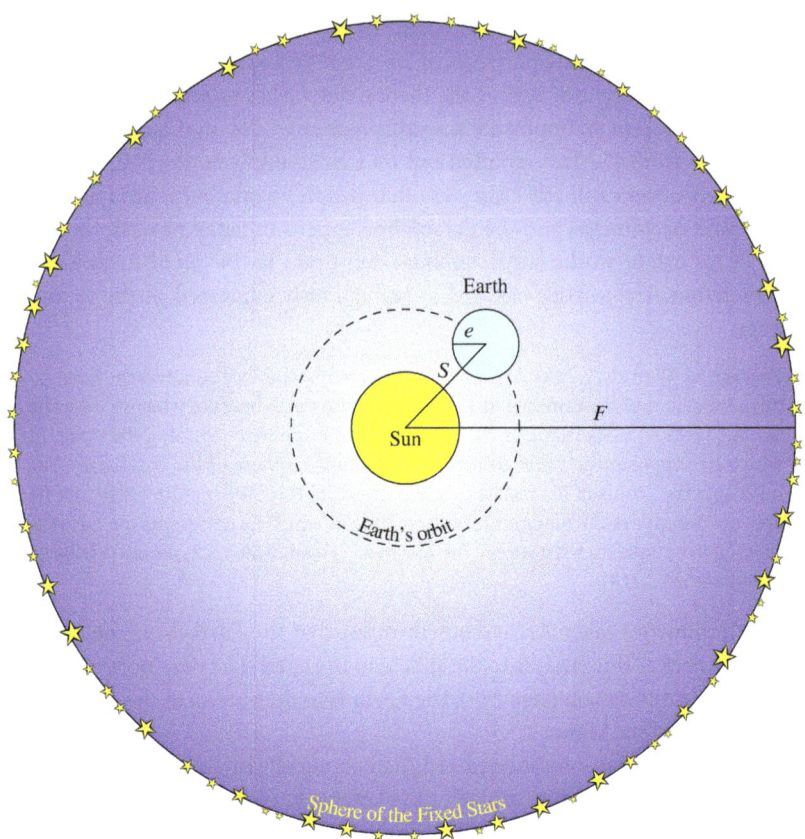

Figure 4.1 In Aristarchus' model of the universe, as understood by Archimedes, the Earth is to its orbit as this orbit is to the sphere of the fixed stars. (Not to scale.)

stars. Whatever this last sentence means has caused much discussion among scholars, with different authors translating it differently in their effort to make sense of it. For example, James of Cremona (1544:155) rendered it as, 'Sphaeram vero stellarum fixarum circa idem centrum cum sole sitam esse'. Meaning, 'The sphere of the fixed stars is situated around the same centre as the Sun'. Heiberg (1881:245) translates it as, 'Sphaeram autem stellarum fixarum circum idem centrum positam, circum quod sol positus sit'. Meaning, 'The sphere of the fixed stars is situated around the same centre around which the Sun lies'. In Stamatis' (1972:219) revision of Heiberg's translation, we find it as, 'Sphaeram autem stellarum fixarum circum idem centrum positam, circum quod sol moveatur'.

Meaning, 'The sphere of the fixed stars is situated around the same centre around which the Sun moves' (which goes against everything the heliocentric theory is about). Erhardt and Erhardt-Siebold (1942:579) give, 'The sphere of the fixed stars is concentric with the Sun' (which is not what the text really says).

According to Boter (2007:428), if we simply delete the words τῷ ἁλίῳ from the original text (in Passage 50), regarding them as a later gloss, then everything falls into place, with the resulting text meaning that 'the sphere of the fixed stars and the orbit of the Earth are concentric, with the Sun as their centre'. (See Passage 53.)

> **Passage 53** The Earth revolves about the Sun in the circumference of a circle, the Sun lying in the middle of the orbit, and the sphere of the fixed stars, situated around the same centre [that is, the Sun], is so great that the circle in which he supposes the Earth to revolve bears such a proportion to the distance of the fixed stars as the centre of this sphere [that is, the Sun] bears to [the] surface [of the sphere in which the Earth moves].

So, according to Boter's interpretation, the Sun is the centre of both the circle in which the Earth moves and the sphere of the fixed stars. It is now clear what was originally meant: 'The Sun is to the circle in which the Earth moves as this circle is to the sphere of the fixed stars'. Or (if we wish to use the definitions on page 129), 'The Sun is to the *cosmos* as the *cosmos* is to the *firmament*'. (See Figure 4.2.) In fact, this is the only interpretation that really makes sense given the small number of spheres available (namely, that of the Earth, that of the Sun, that of the Earth's orbit, and that of the fixed stars).

This can be expressed in mathematical terms by saying that the radius of the Sun s is to the radius of the circle in which the Earth moves S as the radius of this circle is to the radius of the firmament F. That is,

$$\frac{S}{s} = \frac{F}{S}. \tag{4.2}$$

Under this new light, Archimedes' criticism of Aristarchus is out of place. It was always rather suspicious that a person whom we know (from our analysis of his first book *On Sizes*) was the very embodiment of shrewdness, paying attention to the most incredibly tiny mathematical details, a person whose optical propositions (Propositions 1 to 3) are flawless, while some of those in Euclid's *Optics* are flawed (as we saw in Section 2.1), should have made such an 'obvious' mistake as trying to find the ratio of a sizeless point (in the Euclidean sense of the word) to something that has size (let it be what it may). So it turns out that Aristarchus was speaking of the 'centre of the universe' not in the rigorously Euclidean sense of the word, but rather, as Boter (2007:427) says, in the same loose sense in which we speak of the 'centre of the city'.

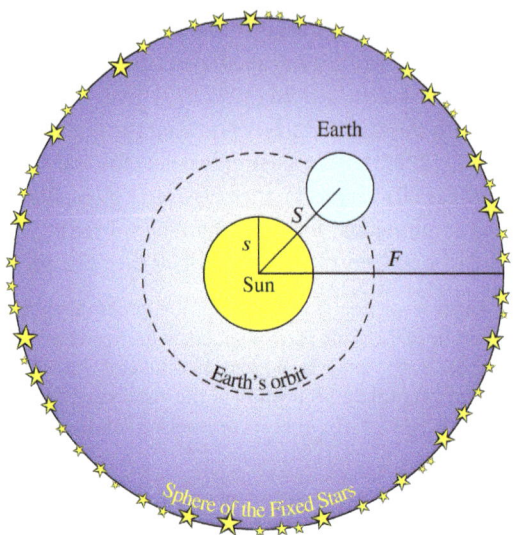

Figure 4.2 In Aristarchus' model of the universe (as inspired by Boter's reading), the Sun is to the Earth's orbit as this orbit is to the sphere of the fixed stars. (Not to scale.)

It follows from either Equations 4.1 and 4.2 or Figures 4.1 and 4.2 that Aristarchus' universe was significantly different from what Archimedes made of it. The former seems smaller than the latter (by a factor of s/e). But, is it? Before venturing an answer, let us first address another question concerning Aristarchus' universe. Was it finite or infinite? Did Aristarchus believe that the stars were all fixed to a celestial sphere, possibly made of some perfectly transparent substance, such as Plato's (1888:212) *aether* or Aristotle's (1929:324, 1922:289^a) *fifth element*? Or did he believe that the Sun was just another fixed star, floating motionless among the myriads of them and looking bigger only because it is closer?

Archimedes' words (in Passage 50), 'The fixed stars and the Sun remain unmoved', highlight what these motionless sources of light have in common by putting them on a level, so that the possibility that they might be the same thing simply suggests itself. A similar suggestion arises, perhaps a little more strongly, from Aetius' words (in *Opinions* 2.24.8 and *Selections* 1.25.3), which run as follows.

> **Passage 54** Aristarchus sets the Sun [motionless] among the fixed stars and holds that the Earth moves round the Sun's circle [that is, the ecliptic] and is put in shadow

according to [the Earth's] inclinations (Heath, 1913:305; see also Mansfeld and Runia, 2009:570; Diels, 1879:355; Wachsmuth and Hense, 1884:212).

So, according to Aetius' reading of Aristarchus, the Sun lies motionless among the fixed stars. These words are enough to trigger anyone's imagination, and, that of the ancient Greeks being so lively, it did not take long after Aristarchus published his second book *On Sizes* for someone, like Posidonius of Apamea (c. 135 – c. 51 BC), as rescued by Cleomedes (*Heavens* 1.11), to speculate as follows.

> **Passage 55** While the Earth has the size demonstrated through the procedures just described, there are several ways of proving that it has the ratio of a point not only to the total size of the cosmos, but also to the height of the Sun, which the sphere that encloses the fixed stars far exceeds. …A [single] pitcher of water would not measure the sea, not even the Nile. So just as the pitcher has no [significant] ratio to the [quantities] mentioned, so too the size of the Earth has no [significant] ratio to the size of the cosmos. …Although the Sun is much larger than the earth and sea combined, it sends out to us …an appearance of being about one foot wide, despite being very bright. We can thus form the notion that the Earth, if we should look toward it from the height of the Sun, would either not be seen at all, or be seen with the size of a minuscule star; but if by hypothesis we were elevated to a distance far beyond the Sun, and right up to the sphere of the fixed stars, the Earth would not be seen by us at all, not even if imagined as having a brightness equal to [that of] the Sun. Hence the [fixed] stars too must be larger than the Earth, in that they are visible from it, whereas the Earth could not be seen from the height of the sphere of the fixed stars. The Earth is certainly far smaller in size than the Sun, since the Sun itself too, if imagined at the height of the fixed stars, will perhaps appear as large as a star (Bowen and Todd, 2004:86-9; see also Ziegler, 1891:102-6).

These words seem directly inspired by a reading of Aristarchus' second book *On Sizes*, where, it appears, the Sun and the Earth were said to be like points compared to the vast size of the sphere of the fixed stars, just as Archimedes' words (in Passage 50) suggest (Collinder, 1964:76).

Arguments had been going on since at least the time of Archelaus, a pupil of Anaxagoras, that the universe was boundless.

> **Passage 56** [Archelaus] declared the Sun to be the largest of the heavenly bodies and the universe to be unlimited (Laertius, 1925a:146).

However, for all the passages read so far, we do not know exactly what Aristarchus said about it. Suggestive though they are, these passages do not provide enough evidence for us to say that he took the universe to be boundless. So, for now and until further proof is found, we shall proceed under the assumption that Aristarchus' universe (like Philolaus' and Aristotle's) was bounded by the sphere of the fixed stars.

But there is one thing that the above passages (and some more we will read later) allow us to do at this point: they allow us to reconstruct some of the hypotheses Aristarchus wrote in his second book *On Sizes*. (The author has carefully chosen the wording of these to faithfully reflect only what the extant sources convey.)

A reconstruction of the hypotheses in *On Sizes 2*

Hypothesis 1 The fixed stars and the Sun stand still (Diels, 1879:355).

Hypothesis 2 The Earth moves in a circle around the Sun (Diels, 1879:355).

Hypothesis 3 The Earth spins on its axis once a day as it revolves around the Sun (Passages 29 and 41).

Hypothesis 4 The circle in which the Earth moves is as wide as (what most astronomers call) the *cosmos* (Passage 52).

Hypothesis 5 The fixed stars and the circle in which the Earth moves are both centred at the Sun (Boter, 2007:428).

Hypothesis 6 The Sun bears such a proportion to the sphere in which the Earth moves as this sphere bears to that of the fixed stars.

Hypothesis 7 The Sun appears to be about 1/720th part of the zodiac circle (Heath, 1897:223; see also Heiberg, 1881:248).

Another thing we can do now is to be more precise about the size of Aristarchus' universe. First, Figure 2.9 (on page 69) gives an idea of the kind of cosmos he imagined in his first book *On Sizes*. In this figure, everything is to scale except the Sun's distance. Should this also be set to scale, it would be clear what a huge *cosmos* he was seeing in his mind. The size of the firmament corresponding to this *cosmos* can be found as follows. Dividing Inequality 2.48 by Inequality 2.35 gives

$$\frac{513/2}{43/6} < \frac{S/e}{s/e} < \frac{4300/9}{19/3}, \tag{4.3}$$

or simply,

$$\frac{1539}{43} < \frac{S}{s} < \frac{4300}{57}. \tag{4.4}$$

The size of the firmament that follows from this and Equation 4.2, is

$$\frac{1539}{43} < \frac{F}{S} < \frac{4300}{57}. \tag{4.5}$$

In other words, the firmament is $35^{34}/_{43}$ to $75^{25}/_{57}$ times as wide as the cosmos. Or, if we want the exact answer that follows from our solution in Chapter 2, we simply divide Equation 2.30 by Equation 2.32 to obtain

$$\frac{S}{s} = \frac{384.2303...}{6.7057...} = 57.2986... \tag{4.6}$$

So, using Equation 4.2 again, the firmament in *On Sizes* 1 is about 57 times as wide as the cosmos. That is,

$$\frac{F}{S} = 57.2986... \tag{4.7}$$

We can also express the size of the firmament (that is, the sphere of the fixed stars) in terms of the Earth's radius. Thus, rearranging Inequality 4.5, we have

$$\frac{1539}{43}S < F < \frac{4300}{57}S.$$

Using Inequality 2.48 gives

$$\frac{1539}{43}\left(\frac{513e}{2}\right) < F < \frac{4300}{57}\left(\frac{4300e}{9}\right),$$

which simplifies to

$$\frac{789507}{86} < \frac{F}{e} < \frac{18490000}{513}. \tag{4.8}$$

So the firmament in *On Sizes* 1 is $9180^{27}/_{86}$ to $36042^{454}/_{513}$ times as wide as the Earth. The exact solution can be found by manipulating Equation 4.2 as follows. Swapping sides gives

$$\frac{F}{S} = \frac{S}{s}.$$

Multiplying through by e/e gives

$$\frac{F/e}{S/e} = \frac{S/e}{s/e}.$$

Multiplying through by S/e gives

$$\frac{F}{e} = \frac{(S/e)^2}{s/e}. \tag{4.9}$$

Substituting Equations 2.30 and 2.32 into Equation 4.9, we obtain

$$\frac{F}{e} = \frac{(384.2303...)^2}{(6.7057...)} \approx 22016. \quad (4.10)$$

So, the firmament in *On Sizes* 1 is about 22016 times as wide as the Earth. Curiously, this is nearly as wide as the real Earth's orbit, which, according to modern standards (Pitjeva and Standish, 2009:370; Luzum et al., 2011:296), is about 23455 times as wide as the Earth.

Should we be able to visualize this along with a fully scaled Figure 2.9, we would get an idea of what huge, mind boggling sizes and distances Aristarchus was seeing in his head and what a quantum leap this was from anything that came before. But even these are just approximations, because, by the time he wrote his second book *On Sizes*, he had made significant changes to the knowledge gained from his earlier work *On Sizes*.

In order to get some insight into the kind of universe he presented in his second book *On Sizes*, we may use Archimedes' report that Aristarchus put the angular size of the Sun (and Moon) at half a degree. This is four times less than the estimate he used in *On Sizes* 1, so the Moon in *On Sizes* 2 is simply four times as far away as that in *On Sizes* 1. That is, four times that in Inequality 2.47, or

$$57 < \frac{M}{e} < \frac{860}{9}. \quad (4.11)$$

Putting this into Inequality 2.12, we obtain a new range of estimates for the Sun's distance. Namely,

$$18 \times 57 < \frac{S}{e} < 20 \times \frac{860}{9},$$

or, simply,

$$1026 < \frac{S}{e} < \frac{17200}{9}. \quad (4.12)$$

Also, multiplying Inequality 4.5 by four gives an estimate of the size of the firmament that Aristarchus may have considered in *On Sizes* 2. Namely,

$$\frac{6156}{43} < \frac{F}{S} < \frac{17200}{57}. \quad (4.13)$$

So, the firmament in *On Sizes* 2 is $143^7/_{43}$ to $301^{43}/_{57}$ times as wide as the cosmos (which, in modern terms, is about three to six times the size of the Kuiper belt, or about once to twice the size of the heliosphere).

Putting Inequality 4.12 into this gives

$$\frac{6156}{43} \times 1026 < \frac{F}{e} < \frac{17200}{9} \times \frac{17200}{57},$$

or, simply,

$$\frac{6316056}{43} < \frac{F}{e} < \frac{295840000}{513}. \tag{4.14}$$

So, the implications of putting the angular size of the Sun and Moon at half a degree (and leaving everything else unchanged in *On Sizes* 1) are that Aristarchus used new estimates for the distance of the Moon, Sun, and firmament in his second book *On Sizes*, and these estimates can be worked out as 57 to 95⅝, 1026 to 1911⅑, and $146887^{137}/_{344}$ to $576686^{82}/_{513}$ Earth radii, each. (The latter firmament is larger than the real orbit of Jupiter, but smaller than that of Uranus.)

There is another question concerning Aristarchus' universe that we may address now: Where are the planets? As Carman (2018:6, 12) points out, quoting Neugebauer (1975:692), there is no mention of them in Aristarchus' first book *On Sizes* or Archimedes' *Sand Reckoner*, and there is no record of anyone before Copernicus saying that Aristarchus had any planet (other than Earth) moving around the Sun. There are, however, a few instances of people saying that Mercury and Venus revolve around the Sun (as in Passages 36 to 39), but they do not mention Aristarchus or extend this type motion to the outer planets.

According to Neugebauer (1975:692), the fact that the ancient sources are silent on this topic suggests it is unlikely that Aristarchus ever developed any planetary theory. For Carman (2018:9, 12), this silence is strong evidence that this is indeed so, and though he knows this goes against the old aphorism that 'absence of evidence is not evidence of absence', he is willing to make an exception here because it seems to him that 'the absence of evidence is eloquent in this case' (Carman, 2018:11). However, the laws of logic apply to all cases and cannot be bent to establish exceptional implications. So, as he himself is aware, it is impossible to prove his point. In fact, a myriad explanations immediately jump to mind that effectively account for the mentioned lack of evidence.[25] Besides, there is (at least) one piece of evidence we may wish to consider. Namely, Philolaus.

---

25 Carman (2018:2) himself quotes the following as possible explanations for the rejection of Aristarchus' theory in antiquity (which might also have reduced the chances of survival of his works): 'The status of crisis that the geocentric paradigm had in Copernicus', but not in Aristarchus' times (Kuhn, 1962:75-6); the great authority that geocentric astronomers like Hipparchus (Heath, 1913:308) and geocentric philosophers like Plato and Aristotle

Philolaus did not think the Earth was so special. What was really special for him was the Centre of the Universe, and whatever happened to be there had everything else moving around it (as in a whirlpool). So, if we put the Earth at the centre, then everything will move around the Earth, but if we put there something else, then everything will move around this 'something else', and this 'everything' includes the planets. Philolaus explicitly said (in Passage 10) that all the planets revolve around the centre of the universe, which he also said was occupied by the Central Hearth (or *Hestia*), and it follows from the way he presents things that they did so at different speeds, otherwise, the Antichthon (or antichthons) could not cause eclipses (as explained in Passages 23 and 24). Thus, the Earth orbited the centre of the universe in one day, the Moon in 29½ days, the Sun in 364½ days (as follows from Passages 15 and 16), and in general, the further out, the slower the orbiting body moved.

The only thing Aristarchus did was to replace Hestia with the Sun at the centre of the universe and slow down the outermost sphere (that of the stars) so much as to make it still. The idea that all planets should move around the Sun in this model suggests itself very strongly, unless, of course, we choose to ignore Philolaus' great contribution and assume that Aristarchus ignored it too. In all likelihood, he did not. By the time he started, the planets had been moving around something other than the Earth for over a century (at least) in Pythagorean circles. Hypothesis 5 clearly states that the Sun is the centre of the universe. Philolaus would have known exactly what these words mean. Would so Aristarchus? A clue to the answer is given by Aristotle ($1922:293^a$) when he says (in *On the Heavens* 2.13) that, for the Pythagoreans, 'the Earth is one of the [wandering] stars, moving [like the other planets] around the centre [of the universe]'.

There is one thing Aristarchus (unlike Philolaus) kept moving around the Earth. Namely, the Moon. His calculations in *On Sizes* showed him that the Moon is 19 ± 1 times closer than the Sun, which makes it impossible for the Moon to orbit anything other than the Earth.

Another piece of evidence we might wish to consider is the fact that Archimedes (in Passage 20) took the planets to be arranged in what looks like a slight variation of the order Ptolemy ascribed to the 'foremost astronomers', which we identified with the Pythagoreans in Section 1.5. As can be seen there (in

---

(Erhardt and Erhardt-Siebold, 1942:595; North, 2008:86) had; or even the good luck that Copernicus had "to be born not only at a time when science was beginning to reach, so to say, a critical mass, but also at a time when scientific works were beginning to be printed" (Gingerich, 1985:41)'.

Table 1.1), both planetary arrangements are identical but for the positions assigned to Mercury and Venus. While Ptolemy (and the Pythagoreans) followed the strict order of increasing sidereal period (which makes sense from both a geocentric and hestiocentric point of view), Archimedes (despite being a geocentrist) placed Venus further from the Sun than Mercury, which is very odd, because this is exactly what might be expected from a heliocentrist trying to explain why the greatest elongation of Venus from the Sun is greater than that of Mercury. Now, do we know of any heliocentrist whose works Archimedes read so intently, and perhaps also admire so much as to influence his own calculations? (Luckily, the list is not very long.)

As Osborne (1983:235) points out, there were two main schools of thought regarding the order of the planets. Whether one of these originated with Philolaus and the second with Plato does not really matter. What matters is that at some point, Aristarchus (who, for some reason, was Strato's most famous pupil) must have faced the question of which system was best. They only differ in the position of Mercury and Venus, and while attempting to understand why these planets never stray too much from the Sun, he might have stumbled upon heliocentrism. The Moon orbits the Earth because it is smaller than the Earth, which in turn orbits the Sun because it is smaller than the Sun. So, apparently, smaller objects orbit bigger ones. So what should Venus and Mercury do? From his calculations in *On Sizes*, Aristarchus knew that if these planets lie between the Moon and the Sun (as Philolaus proposed), then they might be bigger than the Moon, but certainly not bigger than the Sun; otherwise, their apparent size would exceed that of the Sun (and Moon), which is clearly not the case. In fact, simple observation suggests that they are very tiny compared to the Sun. Hence they should orbit the Sun, or, if we prefer, they must orbit the Sun in Aristarchus' model, just as they orbit the Central Hearth in Philolaus', or, more simply, they (as well as all other planets) orbit the centre of the universe in both models.

So, as we said, it is most likely that at some point, Aristarchus braved the Venus and Mercury question. In fact, considering how brilliant minds work (and assuming his was one of those), the mere suggestion that he did not is akin to (subtly) suggesting he was not very bright. But the evidence so far points to the contrary. Would the person Archimedes admired most not be very bright? As history has posed it, this apparently simple yes-no question is still a live subject of debate. Sadly, we do not have Aristarchus' second book to tell us, and in fact, should Copernicus' book have suffered the same fate, we might never have known that he developed any planetary theory either. But fortunately, we do have *The Sand Reckoner*, miraculously snatched from the jaws of time to bless our eyes with a glimpse of the truth.

4.2 Archimedes' assumptions

Archimedes wants to deliberately increase the size of Aristarchus' huge universe in order to impress his readers with an even huger one and show that mathematics can easily deal with such numbers. In order to do so, he makes the following assumptions.

Assumptions in *The Sand Reckoner* (Heiberg, 1881:246; Heath, 1897:222)

Assumption 1 The Earth's girth is about 300 myriad stades and no greater.
Assumption 2 The Sun is larger than the Earth, which is larger than the Moon.
Assumption 3 The Sun is no wider than 30 Moons.
Assumption 4 The side of a regular chiliagon inscribed in the *cosmos* is no wider than the Sun.
Assumption 5 A myriad grains of sand do not take up more space than a poppy seed, and a finger is no wider than 40 poppy seeds.

Let us start with the first of these. Among the estimates for the Earth's circumference available to Archimedes there were (at least) two: namely, the one Aristotle (1922:298^a) reported as 40 myriad stades and the one Archimedes himself reported as 30 myriad stades. Both men ascribed these values to previous mathematicians, without naming them. (This is significant, because if Aristarchus had been the originator of the latter value, Archimedes would surely have said so.) Perhaps this vaguely plural attribution is justified, since, as Collinder (1964:73) says, these values may each have been the result of a collective effort, rather than a single individual's. Others, like Fortenbaugh and Schütrumpf (2001:365), think the extant evidence, though thin, points to Dicaearchus of Messana as the most likely author of the 30-myriad-stade value.

In any case, Cleomedes (Bowen and Todd, 2004:69), without giving attributions either, uses this very value as part of an argument to prove the sphericity of the Earth by showing how absurd the arguments against are. In doing so, he mentions that, if the distance between Lysimachia and Syene is 2 myriad stades, and if the head of Draco is just above Lysimachia when Cancer is just above Syene, and if the distance between the mentioned star signs is 1/15 of the whole celestial circle passing through them, then the length of the whole circle passing thorough Lysimachia and Syene is 15 × (2 myriad stades) = 30 myriad stades. This gives an idea of how the 30-myriad-stade value might have been obtained. The assumptions used above are all very rough, as is the approximation to pi Cleomedes used (namely, $\pi \approx 3$). Dividing this into the Earth's girth, he obtained the Earth's width as 10 myriad stades.

Archimedes chooses the estimate of 30 myriad stades as the length of the Earth's girth, and to make sure he is not falling short of the truth, he arbitrarily increases it tenfold, obtaining 300 myriad stades. He thus makes it clear he is not after accuracy, but after the deliberate exaggeration of existing estimates.

Archimedes' Assumption 2 (that the Moon is smaller than the Earth, which is smaller than the Sun) is simply correct as we know it and as was known to most astronomers of his time. It had been proved beyond doubt in Aristarchus' first book *On Sizes*.

Archimedes' Assumption 3 (that the Sun is about 30 times as wide as the Moon) is most interesting in that Archimedes gives unique data concerning the estimates of three of his favourite astronomers: Eudoxus, Phidias, and Aristarchus. According to these astronomers, the Sun is, respectively, 9, 12, and 19 ± 1 times as wide as the Moon. In saying this, Archimedes quotes the same value we encountered in Proposition 7 and, by extension, Proposition 9 of *On Sizes* (which in the present book, we rendered as Inequalities 2.12 and 2.17, respectively). Then, he arbitrarily goes beyond the largest of these values in order to establish his proposition 'beyond dispute'. Thus, his Sun was taken to be 30 times as wide (and far away) as the Moon.

Archimedes' Assumption 4 states that the Sun is wider than the side of a **chiliagon** (or regular polygon of a thousand sides) inscribed in the greatest circle in the universe (Heath, 1897:223). By this he means the Sun's path (or orbit) around the Earth. That is, the *cosmos* (or greatest circle in Anaximander's universe, as defined on page 129). We can quickly see that this assumption is correct, because the side of a chiliagon is $360°/1000 = 21'36''$ wide, which is less than the Sun's apparent size. At this point, Archimedes makes an astonishing revelation when he says that Aristarchus discovered that the Sun's apparent size goes 720 times into the circle of the zodiac. That is, the Sun appears to be $360°/720 = 30'$ wide. This is the first time in (recorded) history that someone comes up with this virtually perfect value, which could only have been obtained empirically. Archimedes has a reason to admire his hero, but takes nothing for granted, so he repeats the measurement himself by a method he immediately describes.

The method Archimedes describes is potentially dangerous, as one might become blind by accidentally looking into the Sun. So it is recommended to proceed with the utmost care. Performed carefully, it yields upper and lower bounds for the Sun's apparent size. These are obtained as follows.

- Place a white cylinder so that it blocks the rising Sun completely, and place a black cylinder so that some of the white cylinder shows on either side of the black one. The angle between tangents to these cylinders will be an upper bound for the Sun's angular size.
- Place a white cylinder so that some of the Sun shows on either side of it, and place a dark cylinder so that it blocks the white one completely. The angle between tangents to these cylinders will be a lower bound for the Sun's angular size.

Archimedes recommends placing these cylinders upright on a straight, flat board. Hence his choice of a rising Sun, whose horizontal axis (which is what he wants to measure) is not affected by atmospheric refraction, unlike its vertical axis (which is always shrunk, specially when near the horizon).

The mentioned angles θ can be either measured directly (by means of strings and a protractor) or calculated by using the following modern (trigonometric) equation,

$$\theta = 2 \sin^{-1}\left(\frac{R-r}{d}\right), \qquad (4.15)$$

where R is the radius of the white cylinder, r is the radius of the black cylinder, and d is the distance between the centres of these cylinders.[26]

Using this method, Archimedes found the following upper and lower bounds for the Sun's angular size θ.

$$\frac{90°}{200} \leq \theta \leq \frac{90°}{164}. \qquad (4.16)$$

The lower of these bounds is exactly $90°/200 = 27'$, and the upper of these bounds is exactly $90°/164 = 32'55''36'''...$, which is suspiciously close to $33'$. Mathematicians usually have a quick eye for rounding discrepancies and here is a most likely one. In fact, it makes sense to think that Archimedes (who confessed this method is not 'reliable enough') may have chosen his bounds to be $3'$ either side of Aristarchus' $30'$ angular size (to play on the safe side). That is, he seems to have chosen $\theta = 30' \pm 3'$. When converting these sexagesimal bounds into what was then a more conventional language, such as fractions of a quadrant, he chose his denominators to be 200 and 164, since no other whole numbers can do

---

26 Two interactive online illustrations of this experiment can be found by googling the words 'Archimedes' sunwidth measurement' and 'Archimedes' sunwidth reckoning'.

the job best. In particular, the latter is the whole number that gets closest to 33′ when divided into 90 degrees. That is, Inequality 4.16 is close to

$$27' \leq \theta \leq 33'. \tag{4.17}$$

The first time the author thought of this, back in the summer of 2010, while in bed, he immediately realized that Archimedes may have used sexagesimal units to obtain his bounds, and then expressed them into fractions of a quadrant, choosing the best fractional approximations that could possibly be found. Under this light, Inequality 4.17 preceded Inequality 4.16, rather than the other way round.

On June 2, 2011, the author communicated this knowledge to Rawlins, who, inspired by the author's mail, wrote up a short paper on the subject. In this paper, he did not mention his source of inspiration, surely because he genuinely believed he had discovered it himself, as he (2012:3) claims on the very first page of it. But the fact is that the author's words (in the long document attached to his mail), explicitly mentioning the 3′ difference and clearly stating that 'Archimedes' results average the same as Aristarchus', were intended to communicate this discovery. According to Rawlins (and this is genuinely *his* realization), this discovery is strong evidence that the Greeks were starting to adopt the Babylonian way of measuring angles this early in history.

Surely, had there been no mail from the author to Rawlins, there would have been no paper by Rawlins on Archimedes' sunwidth values, because the connection between Inequalities 4.16 and 4.17 does not seem to be immediately obvious to everyone. In fact, Delambre (1817:104) and Shapiro (1975:77) were very close to having made it themselves, yet never saw the 33 minutes of arc hiding behind Archimedes' upper bound. Instead, the latter man even thought it was proof that Archimedes never used sexagesimal degrees (because the match is not perfect), which is the opposite of what the author proposes (because the match is as perfect as it can possibly be).

Neither theory can be proved beyond doubt (nor can, in fact, any theory outside the field of mathematics), but the likelihood of the latter one increases when we note that Archimedes' upper and lower bounds, rather than looking like actual measurements, look artificially chosen to be equidistant from Aristarchus' value. The reason for doing so is that Archimedes, doubting the reliability of the method he describes—as when he says that 'neither vision, hands, nor the instruments required to measure this angle are reliable enough' (Vardi, 1997:3), may have finally decided it was safer to rely on Aristarchus' value and put his own bounds a reasonable distance either side of it.

He then goes on to say (in *Sand Reckoner* 1.17) that assuming Inequality 4.16 is correct, it is possible to prove that the Sun is wider than the side of a chiliagon inscribed in its path. Then, he gives a long geometric proof that this is indeed the case for both his upper and lower bounds (and hence, any value in between, such as Aristarchus' 720th part of a circle). In doing so, Archimedes proves something which is obvious from first principles, even for a child like Gelo, for whom the book thus provides as gentle and inspiring an introduction to the language of mathematical proof as the genius of Archimedes could make it. Thus, the side of a 1000-gon is clearly shorter than the side of a 720-gon (provided they are both regular and inscribed in the same circle).

Now, Inequality 4.16 can be rewritten (in either degrees or radians) as

$$\frac{360°}{800} \leq \theta \leq \frac{360°}{656} \quad \text{or} \quad \frac{2\pi}{800} \leq \theta \leq \frac{2\pi}{656}, \tag{4.18}$$

and the above-mentioned angles (namely, that by Aristarchus, those by Archimedes, and that subtended by a chiliagon's side) compare as follows.

$$\frac{360°}{1000} < \frac{360°}{800} < \frac{360°}{720} < \frac{360°}{656} \quad \text{or} \quad \frac{2\pi}{1000} < \frac{2\pi}{800} < \frac{2\pi}{720} < \frac{2\pi}{656}. \tag{4.19}$$

In other words, the side of a 1000-gon is smaller than that of an 800-gon, which is smaller than that of a 720-gon, which is smaller than that of a 656-gon (when these figures are all regular and of like radius).

4.3 Archimedes' telltale ratio

In his proof that the side of a 1000-gon is smaller than that of a 656-gon, Archimedes argues that the further the Sun is, the smaller it looks. In particular, it looks smaller when seen from the Earth's centre than when seen from a point on the sunlit face of the Earth. In other words, when the Sun is fully above the horizon (as in Figure 4.3), the angle *RED* (between tangents to the Sun and the Earth's centre at *E*) is smaller than the angle *NOT* (between tangents to the Sun and the observer's eye at *O*).

That is, the *RED* angle is smaller than the *NOT* angle. Denoting these, respectfully, by η and θ, we have

$$\eta < \theta, \tag{4.20}$$

provided the Sun's centre is above the horizon.

In fact, the angle *NOT* (in Figure 4.3) is the same as the angle θ (in Inequality 4.18), which is at least 656 times smaller than a whole circle (such as that in which the Sun moves).

Archimedes' telltale ratio

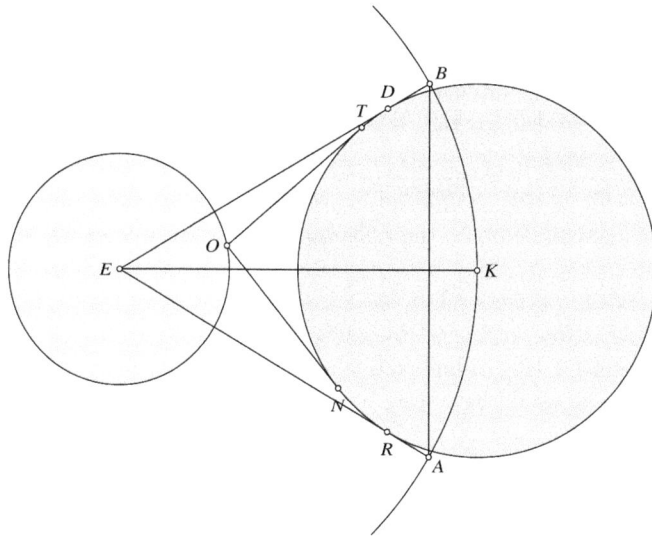

Figure 4.3 Partial illustration of Archimedes' proof that the sides of a 1000-gon are smaller than the sides of a 656-gon.

Thus, combining Inequalities 4.18 and 4.20, we have

$$\eta < \theta \leq \frac{2\pi}{656}. \tag{4.21}$$

Using Archimedes' upper bound for pi (namely, 22/7), the latter inequality becomes

$$\eta < \theta < \left(\frac{2}{656} \times \frac{22}{7} = \frac{11}{1148}\right), \tag{4.22}$$

which is a most interesting ratio, because Archimedes explicitly describes it as the ratio between the Sun's width and distance from Earth. He says so in Passage 57 (reproduced below), after which he gives a proof (not repeated here) that the line segments AB and EK (in Figure 4.3) are equal to the Sun's diameter and distance from Earth.

> **Passage 57** The ratio of AB to EK is less than the ratio of eleven to one thousand one hundred and forty-eight (Heiberg, 1881:258; Vardi, 1997:4).

Not only is Inequality 4.22 a good estimate of the Sun's width-to-distance ratio, but also a good upper bound for the Sun's angular size θ in radians.

Using modern data for the Sun's (photospheric) radius and distance (namely, $s = 695508$ km and $S = 149597870.7$ km, as given by Brown and Christensen-Dalsgaard, 1998:L197 and Luzum et al., 2011:296), we see that the true value of the Sun's width-to-distance ratio is

$$\frac{2s}{S} = \frac{2 \times 695508 \text{ km}}{149597870.7 \text{ km}} = \frac{4636720}{498659569} \approx \frac{11}{1183}, \qquad (4.23)$$

where the approximation on the right-hand side is accurate to five significant figures.

Archimedes' upper bound for the Sun's width-to-distance ratio (in Inequality 4.22) is, therefore, very good, since it is only slightly bigger than the true ratio (in Equation 4.23). Had he bothered to write (in his book) also the lower bound for this ratio, he would have used Inequality 4.18 again and either $\pi \approx 3$ or $\pi \approx 22/7$ to obtain

$$\left(\frac{2}{800} \times 3 = \frac{3}{400}\right) < \left(\frac{2}{800} \times \frac{22}{7} = \frac{11}{1400}\right) < \eta < \theta. \qquad (4.24)$$

Surprisingly, though the concept of *radian* appears to come much later in history, Archimedes is using it here (without giving it a name) to establish a link between an angle (namely, θ) and the lengths of two line segments (namely, AB and EK). Thus, in radians,

$$\frac{3}{400} < \frac{11}{1400} < \theta < \frac{11}{1148}. \qquad (4.25)$$

So, a round object that is as wide as the segment AB and is viewed from a distance equal to the length of the segment EK subtends an angle θ whose measure in radians is equal to the quotient of the mentioned line segments. That is,

$$\theta = \left(\frac{AB}{EK} \text{ radians}\right) = \left(\frac{2s}{S} \text{ radians}\right), \qquad (4.26)$$

where all variables are as before.

Did Archimedes learn this in Aristarchus' second book *On Sizes*? We do not know, but should the answer be 'yes', we might even speculate that Aristarchus' value for the Sun's width-to-distance ratio was somewhere halfway between the bounds in Inequality 4.25. The exact value depends on which approximation to pi Aristarchus used. This is another thing we cannot be sure of, but we may try anyway and start by considering the possibility that he used the same upper bound for pi as Archimedes used in *The Sand Reckoner* (namely, $\pi \approx 22/7$). His solar width-to-distance ratio (as derived from Inequality 4.19) would then be

$$\frac{2s}{S} = \frac{2\pi}{720} < \left(\frac{2}{720} \times \frac{22}{7} = \frac{11}{1260}\right). \qquad (4.27)$$

We know that Aristarchus' book and Archimedes' upper bound for pi both preceded Archimedes' book, but we do not know which of the former two preceded which. If Archimedes found his upper bound for pi very early in his career, there is a chance that this value may have found its way into Aristarchus' second book *On Sizes*. But all things considered, this is an unlikely possibility. It is much more likely that Aristarchus used a different approximation to pi than the one considered above. In *The Sand Reckoner*, Archimedes gives the upper and lower bounds for pi as 22/7 and 3, whereas in his *Measurement of a Circle* (Heath, 1897:93), he gives them as 22/7 and 223/71. This suggests that Archimedes finished *The Sand Reckoner* before he found his lower bound for pi, as Knorr (1978:235) says, but it is also possible that he had already found his lower bound for pi before he started *The Sand Reckoner*, yet decided to keep using the number 3 there for reasons such as convenience or pedagogy. We do not know. But we know that the number 3 was not the only approximation to pi available at the time. Table 4.1 gives some of these early approximations—lone survivors of a bygone age.

Assuming Aristarchus used one of these, the following possibilities arise.

– Using $\pi \approx 3$, Aristarchus' solar width-to-distance ratio becomes

$$\frac{2s}{S} = \frac{2\pi}{720} \approx \left(\frac{2}{720} \times 3 = \frac{1}{120}\right). \tag{4.28}$$

– Using $\pi \approx 25/8$, Aristarchus' solar width-to-distance ratio becomes

$$\frac{2s}{S} = \frac{2\pi}{720} \approx \left(\frac{2}{720} \times \frac{25}{8} = \frac{5}{576}\right). \tag{4.29}$$

Table 4.1: Some pre-Archimedean approximations to pi

source	year (BC)	π
Moscow Mathematical Papyrus (Problem 10)	1897 ± 95	256/81
Ahmes Papyrus (Problems 41-48)	1837 ± 23	256/81
Yale Babylonian Collection No. 7302*	1800 ± 200	3
Susa Mathematical Text No. 3 (line 30)[†]	1700 ± 100	25/8
First Book of Kings 7.23	550 ± 11	3
Second Book of Chronicles 4.2	325 ± 25	3
Cairo Papyri 89137-43 (Problems 32, 33, 36, 38)	3rd cent.	3

*If the numbers in this tablet are read as $45/60 = 9/(4\pi)$, then $\pi = 3$.
[†] As interpreted by Bruins and Rutten (1961:18,33) and Neugebauer (1969:47).

- Using $\pi \approx 256/81$, Aristarchus' solar width-to-distance ratio becomes

$$\frac{2s}{S} = \frac{2\pi}{720} \approx \left(\frac{2}{720} \times \frac{256}{81} = \frac{32}{3645}\right). \tag{4.30}$$

These ratios are all very close to each other, but one of them, the best, is also very close to the value halfway between Archimedes' upper and lower bounds (in Inequality 4.25), and in particular, to the geometric mean of these bounds. Namely,

$$\frac{2s}{S} = \sqrt{\frac{11}{1400} \times \frac{11}{1148}} \approx \frac{4}{461}. \tag{4.31}$$

The reason for choosing the geometric mean in the latter equation is that (unlike the arithmetic mean) it allows both the result and its inverse to be used in all four arithmetic operations. The advantage of this becomes apparent when another way of understanding the above ratios is considered which involves taking their reciprocals. In particular, the reciprocal of Equation 4.29 tells us that the Sun is $576/5 = 115.2$ times as far away as it is wide, and the reciprocal of Equation 4.31 tells us that the Sun is $461/4 = 115.25$ times as far away as it is wide. (In reality, the Sun is about $1183/11 = 107\tfrac{6}{11}$ times as far away as it is wide, so Aristarchus' ratio is only slightly better than the geometric mean of Archimedes' bounds for this ratio). Table 4.2 shows the reciprocals of the above ratios and the approximations to pi on which they are based.

Table 4.2: Reciprocals of the Sun's width-to-distance ratios

author	Inequality	$S/(2s)$	result	π
Aristarchus*	4.28	120/1	120	3
Aristarchus*	4.29	576/5	$115\tfrac{1}{5}$	25/8
Aristarchus*	4.30	3645/32	$113\tfrac{29}{32}$	256/81
Archimedes*	4.24	400/3	$133\tfrac{1}{3}$	3
Archimedes*	4.24	1400/11	$127\tfrac{3}{11}$	22/7
Archimedes†	4.31	461/4	$115\tfrac{1}{4}$	22/7
Archimedes	4.22	1148/11	$104\tfrac{4}{11}$	22/7
Reality	4.23	1183/11	$107\tfrac{6}{11}$	

*Hypothetical reconstruction
†Geometric mean

As we can see, the second and sixth rows in Table 4.2 are virtually the same. (Again, it seems that Archimedes has chosen the denominator of his upper bound in Inequality 4.16 to match Aristarchus' as close as is mathematically possible.)[27]

As said above, the Sun's width-to-distance ratio is also the angle (in radians) subtended by the Sun when seen from Earth, and (at least) to Archimedes, it matters where it is observed from, or rather, when. So (unlike in Figure 4.3, where, for illustrative reasons, the Sun is almost above the observer), Archimedes took his measurements at sunrise of all times. Apart from the advantage that this gives to those trying to keep their cylinders from slipping off the board, Archimedes had a more subtle reason for doing this. Namely, that observers on the Earth's surface watching the Sun rise are virtually as far away from the Sun's centre as are observers at the Earth's centre (though perhaps not as warm). In other words, at sunrise, the lengths of the line segments EK and OK (in Figure 4.3) are virtually the same, and therefore, so are the widths of the angles η and θ (in Inequality 4.20). So, Archimedes' bounds (in Inequality 4.25) are as valid for the angle η as they are for the angle θ. Hence his choice.

But there is more to Archimedes' eleven to eleven hundred and forty-eight ratio for the Sun's width and distance from Earth. Eventually, these numbers were taken to be more than just a ratio or an angle: they were taken to be actual distances measured in Earth radii. So, a Sun that is 11 Earth radii wide and 1148 Earth radii away subtends an angular size of exactly 11/1148 radians, or, as Archimedes put it, the 164th part of a quadrant (assuming $\pi \approx 22/7$). Now, what should Aristarchus' ratio be that inspired Archimedes' ratio? The answer, as given by Equation 4.29, is that the Sun is 10 Earth radii wide and 1152 Earth radii away (since 10/1152 is the expansion of 5/576 that best resembles Archimedes' 11/1148 ratio). Thus, according to this reconstruction, Aristarchus would have put the Sun's angular size at exactly 10/1152 radians, or, as Archimedes reported, the 720th part of a circle (assuming $\pi \approx 25/8$), or, equivalently, that the Sun is $1152/10 = 115.2$ times as far away as it is wide (as said in the paragraph under Equation 4.31).

If we allow for the possibility that Aristarchus used a different approximation to pi than that in Equation 4.29, then his ratio for the Sun's width and distance varies accordingly. Thus, a glimpse at Table 4.2 reveals the range of possible solar

---

27 The reciprocal of the arithmetic mean of these bounds, $2/(11/1400 + 11/1148) = 114.68...$, is different from the arithmetic mean of their reciprocals, $(1400/11 + 1148/11)/2 = 115.81...$, but the problem of choosing between these two different values does not arise when using the geometric mean, which produces a single value that can be inverted at will when needed.

distances that Aristarchus may have used, and in fact, a similar range was used by ancient, medieval, and Early Renaissance astronomers, whose solar distances (as pointed out in Section 3.1) do not venture far from Aristarchus'. For example, Ptolemy's (Toomer, 1984:257) solar distances range from 1160 to 1260 Earth radii, al-Battani's (1116:124v) range from 1070 to 1146 Earth radii, Copernicus' (1543:122v) range from 1105 to 1179 Earth radii, Tycho's (1602:98) range from 1101 to 1182 Earth radii, and so on (Riccioli, 1651:110; Hughes, 2001b:2).

According to this line of reasoning, Archimedes' 11/1148 ratio allows the recovery of Aristarchus' original ratio as 10/1152 to a degree of confidence that will be assessed later. This means that by the time he finished his second book *On Sizes*, Aristarchus had made some changes to the data in his first book *On Sizes*. For example, his reported estimate for the Sun's and Moon's angular size is

$$\theta = \frac{2\pi}{720} \text{ radians} = 30', \tag{4.32}$$

his recovered estimate for the Sun's distance S from Earth (in Earth radii e) is

$$\frac{S}{e} = 1152, \tag{4.33}$$

and his recovered estimate for the Sun's radius s (in Earth radii e) is

$$\frac{s}{e} = 5. \tag{4.34}$$

Assuming Aristarchus did not also change his 87-degree estimate for the angle ε between a half-looking Moon and the Sun, Equation 4.33 allows the recovery (via Equation 2.15) of Aristarchus' ultimate estimate for the Moon's distance M from Earth (in Earth radii e) as

$$\frac{M}{e} = \frac{S}{e} \cos \varepsilon = 1152 \cos(87°) = 60.291..., \tag{4.35}$$

which is virtually perfect. Furthermore, assuming Aristarchus rounded this to 60 Earth radii, and assuming also that the lunar distance of 168 myriad stades which Hippolytus (1851:46; 1921:78; 2016:115) ascribed to Aristarchus is correct, then Aristarchus' value (or one of Aristarchus' values) for the Earth's radius can also be recovered as

$$e = \frac{1680000 \text{ stades}}{60} = 28000 \text{ stades}, \tag{4.36}$$

and from this, it follows that his value for the Earth's circumference was

$$c_\oplus = (28000 \text{ stades}) \times 2 \times \frac{25}{8} = 175000 \text{ stades}, \tag{4.37}$$

which, rounded to the nearest myriad, is one of the values for the Earth's circumference attributed to Posidonius by Strabo in *Geography* 2.2.2 (Howarth, 1917:364).

It is to be noted that it is not attested whether Aristarchus or Archimedes originally meant their solar width-to-distance ratios to be read as actual distances in Earth radii, but this is exactly what they were understood to be by later astronomers. For example, Ptolemy took the distance between the Earth and the Sun to be 1210 Earth radii in *Almagest* 5.15 (Toomer, 1984:257), and in *Planetary Hypotheses* (Goldstein, 1967:7), he specified his upper and lower bounds for it as 1260 and 1160 Earth radii, each, giving enough instructions in *The Almagest* for these bounds to be worked out, as Pappus (Rome, 1931:107) and Neugebauer (1975:110) show. Proclus too quoted these bounds (with errors) in his *Commentary to Plato's Timaeus* (Diehl, 1906:62) and *Hypotyposis* (Manitius, 1974:222, 224).

Curiously, Ptolemy's mean solar distance is very close to the denominator of the fraction in Equation 4.28 expanded by the factor 10 (to resemble Archimedes' ratio), and his upper bound for the Sun's distance is exactly the same as the denominator of the fraction in Inequality 4.27, whose numerator is also the same as Ptolemy's estimate for the Sun's width (in Earth radii), as given in *Almagest* 5.16 (Toomer, 1984:257). We cannot be sure whether or not these are mere coincidences, because Ptolemy does not acknowledge Aristarchus as a source of inspiration.

The same can be said of the Syrian astronomer and mathematician al-Battani (c. AD 858 – 929), whose upper bound for the Sun's distance can be obtained in the same manner by simply using better approximations to pi than those available to Aristarchus, such as Ptolemy's 377/120 (Toomer, 1984:302) or Zu Chongzi's 355/113. Even the younger Aryabhata's 600/191 approximation (which he wrote down a few decades after al-Battani's death) produces the desired result (Datta, 1926:30). Thus, using any of these excellent approximations, say, the latter, we have

$$\frac{2s}{S} = \frac{2\pi}{720} \approx \frac{2}{720} \times \frac{600}{191} = \frac{10}{1146}. \tag{4.38}$$

But this is not how al-Battani allegedly got his solar distance. In fact, even Ptolemy asserts that his own solar distance was obtained by means of a certain approximation, whose derivation, as given in *Almagest* 5.15 (Toomer, 1984:255), is very much along the lines laid out by Hipparchus, as Ptolemy acknowledges in *Almagest* 5.14 (Toomer, 1984:254). This derivation can also be found in the works of Swerdlow (1969:294) and Toomer (1974a:130), as well as in Section C.2 of the present book. This approximation is important enough to receive a name of its own. So, using the fact that it is a **scalar field** (or function of many variables), it will be referred to in this book, whenever it is convenient, as the **solilunar distance field**. Rendered in modern guise (adapted from the cited authors), it is as follows.

$$\frac{S}{e} \approx \left(\left(\frac{\varphi}{\rho_\mathbb{C}} + 1\right)\sin\rho_\mathbb{C} - \frac{e}{M}\right)^{-1}, \tag{C.63}$$

where S is the distance of the Sun (in Earth radii e), M is that of the Moon, $\rho_\mathbb{C}$ is half the angular size of the Moon (to be taken, preferably, when it is equal to that of the Sun $\rho_\odot$), and φ is half the angular size of the Earth's shadow (as cast on the Moon).

Unfortunately, this approximation suffers from a serious drawback. As Swerdlow (1969:294) points out, it is extremely sensitive to small changes in its parameters. To see what this means, we may use Ptolemy's own set of parameters, which are given in *Almagest* 5.13-7 (Toomer, 1984:252, 254, 259; Neugebauer, 1975:106) and repeated here below.

$$\frac{M}{e} = 59 \pm \frac{31}{6} \quad \text{or} \quad \frac{323}{6} \leq \frac{M}{e} \leq \frac{385}{6}, \tag{4.39}$$

$$\rho_\odot = 15'40'', \tag{4.40}$$

$$\rho_\mathbb{C} = 16'40'' \pm 1' \quad \text{or} \quad 15'40'' \leq \rho_\mathbb{C} \leq 17'40'', \tag{4.41}$$

$$\varphi = 43'20'' \pm 2'40'' \quad \text{or} \quad 40'40'' \leq \varphi \leq 46'. \tag{4.42}$$

Putting these into Approximation C.63

– when the Moon is closest to Earth (and therefore, $\rho_\odot < \rho_\mathbb{C}$), we have

$$\frac{S}{e} \approx \left(\left(\frac{46'}{17'40''} + 1\right)\sin(17'40'') - \frac{6}{323}\right)^{-1} \approx -17841; \tag{4.43}$$

– when the Moon is at mean distance from Earth (and therefore, $\rho_\odot < \rho_\mathbb{C}$), we have

$$\frac{S}{e} \approx \left(\left(\frac{43'20''}{16'40''} + 1\right)\sin(16'40'') - \frac{1}{59}\right)^{-1} \approx 1984; \tag{4.44}$$

– and when the Moon is furthest from Earth (and therefore, $\rho_\odot = \rho_\mathbb{C}$), we have

$$\frac{S}{e} \approx \left(\left(\frac{40'40''}{15'40''} + 1\right)\sin(15'40'') - \frac{6}{385}\right)^{-1} \approx 1247. \tag{4.45}$$

Clearly, small changes in one (or more) of the parameters can lead to large changes in the result. Because of this, Approximation C.63 is said to be *ill-conditioned* with respect to M. Also, one of the above results is closer to the truth than the others. Namely, that corresponding to the Moon's mean distance. However, Ptolemy rejected this one and chose the one corresponding to the Moon's furthest distance. The reason why he did so is as follows.

In *Almagest* 5.14 (Toomer, 1984:252), Ptolemy says that his parameters were obtained by means of measurements made with a dioptra and that these measurements showed that the Sun's angular size does not vary appreciably (from the value in Equation 4.40), while that of the Moon does vary (within the range of Equation 4.41) and is never smaller than that of the Sun. This automatically excludes the possibility of annular eclipses, so Ptolemy's statement cannot be true. In reality, the Moon looks sometimes larger and sometimes smaller than the Sun, and it can look smaller by a greater amount than it can look larger. So, if Ptolemy was able to measure the smaller amount by which they differ, then he should have been able to measure the larger amount (if indeed he measured anything). Ptolemy says that in this, he differed from his 'predecessors', who believed that the time of **congruence** (that is, when $\rho_\odot = \rho_\mathbb{C}$) occurred at about mean lunar distance, while he thought it happened at extreme **apogee** (or furthest lunar distance). Clearly, his 'predecessors' were more correct than him on this.

We can now see why Ptolemy chose the result in Approximation 4.45 above all others. The negative one (in Approximation 4.43) was rejected outright. That in Approximation 4.44 is in fact the closer to the truth of the three results above and might have been viewed favourably by his 'predecessors', but Ptolemy preferred the one corresponding to apogee, since according to his own choice of parameters, this is when *congruence* occurs.

But even this one was not entirely pleasing to him. It was still too big. Thus, in order to obtain his 'mean' solar distance (of 1210 Earth radii), Ptolemy had to arbitrarily adjust his lower bounds for either $\rho_\mathbb{C}$ or φ in Approximation 4.45 to either $15'45''$ or $40'45''$, respectively. The former choice involved changing also his estimate for $\rho_\odot$, whereas the latter choice involved adding just one arcsecond to the value of $40'44''$ that results from multiplying his lower bound for $\rho_\mathbb{C}$ by the factor 13/5 (or 2⅗), which is the factor by which he estimated the Earth's shadow to be larger than the Moon's disc, as given in *Almagest* 6.7 (Toomer, 1984:304). That is, $15'40'' \times 13/5 = 40'44''$. So he chose to change φ. Note that Ptolemy's lower bound for φ is $40'40''$, rather than $40'44''$, as it should. Note also that the last of the above approximations is one that should be expected to yield an *extreme* (rather than *mean*) value, which is perhaps why Ptolemy never says in *The Almagest* whether his solar distance of 1210 Earth radii is a mean, minimum, or maximum value, though he calls it a 'mean' in his *Planetary Hypotheses* (Goldstein, 1967:9). Proclus, however, called it a 'maximum' in his *Hypotyposis* (Manitius, 1974:222; Neugebauer, 1975:110).[28]

It can be argued that Ptolemy's maths, as detailed by Swerdlow (1969:295) or Carman (2009:205), led him to his solar distance simply because of a fortuitous accumulation of rounding errors, but it can also be argued that he could have got the right answer (as given by Approximation 4.45) had he wanted to. It was within his grasp. One can only wonder whether his choice of result and method of calculation were both influenced by a secret desire to approach the denominator of the solar width-to-distance ratio in Equation 4.28 (namely, 10/1200) as much as possible by making the most of accumulating rounding errors in a formula he knew was highly sensitive to small changes in the parameters—a formula where the use of rounding can hardly be justified (except, of course, at the very last step). In other words, did the parameters lead to the result, or did the result lead to the parameters? This is not the first time this question has been posed. Toomer (1974a:131), for example, with respect to the mean distance of the Moon, of which we shall speak later, thinks that 'Ptolemy knew in advance approximately what distance he wanted, and selected that observational result which produced it'.

Centuries later, on trying to understand Ptolemy's ways and hopefully improve upon his results, al-Battani concluded that the Sun is at most 1146 Earth radii away. This result eventually became as canonical as Ptolemy's 1210 Earth radii (Swerdlow, 1973:102). The story of how it was found is more illustrative and complex than Inequality 4.38 suggests. Let us just dive into it.

Armed with what he thought was (and indeed is) a more accurate set of parameters than those in *The Almagest*, al-Battani was confident that Ptolemy's value could be improved. This new set of parameters included a variable angular radius for the Sun $\rho_\odot$, which he took from al-Khwarizmi's tables (Björnbo et al., 1914:175-80), as well as that of the Moon $\rho_\mathbb{C}$ and the Earth's shadow φ, which he himself measured from two lunar eclipses he had observed (al-Battani, 1899:84-6; Said and Stephenson, 1997:43). These parameters are neatly given by Swerdlow

---

28 A 3D plot of Approximation C.63 can be viewed online by googling the words 'solilunar distance field'.

(1973:99-100) and repeated here for convenience.

$$\frac{M}{e} = 59 \pm \frac{31}{6} \qquad \text{or} \qquad \frac{323}{6} \leq \frac{M}{e} \leq \frac{385}{6}, \tag{4.46}$$

$$\rho_\odot = 16'15'' \pm 35'' \qquad \text{or} \qquad 15'40'' \leq \rho_\odot \leq 16'50'', \tag{4.47}$$

$$\rho_\mathbb{C} = 16'12''30''' \pm 1'27''30''' \qquad \text{or} \qquad 14'45'' \leq \rho_\mathbb{C} \leq 17'40'', \tag{4.48}$$

$$\varphi = 42'10'' \pm 3'50'' \qquad \text{or} \qquad 38'20'' \leq \varphi \leq 46'. \tag{4.49}$$

Putting these into Approximation C.63, we obtain, respectively, the values -17841, 31864, and -6986 (in round numbers). Two of these numbers are negative (which make no sense as distances) and the other one looks far removed from anything in the vicinity of Ptolemy's 1210 Earth radii. Understandably, al-Battani must have been puzzled by these results (if indeed he got them), but thankfully, one of them is positive. Namely, the one corresponding to a mean lunar distance. So, focusing on it (as Ptolemy's 'predecessors' would), his next move was to find a suitable middle value for the distance of the Moon. Thus, following the requirements of the method, he chose the point at which the Moon's angular size is exactly the same as that of the Sun. This is easy to find. It is just the point of intersection of the linear equations for the angular sizes of the Sun and Moon, which can easily be found by putting al-Battani's parameters into Equations C.68 to C.70 (in Section C.3, where these equations are derived). The resulting linear functions are given below in simplified form and yield angles (in degrees) when lunar distances M (in Earth radii e) are fed into them.

$$\rho_\odot = \frac{947}{2480} - \frac{7}{3720} \times \frac{M}{e}, \tag{4.50}$$

$$\rho_\mathbb{C} = \frac{24449}{44640} - \frac{7}{1488} \times \frac{M}{e}, \tag{4.51}$$

$$\varphi = \frac{3197}{2232} - \frac{23}{1860} \times \frac{M}{e}. \tag{4.52}$$

These functions are valid for the range of lunar distances specified by al-Battani (in Equation 4.46), which is the same as that specified by Ptolemy (in Equation 4.39).

As promised, the point of intersection of Equations 4.50 and 4.51 is easily found (by means of Equation C.72) and is

$$\begin{pmatrix} M/e \\ \rho_\odot = \rho_\mathbb{C} \end{pmatrix} = \begin{pmatrix} 7403/126 \\ 293°/1080 \end{pmatrix}. \tag{4.53}$$

That is, when the distance between the Earth and the Moon is

$$\frac{M}{e} = \frac{7403}{126} = 58.753968..., \tag{4.54}$$

the angular size of the Sun (as given by Equations 4.51 and 4.54) is exactly the same as that of the Moon and has radius

$$\rho_\odot = \rho_\mathbb{C} = \frac{24449}{44640} - \frac{7}{1488} \times \frac{7403}{126} = \frac{293°}{1080} = 16'16''40''', \tag{4.55}$$

and the angular size of the Earth's shadow (as given by Equations 4.52 and 4.54) has radius

$$\varphi = \frac{3197}{2232} - \frac{23}{1860} \times \frac{7403}{126} = \frac{667°}{945} \approx 42'20''57'''. \tag{4.56}$$

However, al-Battani did not have the advantages of modern mathematics to achieve these exact solutions. Hence, he got different results. Namely, a lunar distance of 60;58 (or $60^{29}/_{30}$) Earth radii, a half angular size of $15'40''$ for both the Sun and Moon, and a half angular size of $40'40''$ for the Earth's shadow (which are Ptolemy's lower bounds for $\rho_\mathbb{C}$ and φ exactly). When we put the correct values (as given by Equations 4.54 to 4.56) into Approximation C.63, we get a solar distance of 29667 Earth radii, in round numbers (which is not too far from the truth), but when we put al-Battani's values, we get -63452 Earth radii (which is nonsense). As Swerdlow (1973:97) says, Approximation C.63 is so sensitive to small changes in M, $\rho_\mathbb{C}$, and φ that it is 'exceedingly difficult to use', and al-Battani learned this the hard way. So, after having gone through so much trouble and as if to hush it all up, he never wrote a word of the calculations that in all likelihood led him to eventually abandon the eclipse method and find his solar distance some other way. As Swerdlow (1973:102) explains, 'after puzzling over his strange result, [al-Battani] decided to cut through the Gordian Knot by discarding the correct method and instead adopting Ptolemy's ratio of $18\frac{4}{5}$', which is the ratio of the Sun's distance from Earth to that of the Moon. Ptolemy (Toomer, 1984:257) put this ratio at

$$\frac{S}{M} = \frac{1210}{64\frac{1}{6}} \approx 18\frac{4}{5}, \tag{4.57}$$

though, in fact, the correct answer is $18\frac{6}{7}$.

Thus, as Swerdlow (1973:102) explains, in a desperate, final move, al-Battani multiplied his own *congruent* lunar distance by Ptolemy's lunisolar distance ratio to obtain

$$\frac{S}{e} = 60^{29}/_{30} \times 18\frac{4}{5} \approx 1146. \tag{4.58}$$

This is not a rigorous thing to do, because neither is his lunar distance correct (since, according to his own parameters, the lunar distance at which *congruence* occurs is that in Equation 4.54) nor is his choice of $\rho_{\mathbb{C}}$ and φ correct (since the correct ones are those in Equations 4.55 and 4.56). Furthermore, neither is Ptolemy's distance ratio (in Equation 4.57) reliable (since his numbers do not tally) nor is it clear why al-Battani should use a solar distance that is explicitly described by Ptolemy as 'mean' and a lunar distance that al-Battani would describe as about 'mean' and yet obtain a result that he calls a 'maximum').

Now, in order to see why the *solilunar distance field* (or Approximation C.63) behaves as it does, we may use Equations 4.51 and 4.52 to express it as a function of one variable alone, namely M. This is an interesting thing to do, because it allows a most illuminating analysis of the latter function to be made. Thus, putting the mentioned equations into Approximation C.63 and simplifying, the *solilunar distance field* can be turned into what we will call here the **solilunar distance function based on al-Battani's parameters**. This function automatically adjusts for $\rho_{\odot}$, $\rho_{\mathbb{C}}$, and φ and is as follows.

$$\frac{S}{e} \approx \left(\left(\frac{762M/e - 88389}{210M/e - 24449}\right) \sin\left(\frac{24449}{44640} - \frac{7}{1488} \times \frac{M}{e}\right) - \frac{e}{M}\right)^{-1}. \qquad (4.59)$$

The same can be done with Ptolemy's version of the *solilunar distance field* by proceeding exactly as above. First, we express his parameters (in Equations 4.40 to 4.42) as linear functions (by using again Equations C.68 to C.70). This gives

$$\rho_{\odot} = \frac{47}{180}, \qquad (4.60)$$

$$\rho_{\mathbb{C}} = \frac{653}{1395} - \frac{1}{310} \times \frac{M}{e}, \qquad (4.61)$$

$$\varphi = \frac{3431}{2790} - \frac{4}{465} \times \frac{M}{e}, \qquad (4.62)$$

which are valid for the range of lunar distances specified by Ptolemy (in Equation 4.39). Again, these give angles (in degrees) when lunar distances M are fed into them (in Earth radii e). Putting Equations 4.61 and 4.62 into Approximation C.63 and simplifying, we obtain

$$\frac{S}{e} = \left(\left(\frac{33M/e - 4737}{9M/e - 1306}\right) \sin\left(\frac{653}{1395} - \frac{1}{310} \times \frac{M}{e}\right) - \frac{e}{M}\right)^{-1}. \qquad (4.63)$$

which we will call here the **solilunar distance function based on Ptolemy's parameters**. The one advantage it has over Approximation C.63 is that this function requires only one variable (namely, M, assuming e is always trivially equal to 1), whereas the field requires three (namely, M, $\rho_{\mathbb{C}}$, and φ). (That is, Equation 4.63 gives the Sun's distance as a function of the Moon's distance alone, as promised.)

As said above, Equation 4.63 is much easier to analyse than Approximation C.63. Even so, it is not amenable to analytic methods, but, fortunately, it can be analysed using numerical methods. This means that nice, analytic solutions to this equation are hard (or impossible) to find, but arbitrarily accurate numerical approximations can be found instead. The mathematical analysis performed by the author is not given in this book (as it is typically trivial), but the results are, and they yield information that is relevant to the study in hand, as we shall see. This analysis reveals the following.

Equation 4.63 is not defined when the distance to the Moon is either

$$\frac{M}{e} = 54.245447... \tag{4.64}$$

or

$$\frac{M}{e} = 89.300382... \tag{4.65}$$

(where, as always, the precision displayed is of mathematical interest only).

In fact, it is neither defined for $M/e \approx (1306 + 2790 \times 180 \times n)/9$ (where n is any integer and the number 180 is related to the sexagesimal angle conversion used in the sine function). So, in fact, there are infinite values of M/e for which the function is not defined, but only those in Equations 4.64 and 4.65 need concern us here, since all other breaks fall far beyond the range of any reasonably possible lunar distances we might wish to consider. It is therefore convenient to restrict our analysis to a smaller domain such as, say, from 0 to 55945 (since M/e cannot be less than 0, and 55945 Earth radii is far beyond what is needed).

So, on the specified domain (namely, $0 \leq M/e < 55945$), Ptolemy's function

 - is defined for all values of M/e other than 54.245... and 89.300...,
 - is negative on the interval (0, 54.245...) ∪ (89.300..., 55945),
 - is positive on the interval (54.245..., 89.300...),
 - has a local minimum at

$$\frac{M}{e} = 69.599862..., \tag{4.66}$$

 - is decreasing on (0, 54.245...) ∪ (54.245..., 69.599...),
 - and increasing on (69.599..., 89.300...) ∪ (89.300..., 55945).

In other words, on the specified domain, Equation 4.63

- breaks down at $M/e = 54.245...$ and $M/e = 89.300...$,
- yields negative (and therefore, meaningless) solar distances for values of M/e smaller than 54.245... and greater than 89.300...,
- yields positive solar distances only when M/e lies between 54.245... and 89.300...,
- and yields the smallest (meaningful) solar distance when $M/e = 69.599...$

Putting Equation 4.66 into Equation 4.63 gives

$$\frac{S}{e} \approx \left(\left(\frac{4737 - 33 \times 69.59...}{1306 - 9 \times 69.59...}\right)\sin\left(\frac{1306 - 9 \times 69.59...}{2790}\right) - \frac{1}{69.59...}\right)^{-1}$$
$$= 1114.578147..., \quad (4.67)$$

which is the smallest (meaningful) solar distance Equation 4.63 can possibly yield.

So the local minimum has coordinates

$$N = \begin{pmatrix} M/e \\ S/e \end{pmatrix} = \begin{pmatrix} 69.599862... \\ 1114.578147... \end{pmatrix}. \quad (4.68)$$

All this information allows us to plot a graph of Equation 4.63, such as the one in Figure 4.4.[29]

The local minimum N in Figure 4.4 shows that the smallest (meaningful) solar distance Equation 4.63 can yield is about 1115 Earth radii, but this occurs only when the Moon's distance is set to about 70 Earth radii, which lies beyond the range of lunar distances deemed possible by Ptolemy (as given by Equation 4.39 and highlighted in green in Figure 4.4). Within this range, the Sun's distance cannot be less than about 1247 Earth radii (as given by Approximation 4.45), corresponding to point L (in Figure 4.4), whose coordinates are

$$L = \begin{pmatrix} M/e \\ S/e \end{pmatrix} = \begin{pmatrix} 64.1\dot{6} \\ 1246.525130... \end{pmatrix}. \quad (4.69)$$

However, as we have seen, Ptolemy does something odd at this point: instead of adopting the result (of 1247 Earth radii) that strictly follows from his own function, he comes up with an even smaller result (namely, 1210 Earth radii). We now know that this result comes from Approximation C.63 only if computations

---

29 Interactive 2D plots of this and al-Battani's function can be found online by googling the words 'Ptolemy-based solilunar distance function' and 'al-Battani-based solilunar distance function'.

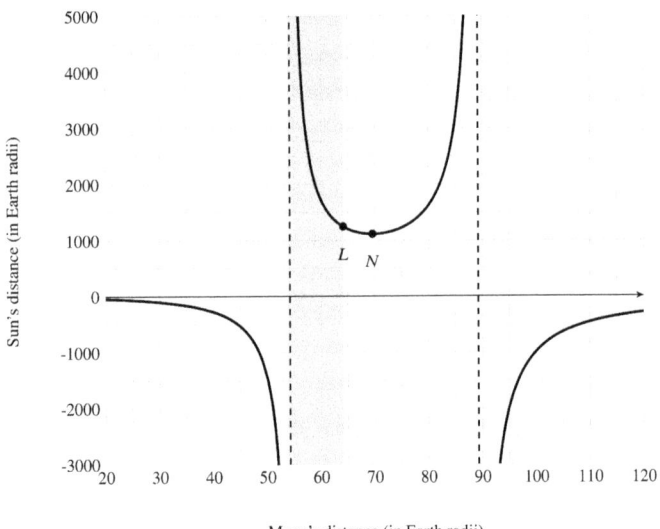

Figure 4.4 Plot of solar against lunar distances as given by the *solilunar distance function* based on Ptolemy's parameters. The range of lunar distances highlighted in grey is that of those deemed possible by Ptolemy.

are done in a certain way. We also saw above that he calls it a 'mean' solar distance only in his *Planetary Hypotheses* (Goldstein, 1967:9), while Proclus called it a 'maximum' in his *Hypotyposis* (Manitius, 1974:222; Neugebauer, 1975:110), but we now know it to be arbitrarily below the minimum strictly allowed by Equation 4.63 (that is, below point L in Figure 4.4).

Be it as it may, Ptolemy relinquished a golden opportunity to get a better solar distance, first, by refusing to use a mean lunar distance (against the advice of his 'predecessors'), and second, by choosing complex sexagesimal calculations rounded to the nearest arcsecond, rather than simple old fractions. Al-Battani, too, missed his chance to get a fairly good measurement of the Sun's distance by failing to use his own parameters correctly. Instead, he chose a number that is exactly the same as the denominator in Aristarchus' solar width-to-distance ratio $2\pi/720$ when a very good approximation to pi is used and the result is rounded to the nearest integer. Ptolemy, too, chose a number that fairly resembles the denominator of the fraction in Equation 4.28 expanded by the factor ten (to match the looks of Archimedes' 11/1148 ratio). Could it be that these men were both using Aristarchus' lunisolar angular size (or its equivalent lunisolar width-to-distance ratio) as a secret reference? We cannot know because neither men explicitly references Aristarchus on this, and because in mathematics, too many coincidences do not make a proof.

As an example of an amazing coincidence that does not make a proof, we may cite a curious book written in 1570 by the German astronomer and astrologer Jofrancus Offusius. In it, Offusius (1570:3r) argues that the most authoritative estimates for the Sun's distance (in Earth diameters) from antiquity to his time (of which he gives Ptolemy's 580, Aratus' 555, and Copernicus' 571 as examples) are all close to the number 576. Curiously, this is exactly the same as the denominator of the fraction in Equation 4.29. Offusius pointed out that, being the square of 24, this number has remarkable properties, such as being the sum of all the right-angled triangles that can be drawn on all the faces of all Platonic solids put together. Because of this, Offusius (1570:7v; Westman, 2011:188) suggested that the distance to the Sun must somehow be related to this 'Quantity of Qualities', as he called it, and concluded that the Sun is $576 \pm 24\frac{1}{2}$ times as far away from the Earth as the Earth is wide (or, if we prefer, that the Sun is 1152 ± 49 Earth radii away). We now know that this figure, as well as those he cites as examples, are gross underestimates, but Offusius's words are not entirely wrong, because the Sun's distance is indeed related to this number in one particular way. Namely, that in Equation 4.29, according to which, the Sun is 576/5 (or 1152/10) times as far away as it is wide, which is a good approximation to the truth (which is 1183/11, to five significant figures, as given by the reciprocal of Equation 4.23).

Eventually, Offusius's book influenced the work of later astronomers, such as Tycho Brahe (1602:472), whose mean solar distance of 1150 Earth radii is virtually the same as Offusius's 576 Earth diameters, and Johannes Kepler (1617:2), who spent much of his life exploring the connection between the Platonic solids and the music of the spheres.

As for the ancients themselves, Plato developed a 'theory of everything' in *Timaeus* based on the two different types of right-angled triangles that, playing the role of elementary particles, tile up the faces of the five Platonic solids: 24 on the tetrahedron, 24 on the cube, 48 on the octahedron, 120 on the icosahedron, and 360 on the dodecahedron (Thorndike, 1941:110; Brown, 2011:20). In all, 576, as Offusius said. However, this number does not appear explicitly mentioned in *Timaeus*. Yet, the number in question was certainly known in antiquity, as it is one of the sides of several **Pythagorean triangles** (that is, those that are right-angled and whose sides are whole numbers). Its square root is also one of the sides of several of these triangles (Agarwal, 2020:17). But apart from these and other wonderful credentials (which need not be repeated here), there is nothing that should connect Offusius's 'Quantity of Qualities' to the denominator in Equation 4.29, apart from its being inspired by Ptolemy's, Aratus', and Copernicus' solar distances, whose connection to Aristarchus' ratio may be coincidental. In any case, Offusius's remarkable insight is no rigorous proof of such a connection.

Returning to the above discussion, we must say that the *solar eclipse method*, with its ancient reversible Approximation C.63 (which we called the *solilunar distance field* on page 154) was humanity's preferred method of estimating the distance to the Sun and Moon for the fifteen centuries that followed the publication of Ptolemy's *Almagest* (Swerdlow, 1969:97). But this is only part of a bigger picture. There is yet another way of estimating these distances that was tried in the past. It uses lunar (rather than solar) eclipses and is called the *lunar eclipse method*. Research on this one produced another reversible approximation whose derivation is given by Carman (2020:179) and can also be found in Section C.1 of the present book, where modern, exact solutions are also found that constitute the general solution to the whole *eclipse method*, both *lunar* and *solar*. These modern, exact solutions, along with the approximations derived in antiquity from both eclipse methods, plus an extra approximation found by the author in Section C.2, are

$$\frac{S}{e} = \left(\sin\rho_\odot + \sin\left(\varphi - \sin^{-1}\left(\frac{e}{M\cos\rho_{\mathbb{C}}}\right)\right)\right)^{-1} \tag{C.40}$$

$$\approx \left(\sin\left(\varphi + \rho_\odot - \sin^{-1}\left(\frac{e}{M}\right)\right)\right)^{-1} \tag{C.9}$$

$$\approx \left(\left(\frac{\varphi}{\rho_{\mathbb{C}}} + 1\right)\sin\rho_{\mathbb{C}} - \frac{e}{M}\right)^{-1} \tag{C.63}$$

$$\approx \left(\sin\varphi + \sin\rho_{\mathbb{C}} - \frac{e}{M}\right)^{-1} \tag{C.65}$$

and

$$\frac{M}{e} = \left(\cos\rho_{\mathbb{C}}\sin\left(\varphi - \sin^{-1}\left(\frac{e}{S} - \sin\rho_\odot\right)\right)\right)^{-1} \tag{C.46}$$

$$\approx \left(\sin\left(\varphi + \rho_\odot - \sin^{-1}\left(\frac{e}{S}\right)\right)\right)^{-1} \tag{C.9}$$

$$\approx \left(\left(\frac{\varphi}{\rho_{\mathbb{C}}} + 1\right)\sin\rho_{\mathbb{C}} - \frac{e}{S}\right)^{-1} \tag{C.64}$$

$$\approx \left(\sin\varphi + \sin\rho_{\mathbb{C}} - \frac{e}{M}\right)^{-1}, \tag{C.66}$$

where all variables are as usual.

In this book, Equations C.40 and C.46 are called the **eclipse method equations**, Approximation C.9 is called the **lunisolar distance field** (as it is derived from the lunar eclipse method), Approximations C.63 and C.64 are called the **solilunar distance field** (as they are derived from the solar eclipse method), and Approximations C.65 and C.66 are called the **neat solilunar distance field** (as they are also derived from the solar eclipse method). In general, the *lunisolar distance field* yields the best approximations, followed by the *neat solilunar distance field*, and the *solilunar distance field*, in that order.

Historically, Approximation C.9 has its origin in Proposition 13 of Aristarchus' first book *On Sizes*, although it is not explicitly developed or used there. In all likelihood, it was developed and used either in Aristarchus' second book *On Sizes*, as hinted at by Vitruvius in *On Architecture* 9.2.3 (Granger, 1934:228; Rowland and Howe, 2001:112), or in one of Hipparchus' own books bearing the same title as those by Aristarchus. Namely, *On the Sizes and distances of the Sun and Moon*. In all, four books apparently bearing this title were written, two by Aristarchus and two by Hipparchus, of which only the first one by Aristarchus is preserved. In the present book, they are all referred to as *On Sizes*, for short, even though, as Toomer (1974a:126) points out, the titles of the missing books are not quite certain, since all that remains is a comment by Pappus of Alexandria (Rome, 1931:68), where Hipparchus' first and second books are both called *On Sizes and Distances*, and another by Theon of Smyrna (Dupuis, 1892:318), where they appear to be called *On the Distances and Sizes of the Sun and Moon*.

Now, in order to understand why Aristarchus' solar width-to-distance ratio (as given by Equation 4.29), which is also an angle in radians, eventually (and wrongfully) came to be understood as a distance ratio, let us now delve into what is currently known about Hipparchus' follow-up books *On Sizes* and how it all connects to the eclipse method above described.

As said above, thanks to the great scholarship and brilliant detective work carried out by Swerdlow (1969), Toomer (1974a), and Carman (2020), the contents of Hipparchus' (lost) books *On Sizes* are (at least partially) recovered. Their reconstruction is based on a comment by Ptolemy in *Almagest* 5.11 (Toomer, 1984:243) and another by Pappus (Rome, 1931:68), in which the following information is given.

- Hipparchus used the *eclipse method* by means of which, if the distance to one of the luminaries is given, then the distance to the other is also given.

- He made two estimates about the distance to the Moon, first assuming that the Sun has no parallax (that is, it is infinitely far away), and then, assuming it has the least perceptible parallax (that is, it is as close as it can possibly be).
- In *On Sizes* 1, he assumes the Sun has no parallax and uses a solar eclipse that was total on the Hellespont, but partial on Alexandria, where 4/5 of the Sun were covered at most, concluding that the Moon is 77 ± 6 Earth radii away.
- In *On Sizes* 2, he assumes the Sun is 490 Earth radii away, concluding that the Moon is 67⅓ ± 5⅓ Earth radii away.

The first to crack the meaning of some of the above points (namely, those concerning book 2 of Hipparchus' *On Sizes*) was Swerdlow (1969). He noted that Ptolemy gives Hipparchus' parameters in *Almagest* 4.9 (Toomer, 1984:205), when he says that, at mean distance from Earth, the Moon's disc goes about 650 times into its own orbit, and the Earth's shadow is twice and a half as wide as this. That is, the angular radius of both the Sun and Moon at mean lunar distance is $\rho_{\mathbb{C}} = 180°/650 = 18°/65$, and that of the Earth's shadow is $\varphi = 2.5 \times 180°/650 = 9°/13$.

Furthermore, in *Almagest* 4.11 (Toomer, 1984:211), Ptolemy gives Hipparchus' ratio of *deferent* radius to *epicycle* radius as either 3144½ to 327⅔ or 3122½ to 247½, as he found from the observation of two different sets of three lunar eclipses each, using the method expounded in *Almagest* 4.6. Toomer (1968:149) showed that Hipparchus used the second of these possibilities in his books *On Sizes*. So, according to Hipparchus, the size of the lunar epicycle is 247½/3122½ = 99/1249. Hence, using this ratio, the recovered set of parameters used by Hipparchus in his second book *On Sizes* is as follows.

$$\frac{M}{e} = \frac{202}{3}\left(1 \pm \frac{99}{1249}\right) \quad \text{or about} \quad 62 \leq \frac{M}{e} \leq \frac{218}{3}, \tag{4.70}$$

$$\rho_\odot = \frac{18°}{65}, \tag{4.71}$$

$$\rho_{\mathbb{C}} = \frac{18°}{65}\left(1 \pm \frac{99}{1249}\right) \quad \text{or about} \quad 15'18'' \leq \rho_{\mathbb{C}} \leq 17'56'', \tag{4.72}$$

$$\varphi = \frac{9°}{13}\left(1 \pm \frac{99}{1249}\right) \quad \text{or about} \quad 38'15'' \leq \varphi \leq 44'50'', \tag{4.73}$$

where, as usual, M is the distance to the Moon (measured in Earth radii e), and $\rho_\odot$, $\rho_{\mathbb{C}}$, and φ are half the angular size of the Sun, Moon, and Earth's shadow, respectively.

Putting these parameters into either Equation C.40 or any of Approximations C.9, C.63 and C.65, gives a number that is very close to the solar distance ascribed to Hipparchus by Pappus. For example, using the neater of these approximations, we have

$$\frac{S}{e} \approx \left(\sin\left(\frac{9°}{13}\right) + \sin\left(\frac{18°}{65}\right) - \frac{3}{202}\right)^{-1} = 484.385... \approx 484, \tag{4.74}$$

which, as Swerdlow (1969:299) says, indicates that the least perceptible solar parallax assumed by Hipparchus is

$$\mu_\odot \approx \sin^{-1}\left(\frac{e}{S}\right) = \sin^{-1}\left(\frac{1}{484}\right) \approx 7', \tag{4.75}$$

as can be found using Equation C.6.

So, Swerdlow proposed that Hipparchus started by assuming that the least perceptible parallax of the Sun is 7 arcminutes, then, he found the solar distance corresponding to this parallax, which is given by either Equation B.2 or an adaptation of Equation C.5 as

$$\frac{S}{e} \approx \frac{1}{\sin \mu_\odot} = \frac{1}{\sin 7'} = 491.107... \approx 491 \approx 490, \tag{4.76}$$

and then used the eclipse method to find the lunar distance corresponding to this solar distance. Thus, using Approximation C.66, which is the simplest of the eclipse method approximations, we have either

$$\frac{M}{e} \approx \left(\sin\left(\frac{9°}{13}\right) + \sin\left(\frac{18°}{65}\right) - \frac{1}{490}\right)^{-1} = 67.226... \approx 67\tfrac{1}{4} \tag{4.77}$$

or

$$\frac{M}{e} \approx \left(\sin\left(\frac{9°}{13}\right) + \sin\left(\frac{18°}{65}\right) - \frac{1}{484}\right)^{-1} = 67.340... \approx 67\tfrac{1}{3}. \tag{4.78}$$

All these values are very close to those in Hipparchus' second book *On Sizes*, as reported by Pappus. Thus, Swerdlow explained them as the result of Hipparchus testing the eclipse method when the Sun is taken to be as close to Earth as a parallax big enough to be visible to the human eye can make it. Toomer (1974a:139) went a step further and realized that Hipparchus also tested the eclipse method when the Sun is taken to be infinitely far away (and therefore has no parallax). In this case, the limit of Approximation C.66 as S/e tends to infinity (or equivalently, as e/S tends to nought) is

$$\frac{M}{e} \approx (\sin \varphi + \sin \rho_\mathbb{C})^{-1} = \left(\sin\left(\frac{9°}{13}\right) + \sin\left(\frac{18°}{65}\right)\right)^{-1} = 59.11578... \tag{4.79}$$

and that of Approximation C.9 is

$$\frac{M}{e} \approx \left(\sin\left(\varphi + \rho_\odot\right)\right)^{-1} = \left(\sin\left(\frac{9°}{13} + \frac{18°}{65}\right)\right)^{-1} = 59.11751..., \tag{4.80}$$

while the exact limit (as derived from Equation C.46) is

$$\frac{M}{e} = \left(\cos\rho_{\mathbb{C}} \sin\left(\varphi + \rho_\odot\right)\right)^{-1} = \left(\cos\left(\frac{18°}{65}\right)\sin\left(\frac{9°}{13} + \frac{18°}{65}\right)\right)^{-1} = 59.11820... \tag{4.81}$$

So, according to Toomer, rounding this to the nearest integer, we obtain a lower bound for the Moon's distance that is likely to have been found by Hipparchus in his second book *On Sizes*. Namely,

$$\frac{M}{e} = 59\left(1 \pm \frac{99}{1249}\right) \quad \text{or about} \quad \frac{163}{3} \leq \frac{M}{e} \leq \frac{191}{3}. \tag{4.82}$$

A proof that this may indeed have been so is that, while Hipparchus would have regarded these as a single lower bound made up of three numbers (namely, 54⅓, 59, and 63⅔) whose *mean* is 59, Ptolemy's *mean* lunar distance is exactly 59. However, we cannot be sure of this because Ptolemy does not acknowledge Hipparchus as the ultimate source of this number, but claims instead, in *Almagest* 5.15 (Toomer, 1984:257), that he found it himself. Should he be right, this would be one of those amazing coincidences that can never be proved to be anything other than mere coincidences.

So, the upper and lower bounds for the Moon's distance found by Hipparchus in his second book *On Sizes* are each likely to have been a set of three numbers. Namely, those in Equation 4.70, which correspond to the smallest possible distance of the Sun, and those in Equation 4.82, which correspond to the greatest possible distance of the Sun. These sets are

$$59\left(1 \pm \frac{99}{1249}\right) \leq \frac{M}{e} \leq \frac{202}{3}\left(1 \pm \frac{99}{1249}\right), \tag{4.83}$$

which, to the nearest third of an Earth radius, can be rendered as 54⅓, 59, 63⅔ and 62, 67⅓, 72⅔.

The next step forward in the effort to understand the meaning of Ptolemy's and Pappus' reports on Hipparchus was given by Toomer (1974a), who this time focused on the data contained in Hipparchus' first book *On Sizes*. In this book, Hipparchus used a solar eclipse that was total as seen from the regions around the Hellespont, but partial as seen from Alexandria, where only 4/5 of the Sun were covered. As Toomer (1974a) explains, Hipparchus used this eclipse to estimate a lower bound for the Moon's distance. According to the eclipse method, a closer Moon corresponds to a further Sun, so on this occasion, Hipparchus considered the possibility that the Sun is infinitely far away.

Now, the latitudes of Alexandria and the Hellespont were both known to Hipparchus, as can be inferred from Strabo's comments in *Geography* 2.5.34 and 2.5.40 (Howarth, 1917:504, 512), according to the first of which, Hipparchus took 700 stades as the equivalent of one degree of latitude, and according to the second, the region of the Troad (where the longest day is 15 hours) is 7000 stades north of Alexandria and 28800 stades north of the Equator. Therefore, the latitude of the Troad is $\phi_H = 28800/700 \approx 41°$ and that of Alexandria is $\phi_A = (28800 - 7000)/700 \approx 31°$. This is confirmed by Hipparchus himself in *On Aratus* 1.3.7 (Manitius, 1894:26), his only preserved work, when he says that the latitude of regions like the Hellespont, where the longest day is 15 hours, is 41 degrees.

On this memorable occasion, Hipparchus assumed the Sun had no parallax, and so, that of the Moon is simply

$$\mu_{\mathbb{C}} \approx \frac{1}{5} \times \frac{360°}{650} = \frac{36°}{325}, \tag{4.84}$$

where 1/5 is the part of the Sun that was never covered, and 360°/650 is Hipparchus' estimate for the radius of the solar disc, as revealed by Ptolemy in *Almagest* 4.9 (Toomer, 1984:205).

With all these data, Hipparchus was able to find a lower bound for the Moon's distance by means of the method explained in Section C.4, which is based on Toomer's (1974a) and Carman's (2020) reconstructions. There is just one more datum that is needed to achieve the stated goal. Namely, the solar declination $\delta_\odot$. In order to find it, Toomer (1974a:135) checked all the possible solar eclipses that Hipparchus may have used and found that only one of them gives a lunar distance that coincides with one of those transmitted by Pappus. Namely, the solar eclipse of March 14, 190 BC. At this time, the Sun's declination is closer to −4 degrees than it is to −3 degrees, but assuming, as Carman (2020:189) does, that Hipparchus' take on this occasion was the latter,[30] and putting all these data

---

30 On March 14, 190 BC, at the time of greatest eclipse, 6:57:31 UT, the Moon's apparent declination was −4°22′44″, as seen from the centre of the Earth. However, the tables in *Almagest* predict a geocentric lunar declination of −2°58′36″, which supports Carman's assumption. Alternatively, at the time of greatest eclipse as seen from Alexandria, 8:16:44 UT, the Moon's apparent declination was −3°59′48″, whereas the tables in *Almagest* predict a geocentric lunar declination of −2°35′58″, which also supports Carman's assumption.

into Approximation C.85, we have

$$\frac{M}{e} \approx 1 + \frac{\sin\left(41° - 31°\right)\cos\left(\frac{41° + 31°}{2} - (-3°)\right)}{\sin\left(\frac{36°}{325}\right)\cos\left(\frac{41° - 31°}{2}\right)} \approx 71. \tag{4.85}$$

At this time, the Moon was rather close to Earth, as Hipparchus surely knew by methods such as those employed in *The Almagest*,[31] so the number 71 is the lowest of the associated threesome, and therefore, the whole set can be reconstructed as

$$77\left(1 \pm \frac{99}{1249}\right) \leq \frac{M}{e}, \tag{4.86}$$

which matches Pappus' report exactly when rounded to the nearest whole number.

Toomer (1974a:140) also noticed that the lower bound found in *On Sizes* 1 (as given by Equation 4.86) is greater than the upper bound found in *On Sizes* 2 (as given by Equation 4.83), which, according to him, shows 'the complete honesty with which Hipparchus reveals his two discrepant results'.

There are a few more data that have been preserved concerning Hipparchus' estimates for the size of the Sun, Earth, and Moon. These are given by Cleomedes and Theon of Smyrna. If their reports are correct, then two extra solar distances and one extra lunar distance can be recovered.

Thus, according to Cleomedes (Ziegler, 1891:152; Bowen and Todd, 2004:118), Hipparchus said that the Sun is 1050 times the size of the Earth. That is, the volume of the Sun $V_\odot$ and the volume of the Earth $V_\oplus$ (if indeed the word 'size' means 'volume' here) are in the ratio

$$\frac{V_\odot}{V_\oplus} = \frac{4\pi s^3/3}{4\pi e^3/3} = \frac{s^3}{e^3} = 1050. \tag{4.87}$$

Hence, in terms of the Earth's radius e, the Sun's radius s is

$$\frac{s}{e} = \sqrt[3]{1050} \approx 10\frac{1}{6}, \tag{4.88}$$

---

31 *The Almagest* predicts a lunar distance of 56;20,55 Earth radii at the time of greatest eclipse (versus 57;23,26 Earth radii in reality), and a lunar distance of 56;26,17 Earth radii at the time of greatest eclipse as seen from Alexandria (versus 57;24,58 Earth radii in reality).

and (by basic trigonometry) the Sun's distance S from Earth is

$$\frac{S}{e} = \frac{s/e}{\sin \rho_\odot} = \frac{\sqrt[3]{1050}}{\sin\left(\frac{180°}{650}\right)} \approx 2103. \tag{4.89}$$

Alternatively, according to Theon of Smyrna (Dupuis, 1892:318), Hipparchus said that the Sun is 1880 times the size of the Earth, and the Earth, 27 times the size of the Moon. Assuming again that the word 'size' means 'volume' here, we have

$$\frac{V_\odot}{V_\oplus} = \frac{4\pi s^3/3}{4\pi e^3/3} = \frac{s^3}{e^3} = 1880 \quad \text{and} \quad \frac{V_\oplus}{V_\mathbb{C}} = \frac{4\pi e^3/3}{4\pi m^3/3} = \frac{e^3}{m^3} = 27. \tag{4.90}$$

Hence, in terms of the Earth's radius e, the radius s of the Sun and the radius m of the Moon are, respectively,

$$\frac{s}{e} = \sqrt[3]{1880} \approx 12\tfrac{1}{3} \quad \text{and} \quad \frac{m}{e} = \frac{1}{\sqrt[3]{27}} = \frac{1}{3}. \tag{4.91}$$

Hence (by basic trigonometry or any of Equations 2.1 to B.2), the distance to the Sun and Moon are, respectively,

$$\frac{S}{e} = \frac{s/e}{\sin \rho_\odot} = \frac{\sqrt[3]{1880}}{\sin\left(\frac{180°}{650}\right)} \approx 2554 \quad \text{and} \quad \frac{M}{e} = \frac{m/e}{\sin \rho_\mathbb{C}} = \frac{1/3}{\sin\left(\frac{180°}{650}\right)} \approx 69. \tag{4.92}$$

However, the latter distances do not seem to have been obtained by the eclipse method (since they do not yield each other when put into any of the relevant approximations). So, if indeed they are correctly reported, they may have been obtained by a different, so far unidentified method. This was to be expected. After Aristarchus opened the Pandora's box of cosmic sizing, there must have been a number of hopeful attempts, and in the case of Hipparchus, the resulting solar distances ranged from one of his lower bounds to infinity. Eventually, the dust started to settle around one value. Namely, the one that was viewed as the most unassailable one among the plethora of infinite possibilities. This is the value that Offusius identified as the 'Quantity of Qualities', the number which (in Earth radii) seems to have guided the choices of Ptolemy, al-Battani, Copernicus, Tycho, and so many others, and which we have assumed was the brainchild of Aristarchus.

The influence of the man who moved the Earth without pulling a lever can be felt perhaps most strongly on Hipparchus, whose books are a sequel of Aristarchus', and whose choice of lower bound for the Moon's distance in his first book *On Sizes* (namely, 77 Earth radii) yields a solar distance (by the eclipse method) that is uncannily close to the lower bound for the Sun's distance in Aristarchus' first book *On Sizes* (namely, 256½ Earth radii), and whose choice of upper bound for the Moon's distance in his second book *On Sizes* (namely, 67⅓ Earth radii) yields a solar distance (also by the eclipse method) that is uncannily close to the upper bound for the Sun's distance in Aristarchus' first book *On Sizes* (namely, 477⅞ Earth radii), as the reader may easily check. Thus, for example, using the neater of the eclipse method approximations (which is perhaps the most pleasant to the eye), we see that when the Moon is 77 Earth radii away, the Sun's distance is

$$\frac{S}{e} \approx \left(\sin\left(\frac{9°}{13}\right) + \sin\left(\frac{18°}{65}\right) - \frac{1}{77}\right)^{-1} = 254.521..., \tag{4.93}$$

and when the Moon is 67⅓ (or 202/3) Earth radii away, the Sun's distance is

$$\frac{S}{e} \approx \left(\sin\left(\frac{9°}{13}\right) + \sin\left(\frac{18°}{65}\right) - \frac{3}{202}\right)^{-1} = 484.385..., \tag{4.94}$$

which are results very close to those in Aristarchus' implicit Inequality 2.48. Hipparchus' otherwise arbitrary choice of solar parallax in Equation 4.75 is thus explained.

Furthermore, if we take the numbers 1050 and 1880 in Equations 4.87 and 4.90 to mean, not the size of the Sun (as Cleomedes and Theon thought), but the size of the Sun's orbit (in Earth diameters), then Hipparchus' best estimates for the Sun's distance are close to those assumed for Aristarchus in Inequality 4.12.

Ptolemy, too, either had a secret thing about Aristarchus or he was extraordinarily lucky in his choice of parameters, because, if instead of making his estimates when the Moon is very far away, as he prescribed, we use his *mean* (rather than *apogean*) values for $\rho_☾$ and φ (as given by Equations 4.41 and 4.42 and as his 'predecessors' would advise), all while assuming that $\rho_☉ = \rho_☾$ (as the method requires and as Hipparchus would applaud because his value of $\rho_☉$, in Equation 4.71, is virtually the same as Ptolemy's mean $\rho_☾$), then the eclipse method yields virtually the same lunar distance assumed for Aristarchus in Equation 4.35. As a check, using the neat eclipse method approximation, we have

$$\frac{M}{e} \approx \left(\sin(43'20'') + \sin(16'40'') - \frac{1}{1152}\right)^{-1} = 60.295868..., \tag{4.95}$$

or, working backwards,

$$\frac{S}{e} \approx \left(\sin(43'20'') + \sin(16'40'') - \frac{1}{60.296}\right)^{-1} = 1152.317232..., \qquad (4.96)$$

which is very close to the solar distance envisioned by Offusius.

This happens only when Ptolemy's *mean* parameters are used and the Sun's angular size is equated to the Moon's (as the method requires). We cannot obtain this nearly perfect match using Hipparchan or al-Battanian parameters. In fact, the match is all the more remarkable when we note that the eclipse method approximations are *ill-conditioned* with respect to some of their parameters, which makes it nearly impossible to achieve this match by mere chance. So, at this point, and since it is impossible to prove anything historical, we are left with two possibilities: either these are just mere coincidences or they are the product of human purpose. If we assume the former, then none of our previous assumptions about Aristarchus' possible use of the fraction 25/8 as an approximation to pi is affected. However, if we assume the latter (that is, if we assume that both Hipparchus and Ptolemy adjusted their parameters to yield results that could never be judged to be worse than Aristarchus'), then all our previous assumptions, in particular that in Equation 4.29, are suddenly confirmed, in so far as historical research allows.

We can now see why the top and bottom of Aristarchus' solar width-to-distance ratio (namely, 10/1152), which is both a ratio and an angle (in radians), eventually came to be regarded as actual distances, which they are not. Strictly speaking, it is correct to say that the Sun and Moon are both about 1152/10 = 115.2 times as far away as they are wide (versus about 107.5 and 110.6 times, each, in reality) and it is also correct to say that both luminaries subtend an angle of about 10/1152 radians when seen from Earth (versus about 10/1075 and 10/1106 radians, each, in reality), but it is incorrect to say that either the Sun or the Moon are 1152 Earth radii away or 10 Earth radii wide. Yet, this is what history made of Aristarchus' ratio when it was put into the eclipse method (along with Ptolemy's mean parameters).

We do not know whether Aristarchus intended things to turn out this way, or whether it was someone else who turned his width-to-distance ratio into an actual distance dictum, but this dictum seems to be based on the *eclipse method* (as Equations 4.95 and 4.96 strongly suggest), and at least part of this method (in particular, the *lunar eclipse method*) follows straight from Proposition 13 of his first book *On Sizes*, and so it is likely that at least this part of the method is due to him. Alternatively, it is also possible that Hipparchus complimented Aristarchus' work by devising the *solar eclipse method*. After all, Strabo reports in *Geography* 1.1.12 (Howarth, 1917:24) that Hipparchus advocated the use of both lunar and solar eclipses as a complement to each other.

We cannot know for sure who devised the *eclipse method*, or whether more than one person was involved, but Ptolemy's words give a few clues as to the origin of this remarkable method when he speaks of his 'predecessors' in the plural, meaning that Hipparchus was not alone in the belief that *congruence* was achieved at *mean* distances, and presumably not alone in his preference for *mean* parameters to use in this method, and should Ptolemy's other 'predecessor' or 'predecessors' precede also Hipparchus, then the mere suggestion that they preferred *mean* parameters means they were using the eclipse method (or part of it) before Hipparchus. Another clue is Ptolemy's failure to clearly and unambiguously attribute the method to Hipparchus. Instead, Ptolemy (Heiberg, 1898:421; Toomer, 1984:254) says that Hipparchus 'followed' (ἠκολούθησεν) a demonstration that is simply repeated in *Almagest* 5.15, but whether Hipparchus devised the method or simply 'followed' someone else's is something suggested but not confirmed by Ptolemy's words. So, despite these clues, we cannot be sure who developed the eclipse method, and saying that Aristarchus developed part of it and Hipparchus developed the other part is just one of several possibilities.

Fortunately, there is less uncertainty as to whether the concept of *radian* as a natural unit of angle measurement (which has often been associated with Aristarchus' ratio in the present book) was used this early in history, because, in fact, it unequivocally played a central role (at least) in Hipparchus' chord table, as reconstructed by Toomer (1974b:8). So, it is possible that Aristarchus too may have been aware of its advantages.

As it happened, his half-Moon method, along with the eclipse method he might also have devised, was the best that humanity could afford for the nearly two thousand years that span Aristarchus' first use of it and the time when Jeremiah Horrocks tried the first truly new alternative method ever. Namely, that of observing the transits of Venus. Just a few decades before this momentous event, Kepler had started to lose faith in the eclipse method, and by 1617, he had completely abandoned it in favour of Aristarchus' half-Moon method (Helden, 1985:80). But perhaps this was a hasty decision, because, in the author's opinion, the true potential of the eclipse method is yet to be discovered. We got the first hint of why this is so in Section 4.3 (just after al-Battani's observational parameters are given in Equations 4.46 to 4.49), when we saw that if we put the mean values of these parameters into Approximation C.63 (that is, into the *solilunar distance field*), we obtain the solar distance

$$\frac{S}{e} \approx \left(\left(\frac{42'10''}{16'12''30'''} + 1\right)\sin(16'12''30'') - \frac{1}{59}\right)^{-1} \approx 31864, \tag{4.97}$$

Archimedes' telltale ratio 175

and if we put al-Battani's mean parameters into Approximation C.9 (that is, into the *lunisolar distance field*), we obtain

$$\frac{S}{e} \approx \left(\sin\left(42'10'' + 16'15'' - \sin^{-1}\left(\frac{1}{59}\right)\right)\right)^{-1} \approx 23389. \tag{4.98}$$

Clearly, these numbers are no longer close to each other. This is because one of these approximations is better than the other, but this only becomes apparent when the eclipse method is pushed to limits such as these. Only then does the superiority of the *lunar* approximation over the *solar* one become apparent, in that the *solar* approximation starts to depart from the truth, whereas the *lunar* approximation continues to yield results that are fairly close to the true solution, which, as given by Equation C.40, is

$$\frac{S}{e} = \left(\sin(16'15'') + \sin\left(42'10'' - \sin^{-1}\left(\frac{1}{59} \times \sec\left(16'12''30'''\right)\right)\right)\right)^{-1}$$
$$\approx 23493. \tag{4.99}$$

At some point in his calculations, al-Battani's eyes must have seen at least one of the numbers in Approximations 4.97 and 4.98 (most likely, the first one, since it is based on the solar eclipse method promoted in *The Almagest*), but he discarded it because it seemed too unbelievably remote from Ptolemy's modest figure.

Strikingly, the latter two solar distances (that is, the ones in Approximation 4.98 and Equation 4.99) are virtually the same as the modern value of 23455 Earth radii. Of course, al-Battani had no access to this knowledge or to the exact solution given by Equation 4.99 (since Equation C.40 had not yet been developed). However, had he used the *lunisolar distance field* (rather than the *solilunar* one promoted in *The Almagest*), he would have gone down in history as the man who pinned down the Sun's distance to within $23455 - 23389 = 66$ Earth radii, which is less than 0.3% short of the truth (even though back then nobody could have checked).

But let us not jump to conclusions. Things are rarely so simple. Thus, if we want to be mathematically rigorous, it is not the above parameters that must be used, but those in Equations 4.54 to 4.56 (that is, those corresponding to *lunisolar congruence* as derived from al-Battani's parameters). Using these (in the best of the approximations), we have

$$\frac{S}{e} \approx \left(\sin\left(\frac{667°}{945} + \frac{293°}{1080} - \sin^{-1}\left(\frac{126}{7403}\right)\right)\right)^{-1} \approx 30349, \tag{4.100}$$

which is not as good an estimate as that in Approximation 4.98.

Furthermore, the equations and approximations in the eclipse method are all purely geometric. That is, they do not take into account the strange fact, first discovered by la Hire (1687:73) and explained by Cassini (1740:34), that the Earth's atmosphere makes the Earth's shadow look larger than it should by a factor of 1/40, as la Hire put it; or 1/50, as Chauvenet (1891:542) put it; or 1/100, as Danjon (1951:53) put it. If we take this effect into account, then we must note that the angular size of the Earth's shadow measured by al-Battani is the apparent one (that is, the one increased by the Earth's atmosphere), and therefore must be reduced by one of the above factors. Using Chauvenet's 1/50, for example, Equation C.40 yields

$$\frac{S}{e} = \left(\sin\left(\frac{293°}{1080}\right) + \sin\left(\frac{667°}{945} \times \frac{50}{51} - \sin^{-1}\left(\frac{126}{7403} \times \sec\left(\frac{293°}{1080}\right)\right)\right)\right)^{-1}$$
$$\approx -4790, \qquad (4.101)$$

which is a meaningless result. So, once all the necessary corrections have been properly made, al-Battani's parameters lead to no result at all, and so it is a small wonder that he looked back on Aristarchus' ratio as the only firm ground on which to tread. This is a consequence of the equation being so sensitive to small changes in its parameters. But this need not deter us, because, as Equation 4.99 promises, the eclipse method has a great potential. We just need to supply the right parameters and correct for the distorting effects of the Earth's atmosphere. But which are the correct parameters? Which is the best correcting factor to use? Chauvenet's or Danjon's, or otherwise? Can the advantages of an *ill-conditioned* method help us decide between the two or find an even better one? This is a topic of research outside the scope of the present book which the author leaves the reader to enthusiastically undertake and hopes thou will come out more victorious than al-Battani or Kepler did.[32]

To conclude this section, we have seen what Archimedes' apparently unassuming 11 to 1148 ratio in *The Sand Reckoner* can reveal, which is one of the reasons why this little book can be regarded as one of the most heavily pregnant ever written in the history of science. Let us now continue our analysis of it.

---

32 In order to help the reader explore the possibilities of the eclipse method, the author has created a series of free online interactive illustrations where the approximations and equations of the method, both ancient and modern, are already preset to Hipparchan, Ptolemean, or al-Battanian parameters. Just google the words 'Eclipse method (equations)' and choose the one you wish to explore.

4.4 Archimedes' big numbers

After proving that the side of a regular chiliagon inscribed in the *cosmos* (as defined in Passage 49) is less than the Sun's diameter, Archimedes started to explore where his deliberately exaggerated assumptions led him to. He reasoned that the side of such a chiliagon is narrower than the Sun (by Assumption 4), which is narrower than 30 Moons (by Assumption 3), which are narrower than 30 Earths (by Assumption 2). That is,

$$k < 2s \leq 30(2m) < 30(2e), \tag{4.102}$$

where k is the length of the side of a regular chiliagon inscribed in the cosmos, and s, e, and m are the radii of the Sun, Earth, and Moon, respectively.

Multiplying through by 1000, we have

$$1000k < 1000(2s) < 30000(2e), \tag{4.103}$$

where $1000k$ is the perimeter of the chiliagon in question, which is more than $\pi \approx 3$ (but less than $\pi \approx 22/7$) times the radius S of the cosmos. That is,

$$3(2S) < 1000k < 30000(2e). \tag{4.104}$$

Hence,

$$2S < 10000(2e) \quad \text{or} \quad \frac{S}{e} < 10000. \tag{4.105}$$

That is, the cosmos is less than a myriad times as wide as the Earth, or, as Archimedes put it (in *Sand Reckoner* 2.1),

Passage 58 The diameter of the *cosmos* is less than a myriad times the diameter of the Earth (Vardi, 1997:5; see also Heiberg, 1881:262).

So, Archimedes' convenient simplifications and exaggerated Assumption 3 lead to a cosmos that is at least $10000/(17200/9) \approx 5$ times, and at most $10000/1026 \approx 10$ times as large as that in Aristarchus' second book *On Sizes*, as reconstructed in Inequality 4.12.

Archimedes then goes on to reason that the circumference of the Earth $c_\oplus$ is greater than $\pi \approx 3$ times its diameter, which (by Assumption 1) is less than 300 myriad stades. That is,

$$3(2e) < c_\oplus < 300{,}0000 \text{ stades}. \tag{4.106}$$

Dividing through by 3, we have

$$2e < 100,0000 \text{ stades}. \tag{4.107}$$

Substituting this into Inequality 4.105 gives

$$2S < 10000(2e) < 10000 \times (100,0000 \text{ stades}), \tag{4.108}$$

which simplifies to

$$2S < 100,0000,0000 \text{ stades}. \tag{4.109}$$

So, the diameter of the cosmos is less than a hundred myriad myriad stades.

But we also know that, According to Archimedes' report (as illustrated in Figure 4.1), the firmament is to the cosmos as the cosmos is to the Earth. So, combining Equation 4.1 and Inequality 4.105, we have

$$\frac{2F}{2S} = \frac{2S}{2e} < 10000. \tag{4.110}$$

That is,

$$\frac{F}{S} < 10000. \tag{4.111}$$

So, the firmament is not as wide as a myriad cosmos, or, as Archimedes put it (in *Sand Reckoner* 4.17),

> **Passage 59** The diameter of the sphere of fixed stars is smaller than a myriad times the diameter of the *cosmos* (Vardi, 1997:9; see also Heiberg, 1881:288).

Substituting for $2S$ from Inequality 4.109 into Inequality 4.111, we have

$$2F < 10000 \times (100,0000,0000 \text{ stades}), \tag{4.112}$$

which can be rewritten as

$$2F < 100,0000,0000,0000 \text{ stades} = 10^{14} \text{ stades}. \tag{4.113}$$

So, the firmament is not as wide as a hundred myriad myriad myriad stades.

Then, Archimedes explains his proposal of a new method for naming big numbers. He starts by noting that the Greek number system can name numbers up to a myriad myriads (that is, $(10^4)^2 = 10^8$, or an **octad**, as he calls it), but this is not nearly enough for *The Sand Reckoner*'s purpose. The system he proposes, however, does not seem to have been used by anyone outside the context of *The Sand Reckoner*, despite Archimedes having sent an earlier version of his book to Zeuxippus, who presumably was a fellow mathematician in Alexandria (as we saw in Section 4.1). (Perhaps no one saw much need for it.)

Archimedes' system has been described by Vardi (1997:27) as 'sub-optimal' in the sense that better systems could have been devised, such as one he says could be based on the work of the Alexandrian mathematician Diophantus (c. AD 200 – c. 284), as described by Heath (1910:47).

One way of understanding Archimedes' system is by means of the expression

$$\left(10^8\right)^{10^8(p-1)+d}, \tag{4.114}$$

where p stands for 'period' and d stands for 'order' (in Heath's translation), so that numbers can be succinctly referred to as (p,d)-numbers. Thus, for example, what Archimedes calls 'second-period third-order numbers', or simply $(2,3)$-numbers, are those from

$$\left(10^8\right)^{10^8(2-1)+3} \quad \text{to} \quad \left(10^8\right)^{10^8(2-1)+4} - 1. \tag{4.115}$$

That is, those from

$$\left(10^8\right)^{10^8+3} \quad \text{to} \quad \left(10^8\right)^{10^8+4} - 1. \tag{4.116}$$

Working the other way round, we can say that any number from

$$\left(10^8\right)^4 \quad \text{to} \quad \left(10^8\right)^5 - 1 \tag{4.117}$$

is a 'first-period fourth-order number', or $(1,4)$-number, since, in this case, $p=1$ and $d=4$.

In this system, the biggest number that can be named is the 'myriad-myriadth-period myriad-myriadth-order number', or, more succinctly, the $(10^8, 10^8)$-number. That is,

$$\left(10^8\right)^{10^8(10^8-1)+10^8} = \left(10^8\right)^{(10^8)^2-10^8+10^8} = \left(10^8\right)^{10^{16}}. \tag{4.118}$$

That is, a 1 followed by 80 quadrillion noughts.

But Archimedes does not need a number this big to fill up the universe with grains of sand. As he says in *Sand Reckoner* 2.4 (Heiberg, 1881:266; Vardi, 1997:5), the number of grains of sand that fit into a sphere the size of a poppy seed is less than a myriad, and forty of these poppy seeds are wider than a finger (when placed side by side in a straight line). Now, an ancient Greek finger was about 19.3 mm, so, according to Archimedes, the mean breadth of a grain of sand is (19.3 mm)/40 ≈ 0.5 mm (which corresponds to a *medium* type of sand, according to the ISO 14688-1 classification).

Denoting the width of a poppy seed by p, the width of a finger by f, the radius of the Earth by e, the radius of the cosmos by S, and the radius of the firmament (or sphere of fixed stars) by F, we see that the radii and volumes of the spheres bound by these bodies (measured in grains of sand) can be found by reasoning as follows.

First, the sphere of a poppy seed has volume

$$V_{\text{poppy seed}} = \frac{4}{3}\pi \left(\frac{p}{2}\right)^3. \tag{4.119}$$

Combining this and Assumption 5, we obtain

$$\frac{4}{3}\pi \left(\frac{p}{2}\right)^3 \approx 10^4 \text{ grains of sand.} \tag{4.120}$$

Making p the subject of this approximation, we have

$$p \approx 2\sqrt[3]{\frac{3 \times 10^4}{4\pi}} \text{ grains of sand} < 27 \text{ grains of sand.} \tag{4.121}$$

So, a poppy seed containing a myriad grains of sand has a diameter slightly less than 27 grains of sand. The radius of this poppy seed is

$$\frac{p}{2} \approx \sqrt[3]{\frac{3 \times 10^4}{4\pi}} \text{ grains of sand} \tag{4.122}$$

and can be used to find the number of grains of sand contained in a sphere a finger wide, since (by Assumption 5) the radius of this sphere is about 40 times that of a poppy seed.

Archimedes also rounds the number of fingers in a stade up to a myriad, since, in fact, there were about 9600 fingers in a stade (depending on the type of stade considered). Thus, under this convenient assumption, the radius of a sphere one stade wide is a myriad times that of a sphere a finger wide, and so we can find the number of grains of sand in the larger sphere by using what we know about the smaller one.

Thus, Equation 4.122 can be used again to find the number of grains of sand contained in any relevant sphere, such as that of the Earth, the cosmos, or the whole firmament. We only need to use the relations between these spheres, which

Table 4.3: Archimedes' cosmic ladder

A sphere as wide as one	has width	and volume (in grains of sand)	
poppy seed	$1p$	$\frac{4}{3}\pi\left(\sqrt[3]{\frac{30000}{4\pi}}\right)^3$	$= 10^4$
finger	$1f$	$\frac{4}{3}\pi\left(40\sqrt[3]{\frac{30000}{4\pi}}\right)^3$	$= 6.4 \times 10^8$
stade	$1\,\mathrm{st}$	$\frac{4}{3}\pi\left(10^4 \times 40\sqrt[3]{\frac{30000}{4\pi}}\right)^3$	$= 6.4 \times 10^{20}$
Earth	$2e$	$\frac{4}{3}\pi\left(10^6 \times 10^4 \times 40\sqrt[3]{\frac{30000}{4\pi}}\right)^3$	$= 6.4 \times 10^{38}$
cosmos	$2S$	$\frac{4}{3}\pi\left(10^4 \times 10^6 \times 10^4 \times 40\sqrt[3]{\frac{30000}{4\pi}}\right)^3$	$= 6.4 \times 10^{50}$
firmament	$2F$	$\frac{4}{3}\pi\left(10^4 \times 10^4 \times 10^6 \times 10^4 \times 40\sqrt[3]{\frac{30000}{4\pi}}\right)^3$	$= 6.4 \times 10^{62}$

are given by Inequalities 4.105, 4.107 and 4.111. Using these, the number of grains of sand in each of the mentioned spheres is as given in Table 4.3.

Thus, the number of grains of sand needed to fill Archimedes' exaggerated version of Aristarchus' already huge universe is somewhat less than 10^{63}, which—if classified in Archimedes' big number naming system—is a mere $(1, 7)$-number. That is, a first-period seventh-order number—a very tiny one, compared to what this remarkable system can cope with. Archimedes has thus given an answer to the kind of question that is guaranteed to fascinate any child walking along any beach at any time. Even today, questions such as this keep captivating the attention of many scientists, who currently put the number of particles (that is, protons, neutrons, neutrinos, and electrons) in the observable universe at 10^{80}—a mere $(1, 10)$-number.

4.5 Conclusion

In this chapter, we have seen how Aristarchus may have found the Moon's distance from Earth by means of a simple equation relating the Moon's elongation from the Sun at half phase (as given in *On Sizes* 1) to the Moon's angular size (as reportedly given in *On Sizes* 2). This equation (namely, Equation 4.35, or rather, a geometric version of it, as Aristarchus may have used it, if at all) is based on a chain of assumptions, of which always the most likely one is chosen. One of them is that Aristarchus used the best of the (extant) Babylonian approximations to pi (namely, $\pi \approx 25/8$). The main evidence supporting this assumption is that the solar width-to-distance ratio resulting from using this approximation is virtually the same as the geometric mean of Archimedes' upper and lower bounds for this ratio (as shown in Table 4.2). The reason why Archimedes should have chosen the geometric mean is that it produces a single value (which can be inverted in calculations involving reciprocals), whereas the arithmetic mean produces two (namely, the reciprocal of the arithmetic mean and the arithmetic mean of the reciprocals), and therefore also a conflict of choice between them. Another piece of evidence supporting the above assumption is that the mentioned equation fits almost exactly the results of the eclipse method when using Ptolemy's observational parameters in it, which may or may not be coincidental.

Also, in the appendix corresponding to this chapter, the eclipse method is explained, its approximations derived, and in the process, the exact solution to the whole method is found along with an extra, neat approximation.

The dimensions of Aristarchus' cosmos and firmament have also been assessed within the frame of the extant data. When expressed in terms of the Sun's distance, Aristarchus' firmament is about once to twice the size of the heliosphere (as follows from Inequality 4.13). However, when expressed in terms of the Earth's radius, it does not extend beyond the modern value for the orbit of Neptune (as follows from Inequality 4.14).

The current theories on the origin of Hipparchus' solar distances, as well as Ptolemy's and al-Battani's, have also been discussed and evidence has been found that they may all have ultimately been inspired by Aristarchus. In particular, the upper and lower bounds for the Sun's distance in *On Sizes* 1 (Inequality 2.48) may have inspired those by Hipparchus (in Approximations 4.93 and 4.94), and the distances in *On Sizes* 2 (Inequality 4.12) may have inspired those by Hipparchus as conveyed by Cleomedes and Theon of Smyrna (in Equations 4.87 and 4.90), as well as those by Ptolemy and al-Battani (as based on Inequalities 4.27 and 4.38 and Equation 4.28).

In the next chapter, we will investigate whether there is a more empirical dimension to Equation 4.35. That is, the one relating the Moon's elongation (of 87 degrees) and angular size (of 30 minutes of arc), the one that seems to be encoded in Ptolemy's version of the eclipse method.

5 On shadows, time, and light

Abstract: In this chapter, we shall consider whether Aristarchus ever measured the Moon's distance, and what kind of improvements (if any) he might have introduced in his (lost) second book *On Sizes* (assuming he ever wrote it). His estimate for the length of the year and its connection to those of others will also be studied.

5.1 The Moon's distance

Many are the theories that have been advanced, especially in recent years, trying to explain how Aristarchus measured the Moon's distance (if indeed he ever did). Take Rogers' (1960:234), for one. However, nobody knows for sure how he did it, or indeed whether he ever improved upon the answers he gave in *On Sizes* 1, but Archimedes' report (in *Sand Reckoner* 1.10) indicates that he did at least four times better in his second book than he did in his earlier one. In this section we are going to explore the possibility that he used lunar eclipses for this purpose.

Apart from wondering at nature's bountiful display of beauty during such events, we might also wonder whether there is anything out there that can be measured. Immediately, two things jump to mind. Namely, the angular size of the Moon and the angular size of the Earth's shadow, which we will denote, respectively, by $2\rho_{\mathbb{C}}$ and 2φ. In fact, the Canadian astronomer Martin Beech (2008:99) found an approximation to the Moon's distance M based on just these two parameters (measured in radians). Namely,

$$\frac{M}{e} \approx \frac{1}{\rho_{\mathbb{C}}}\left(1 - \frac{1+\rho_{\mathbb{C}}}{1+\rho_{\mathbb{C}}/\varphi}\right), \qquad (5.1)$$

which can be simplified to

$$\frac{M}{e} \approx \frac{1-\varphi}{\rho_{\mathbb{C}}+\varphi}, \qquad (5.2)$$

where e is the Earth's radius.

Another thing we might wonder about is the size and shape of the Earth's shadow, which is a topic to which Aristarchus dedicated a whole proposition in his first book *On Sizes*. This proposition cannot be found in the extant Greek manuscripts, but has been preserved in the Arabic translations of Thabit ibn Qurra and Nasir al-Din al-Tusi (as we saw in Section 1.2).

In these medieval Arabic translations, Aristarchus says that the Earth's shadow has the shape of a 'cone'. This was known from at least the time of Aristotle, who mentions such a shape in *Meteorology* 1.8, when he says,

Passage 60 Astronomical researches have now shown that the size of the Sun is greater than that of the Earth and that the stars are far farther away than the Sun from the Earth, just as the Sun is farther than the Moon from the Earth. Therefore, the vertex of the cone formed by the rays of the Sun will not fall very far from the Earth, nor will the Earth's shadow (which we call night) reach the stars (Aristotle, 1952:60).

In Chapter 2, we saw how the upper and lower bounds for the Sun's and Moon's distance and the length of the Earth's shadow in *On Sizes* 1 can all be expressed in terms of the Earth's radius (as in Inequalities 2.47 to 2.49), even though such expressions cannot be found in Aristarchus' extant work. Also, in Section 4.1, we saw how setting the angular size of the Sun and Moon to half a degree and assuming no further changes in *On Sizes* 1 lead to a lunar distance that is four times that in *On Sizes* 1. In other words, the lunar distance in *On Sizes* 2 is four times the lunar distance in *On Sizes* 1, and by extension, so is the length of the Earth's shadow W. Thus, multiplying Inequality 2.49 by four gives

$$\frac{6156}{37} < \frac{W}{e} < \frac{1075}{3}, \tag{5.3}$$

which is the reconstructed length of the Earth's shadow (under the given assumptions). According to this reconstruction, the cone of the Earth's shadow in *On Sizes* 2 is $166\frac{14}{37}$ to $358\frac{1}{3}$ Earth radii long.[33]

Now, using purely geometric calculations, such as those by Meeus (2007:365) or those leading to Equation B.19 (on page 242 of the present book), the length of the Earth's shadow is given by

$$\frac{W}{e} = \frac{S}{s-e}, \tag{5.4}$$

where S is the distance from the Earth to the Sun, s is the radius of the Sun, and e is the equatorial radius of the Earth (though, in fact, the volumetric radius should be more appropriate in this case).

Taking $S = 149597870.7\,\text{km}$ (from Pitjeva and Standish, 2009:370, or Luzum et al., 2011:296), $s = 695658\,\text{km}$ (from Haberreiter et al., 2008:L55), and $e = 6378.1366\,\text{km}$ (from Groten, 2004:726), the length of the Earth's shadow when the Earth is at perihelion, mesohelion, and aphelion, is, respectively,

$$\frac{W_{\min}}{e} = \frac{S_{\min}}{s-e} = \frac{(147003897\,\text{km})}{(695658\,\text{km}) - (6378.1366\,\text{km})} = 213.27..., \tag{5.5}$$

---

[33] By googling the words 'Sizes in *On Sizes* 2', the reader may check an online interactive illustration of the updated contents of *On Sizes* 1 when no change is considered other than setting the angular size of the Sun and Moon to half a degree.

$$\frac{W_{\text{mean}}}{e} = \frac{S_{\text{mean}}}{s-e} = \frac{(149597870.7\,\text{km})}{(695658\,\text{km}) - (6378.1366\,\text{km})} = 217.03..., \quad (5.6)$$

$$\frac{W_{\text{max}}}{e} = \frac{S_{\text{max}}}{s-e} = \frac{(152154293.5\,\text{km})}{(695658\,\text{km}) - (6378.1366\,\text{km})} = 220.74..., \quad (5.7)$$

where S_{min} and S_{max} are the shortest and longest gaps between the Sun and Earth occurring between the years 1000 and 3000, say, and measured at the time of maximum lunar eclipse, such as those of December 9, 1071, and June 14, 1025.

Alternatively, we may use Equation B.20 (on page 243), according to which

$$\frac{W}{e} = \frac{1}{\sin \zeta} = \frac{S}{s-e}, \quad (5.8)$$

where ζ is half the angular size of the Earth as seen from the tip of its own shadow, and everything else is as before. Hence,

$$\zeta = \sin^{-1}\left(\frac{s-e}{S}\right). \quad (5.9)$$

It is easy to deduce from this equation that the *geometric* minimum, mean, and maximum values of this half shadow-tip angle are, respectively,

$$\zeta_{\text{max}} = \sin^{-1}\left(\frac{(695658\,\text{km}) - (6378.1366\,\text{km})}{(147003897\,\text{km})}\right) \approx 16'7.15''. \quad (5.10)$$

$$\zeta_{\text{mean}} = \sin^{-1}\left(\frac{(695658\,\text{km}) - (6378.1366\,\text{km})}{(149597870.7\,\text{km})}\right) \approx 15'50.38''. \quad (5.11)$$

$$\zeta_{\text{min}} = \sin^{-1}\left(\frac{(695658\,\text{km}) - (6378.1366\,\text{km})}{(152154293.5\,\text{km})}\right) \approx 15'34.41''. \quad (5.12)$$

As a check, we have

$$\frac{W_{\text{min}}}{e} = \frac{1}{\sin(16'7.15'')} = 213.27..., \quad (5.13)$$

$$\frac{W_{\text{mean}}}{e} = \frac{1}{\sin(15'50.38'')} = 217.03..., \quad (5.14)$$

$$\frac{W_{\text{max}}}{e} = \frac{1}{\sin(15'34.41'')} = 220.74..., \quad (5.15)$$

which are the same results as in Equations 5.5 to 5.7.

As Meeus (2007:366) points out, the length of the Earth's shadow is continuously changing because the Earth's distance to the Sun is also continuously changing, and (to a lesser extent) also because neither the Earth nor the Sun are perfectly spherical bodies, nor does the Sun always lie exactly in the plane of the equator. It is for these reasons that the extreme shadow lengths in the period considered are not those of December 9, 1071 (the shortest), and June 14, 1025 (the longest), as stated above, but those of December 12, 1163 (the shortest), and June 16, 1155 (the longest). (The difference, however, is negligible.) He also points out that Equation 5.4 is purely geometric, so it neglects the presence of the Earth's atmosphere.

In order to account for this effect, we must first reflect upon the fact that an object's shadow is shorter the closer it is to the Sun, and longer the further it is from the Sun. Thus, the Moon's shadow is shorter during a solar eclipse than it is during a lunar eclipse. However, the angle by which the Moon's shadow tapers off is always bigger than the angle by which the Earth's shadow tapers off, even when the Moon is furthest from the Sun. This is because smaller objects cast shorter shadows, and in this case, the Moon is a smaller body than the Earth. As a check, half the angle by which the Earth's *geometric* shadow tapers off (as given by Equation 5.11) is

$$\zeta_\oplus = \sin^{-1}\left(\frac{s-e}{S}\right) \approx 15'50.38'',$$

where s and S are the Sun's radius and distance, and e is the Earth's radius; and half the angle by which the Moon's shadow tapers off varies between the extremes

$$\sin^{-1}\left(\frac{s-m}{S+M}\right) \leq \zeta_\mathbb{C} \leq \sin^{-1}\left(\frac{s-m}{S-M}\right), \tag{5.16}$$

where m and M are the Moon's radius and distance. When appropriate values for these parameters are provided, we have

$$15'54.32'' \leq \zeta_\mathbb{C} \leq 15'59.24'', \tag{5.17}$$

which, as said above, is always bigger than $\zeta_\oplus$.

Alternatively, we may use Equation 5.8 and the Cosine Rule to deduce a similar one for the length of the Moon's shadow. Namely,

$$\frac{L}{m} = \frac{1}{\sin\zeta_\mathbb{C}} = \frac{S}{s-m} = \frac{\sqrt{S^2+M^2-2SM\cos\varepsilon}}{s-m}, \tag{5.18}$$

where m is the Moon's radius, ε is the Moon's elongation from the Sun, and everything else is as before. Hence, a more general equation for $\zeta_\mathbb{C}$ than those used in Inequality 5.16 is

$$\zeta_\mathbb{C} = \sin^{-1}\left(\frac{s-m}{\sqrt{S^2 + M^2 - 2\,SM\cos\varepsilon}}\right). \tag{5.19}$$

Now, should the Sun be infinitely far away and yet preserve its angular size, then the Sun's rays tangent to the Earth and Moon would be parallel to each other at all times, and therefore, the angles $\zeta_\oplus$ and $\zeta_\mathbb{C}$ would always be the same. Of course, the Sun cannot be infinitely far away, which is why we were able to find the value of $\zeta_\oplus$ that satisfies all of Aristarchus' requirements in *On Sizes* 1 (namely, that in Equation B.25), as well as the corresponding solar and lunar distances (namely, those in Equations 2.29 and 2.30).

Now, let us consider the following simplifying assumptions.

- The Earth and Moon are perfect spheres moving in perfectly circular, coplanar orbits.
- The Sun is infinitely far away and yet preserves its angular size.
- The Sun and Moon have exactly the same angular size (as stated in Propositions 3 and 8 of *On Sizes* 1).
- The tip of the Moon's shadow is at our eye during a solar eclipse (as stated in Propositions 3 and 8 of *On Sizes* 1).
- 'The Earth is in the relation of a point and centre to the sphere in which the Moon moves' (as stated in Hypothesis 2 of *On Sizes* 1), which is Aristarchus' way of saying that the observer is assumed to be at the Earth's centre.

These simplifying assumptions are based on the fact that all shadows taper off by about half a degree (provided the objects casting them are about the same distance from the Sun as the Earth-Moon system is). By these assumptions, the angle $\zeta_\mathbb{C}$ (which is half the angular size of the Moon as seen from the tip of its shadow) is the same as the angle $\rho_\mathbb{C}$ (which is half the angular size of the Moon as seen from Earth during a lunar eclipse). At the same time, the Earth itself casts a shadow that tapers off by about half a degree. Hence, the angle $\zeta_\oplus$ (which is half the angular size of the Earth as seen from the tip of its shadow) is about the same as the angles $\zeta_\mathbb{C}$ and $\rho_\mathbb{C}$.

Under these assumptions, and taking Aristarchus' value for the angular radius of the Sun $\rho_\odot$ and Moon $\rho_\mathbb{C}$ to be as reported by Archimedes (in *Sand Reckoner* 1.10), we have

$$\zeta = \zeta_\oplus = \zeta_\mathbb{C} = \rho_\odot = \rho_\mathbb{C} = 15' \tag{5.20}$$

(versus $\zeta = \zeta_\oplus \approx \zeta_\mathbb{C} \approx \rho_\odot \approx \rho_\mathbb{C} \approx 15'$, in reality).

It is important to note that the angle ζ in Equation 5.20 is an *apparent* one (that is, one affected by the Earth's atmosphere, and therefore corresponding to the real thing out there) and is significantly smaller than the *geometric* value (in Equation 5.11).

This realization allows us to establish a curious connection between the Moon and the Earth's shadow. To see what it is, we may start by looking at a simplified diagram of a lunar eclipse, such as that in Figure 5.1.

This diagram differs from that in Figure 2.8 (of Proposition 13) in that the Sun is now infinitely far from Earth, so that all shadow-tip angles are the same (as in Equation 5.20), and also that the arc OMN is now part of the orbit of the Moon's centre, rather than part of the orbit of the Moon's disc.

The length of the line segment NR (in Figure 5.1) can be deduced in two ways depending on whether we regard it as one of the legs of the right-angled triangle ERN, or one of the legs of the right-angled triangle ZRN. Hence, by simple trigonometry,

$$RN = EN \sin \varphi = ZN \sin \zeta, \tag{5.21}$$

where ζ is defined as above, and φ is half the angular size of the Earth's shadow where it cuts the orbit of the Moon's centre.

The other legs of the above mentioned triangles are, respectively,

$$ER = EN \cos \varphi, \tag{5.22}$$

and

$$ZR = ZN \cos \zeta. \tag{5.23}$$

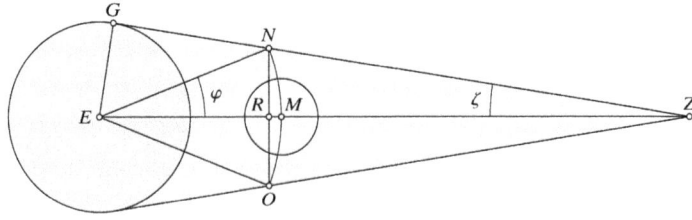

Figure 5.1 Simplified lunar eclipse diagram

Now, the length of the Earth's shadow is

$$W = ER + ZR \quad \text{(by inspection of the diagram)}$$
$$= EN \cos \varphi + ZN \cos \zeta \quad \text{(by Equations 5.22 and 5.23)}$$
$$= EN \cos \varphi + \left(\frac{EN \sin \varphi}{\sin \zeta}\right) \cos \zeta \quad \text{(by Equation 5.21)}$$
$$= \frac{e}{\sin \zeta}. \quad \text{(by Equation 5.8)}$$

Multiplying through the latter equality by $\sin \zeta$ gives

$$EN \cos \varphi \sin \zeta + EN \sin \varphi \cos \zeta = e,$$

which can be rearranged as

$$\frac{EN}{e} = \frac{1}{\cos \varphi \sin \zeta + \sin \varphi \cos \zeta}.$$

Now, EN is equal to the distance between the Earth and the Moon. Denoting this by M and using a common trigonometric addition identity, this reduces to

$$\frac{M}{e} = \frac{1}{\sin(\zeta + \varphi)}, \tag{5.24}$$

which (using Equation 5.20) can be written as

$$\frac{M}{e} = \frac{1}{\sin(\rho_\mathbb{C} + \varphi)}. \tag{5.25}$$

This equation is both simple and exact (under the given assumptions), and it is perhaps the most important of all those in this book. It gives the Moon's distance M in terms of the Earth's radius e as a function of just two variables: namely, half the angular size of the Moon $\rho_\mathbb{C}$ and half the angular size of the Earth's shadow φ. These are the two parameters mentioned at the beginning of this section that can be measured directly (the first one, whenever the Moon is visible; the second, only during lunar eclipses). The difference between Equations 2.28 and 5.25 is that the latter assumes the Sun to be infinitely far away and the former does not.

In his first book *On Sizes* (Hypotheses 5 and 6), Aristarchus assumed $\rho_\mathbb{C} = 1°$ and $\varphi = 2°$. (The sum of these is 3°, which may or may not explain the origin of his 87-degree angle.) Putting these into Equation 5.25, we obtain

$$\frac{M}{e} = \frac{1}{\sin(1° + 2°)} = 19.107..., \tag{5.26}$$

which is less than the result in Equation 2.29 (for the reasons explained above).

However, using the value of $\rho_{\mathbb{C}}$ that Archimedes attributes to Aristarchus (which is less than one arcminute short of the truth) and assuming he also measured the angular size of the Earth's shadow to a similar level of accuracy, putting it, say, at 'the 257th part of a circle' (or about $\varphi \approx 42'1''$), then Equation 5.25 yields

$$\frac{M}{e} = \frac{1}{\sin(15' + 42'1'')} = 60.296..., \tag{5.27}$$

which compares well with the value in Equation 4.35, which (as we saw in Section 4.3) shows up in the eclipse method when using Ptolemy's mean observational parameters. The reason for the above choice, namely,

$$\varphi = \frac{180°}{257} \approx 42'1'', \tag{5.28}$$

is that this is the smallest, neat value of φ that yields a meaningful solar distance when used in the eclipse method along with Aristarchus' $\rho_{\mathbb{C}} = 15'$.

Again, it is important to note that both ζ and φ are *apparent* angles, since it is only the real shadow that can be measured empirically (not the geometric one). So, rearranging Equation 5.21 to make φ the subject and denoting the line segment RN in this equation by the letter U, we have

$$\varphi = \sin^{-1}\left(\frac{U}{M}\right), \tag{5.29}$$

where U and M are both measured in the same unit of length (say, Earth radii e).

Also, rearranging Equation 5.24 to make ζ the subject and substituting for φ from Equation 5.29, we have

$$\zeta = \sin^{-1}\left(\frac{e}{M}\right) - \varphi \tag{5.30}$$

$$= \sin^{-1}\left(\frac{e}{M}\right) - \sin^{-1}\left(\frac{U}{M}\right). \tag{5.31}$$

Hence, combining Equations 5.8 and 5.31, the length of the Earth's *apparent* shadow is given by

$$\frac{W}{e} = \left(\sin\left(\sin^{-1}\left(\frac{e}{M}\right) - \sin^{-1}\left(\frac{U}{M}\right)\right)\right)^{-1}. \tag{5.32}$$

Now, the values of M and U can be found using astronomical algorithms, such as those by Meeus (1998:382). Using these algorithms to analyse all the lunar eclipses occurring within a sufficiently large period of time, such as, say, that between the years 1000 and 3000, we obtain the extreme shadow lengths in Table 5.1.

As we can see, it matters whether we use Chauvenet's or Danjon's increasing factors. Bearing this in mind, the length of the Earth's *apparent* shadow (as averaged out of the 5754 lunar eclipses occurring during the period of time considered) is 227.520... Earth radii, when using Chauvenet's factor, and 225.028... Earth radii, when using Danjon's (at a mean lunar distance of about 59¾ Earth radii).

Using Aristarchus' value for ζ (which Equation 5.20 equates to 15′), Equation 5.8 yields the length of the Earth's *apparent* shadow as

$$\frac{W}{e} = \frac{1}{\sin 15'} = 229.18..., \tag{5.33}$$

which is closer to the *apparent* mean shadow length based on Chauvenet's factor than it is to Danjon's, and all are, of course, larger than the *geometric* values in Equations 5.5 to 5.7. So, the assumptions on page 187 lead to a fairly good approximation to the length of the Earth's *apparent* shadow.

Let us now see how Aristarchus may have measured the breadth of the Earth's shadow. This can only be done by observing lunar eclipses. The main characters in these amazing celestial displays are the Earth's shadow and the Moon's disc, and both move against the background of stars at measurable speeds. The Earth's shadow moves at a rate of one zodiac per year (or about one in 365¼ parts of the zodiac circle per day), and the Moon moves at a rate of one zodiac per sidereal month (or about one in 27⅓ parts of the zodiac circle per day). So the Moon moves about $365.25/27.3 \approx 13.4$ times faster than the Earth's shadow, catching up with it and overtaking it in a period of time that can also be measured. Aristarchus may have used this knowledge to assess the angular size of the Earth's shadow (when it is cast on the Moon). If so, this is how he might have proceeded.

The Earth's shadow takes a year to cross the zodiac (that is, it takes $t_\odot \approx 365¼$ days to sweep 360 degrees of the sky), and it sweeps an angle δ of the sky during the time t that the centre of the Moon's disc remains eclipsed. Hence, by a rule of three,

$$\frac{\delta}{360°} = \frac{t}{t_\odot},$$

which can be rearranged as

$$\delta = 360° \times \frac{t}{t_\odot}. \tag{5.34}$$

Table 5.1 Earth's extreme and mean shadow lengths W, as functions of the Moon's distance M and the umbral radius φ (in degrees) and U (in terms of the Earth's radius e), as given by Equation 5.32, using either Chauvenet's (1891:542) or Danjon's (1951:53) increasing factors.

	M/e	Chauvenet $\varphi(°)$	Chauvenet U/e	Chauvenet W/e	Danjon $\varphi(°)$	Danjon U/e	Danjon W/e
max	63.71346715	0.7841599	0.7592290	232.55614	0.7739253	0.7563290	229.78986
mean	59.74939018	0.7161126	0.7373351	227.52009	0.7065026	0.7344351	225.02867
median	59.66634851	0.7154853	0.7378877	227.51437	0.7058517	0.7349877	225.01822
min	55.90146935	0.6486406	0.7135229	222.36681	0.6396468	0.7106229	220.13844

At the same time, the Moon takes a sidereal month to cross the zodiac (that is, it takes $t_{\mathbb{C}} \approx 27\frac{1}{3}$ days to sweep 360 degrees of the sky), and during the time t the centre of the Moon's disc remains eclipsed, not only does it sweep the breadth of the Earth's shadow, but also the above-mentioned angle δ. That is, during the time t the Moon's centre remains eclipsed, it sweeps an angle equal to $2\varphi + \delta$, where φ is half the angular size of the Earth's shadow. Hence, by a rule of three,

$$\frac{2\varphi + \delta}{360°} = \frac{t}{t_{\mathbb{C}}},$$

which can be rearranged as

$$2\varphi = 360° \times \frac{t}{t_{\mathbb{C}}} - \delta. \tag{5.35}$$

Substituting for δ (from Equation 5.34) into this and simplifying, we have

$$\varphi = 180°t \times \left(\frac{1}{t_{\mathbb{C}}} - \frac{1}{t_{\odot}}\right). \tag{5.36}$$

Now, the time t the centre of the Moon's disc remains eclipsed is to be measured directly, and it turns out that the best lunar eclipses for this purpose are those classified as **central**, or **nearly central** (that is, those in which the centres of the solar and lunar discs get very close to each other at the time of maximum eclipse). Other types of lunar eclipses would make us believe that the Earth's shadow is narrower than it really is, because the Moon's centre remains eclipsed for a shorter period of time.

Aristarchus may have learned about the duration of such events from Babylonian records, which, we know, had fallen into Macedonian hands and had been (or were being) translated into Greek at the wise request of Alexander's (and Strato's) tutor, Aristotle and his nephew Callisthenes of Olynthus, as we learn from Simplicius' *On the Heavens* 2.12 (Heiberg, 1894:506). So, assuming Aristarchus had access to these records (as is most likely the case), he may then have learned about the duration of at least a few (nearly) central eclipses of the Moon that may have been recorded by the Babylonians. One of them occurred only a century earlier, on June 6, 381 BC, and is one of the longest lasting lunar eclipses ever. Now, the Babylonians used to record the duration of the partial and total phases of most lunar eclipses, rounding their numbers to the nearest four minutes of time. However, errors of up to half an hour were frequent, presumably due to the inaccuracy of their primitive clocks (Stephenson and Fatoohi, 1993:261).

The good thing about central or nearly central eclipses is that (because the centre of the Moon's disc goes through that of the Earth's shadow) it is possible to estimate the time t that the centre of the Moon's disc remains eclipsed by simply averaging the times of total and partial phases of the eclipse—just the kind of information that the Babylonians used to record. Thus, for example, according to Espenak and Meeus (2006*b*:196), on June 6, 381 BC, the total phase lasted 106 minutes, and the partial phase lasted 235 minutes. Hence, the centre of the Moon's disc remained eclipsed for $t \approx (106 + 235)/2 = 170.5$ minutes. Another such eclipse was that of May 29, 334 BC. Espenak and Meeus (2006*b*:202) give the times of total and partial phases of this eclipse as 100 and 214 minutes, each. So, in this case, the required time was $t \approx (100 + 214)/2 = 157$ minutes.

Aristarchus was very lucky that one of these rare events took place in his own lifetime. Namely, the (nearly) central lunar eclipse of May 20, 287 BC. Assuming weather and circumstances allowed him to observe this one, and assuming also his timekeeper was good enough, he should have recorded something in the vicinity of 101 and 221 minutes as the times of total and partial phases of this eclipse (as estimated by Espenak and Meeus, 2006*b*:207). Hence, in this case, $t \approx (101 + 221)/2 = 161$ minutes. Putting this into Equation 5.36, along with the above estimate of half-an-hour uncertainty, gives

$$\varphi = 180° \times \frac{(161 \pm 30 \text{ minutes})}{\left(\frac{24 \text{ hours}}{1 \text{ day}}\right) \times \left(\frac{60 \text{ minutes}}{1 \text{ hour}}\right)} \times \left(\frac{1}{27\frac{1}{3} \text{ days}} - \frac{1}{365\frac{1}{4} \text{ days}}\right)$$

$$\approx 40'52'' \pm 7'37'', \qquad (5.37)$$

which compares well with both reality and the guesstimate in Equation 5.28. In fact, a statistical study of all the lunar eclipses occurring between the years 1000 and 3000 based on Meeus' algorithms (which use Danjon's factor) reveals that the angular radius of the Earth's shadow is

$\varphi_{\min} \approx 38'22.73''$ (as on December 12, 1163), $\qquad$ (5.38)

$\varphi_{\mathrm{mean}} \approx 42'23.41''$, $\qquad$ (5.39)

$\varphi_{\max} \approx 46'26.13''$ (as on June 30, 1730). $\qquad$ (5.40)

(These values are close to al-Battani's in Equation 4.49.)

It is to be noted that the Earth's shadow is slightly wider at the distance of the Moon's disc than it is at the distance of the centre of the Moon's spherical body. Aristarchus used the former value in *On Sizes* 1 (Figures 2.8 and C.1) and we used the latter value in the present section (Figure 5.1). We may find the latter value by rearranging Equation 5.24, thus,

$$\varphi_{\mathrm{centre}} = \sin^{-1}\left(\frac{e}{M}\right) - \zeta, \qquad (5.41)$$

and the former value, by combining Equations B.3 and 5.41, thus,

$$\varphi_{\text{disc}} = \sin^{-1}\left(\frac{e}{M \cos \rho_{\mathbb{C}}}\right) - \zeta, \tag{5.42}$$

where all variables are as usual.

Finally, an estimate of the upper and lower bounds for φ which Aristarchus may have used in *On Sizes* 2 can be derived from Inequality 4.11 and Equation 5.24. Combining these, we have

$$57 \leq \frac{1}{\sin(\rho_{\mathbb{C}} + \varphi)} \leq \frac{860}{9}, \tag{5.43}$$

which can be rearranged as

$$\sin^{-1}\left(\frac{9}{860}\right) - \rho_{\mathbb{C}} \leq \varphi \leq \sin^{-1}\left(\frac{1}{57}\right) - \rho_{\mathbb{C}}. \tag{5.44}$$

Substituting for $\rho_{\mathbb{C}} = 15'$, this becomes

$$20'58.62\ldots'' \leq \varphi \leq 45'18.87\ldots'', \tag{5.45}$$

which is a range of values containing those assumed above for Aristarchus.

To recap, we have seen a method by which Aristarchus may have estimated the Moon's distance to the accuracy conveyed by Equations 4.35, 4.95 and 5.27. This method uses lunar eclipses of long duration. He may have timed the partial and total phases of one of these himself or used data from Babylonian records, or both, to obtain an accurate value for φ, which yields the required distance when used along with his accurate value for $\rho_{\mathbb{C}}$.

His upper and lower bounds for the distances of the Sun and Moon, along with the size of the sphere of the fixed stars and the length of the Earth's shadow, as may have been given in his second book *On Sizes*, were also found (in this and Section 4.1) by assuming that everything is left unchanged in his first book *On Sizes*, except Hypothesis 6, which, when updated to reflect Archimedes' report that Aristarchus put the angular size of the Sun (and Moon) at half a degree, gives the results in the summary box below.

A summary of the main results possibly obtained in *On Sizes* 2, assuming no changes in *On Sizes* 1 other than a half-a-degree wide Sun and Moon

- The ratio between the Moon's distance and the Earth's radius is

$$57 < \frac{M}{e} < \frac{860}{9}. \tag{4.11}$$

- The ratio between the Sun's distance and the Earth's radius is

$$1026 < \frac{S}{e} < \frac{17200}{9}.\qquad(4.12)$$

- The ratio between the sizes of the firmament and the cosmos is

$$\frac{6156}{43} < \frac{F}{S} < \frac{17200}{57}.\qquad(4.13)$$

- The ratio between the sizes of the firmament and the Earth is

$$\frac{6316056}{43} < \frac{F}{e} < \frac{295840000}{513}.\qquad(4.14)$$

- The ratio between the length of the Earth's shadow and the Earth's radius is

$$\frac{6156}{37} < \frac{W}{e} < \frac{1075}{3}.\qquad(5.3)$$

Let us now turn our attention to some of Aristarchus' other concerns.

5.2 The length of the year

The only event in Aristarchus' life that can be dated with some precision is his reading of the summer solstice in 280 BC, which Ptolemy has preserved along with four other like observations from antiquity plus enough information to identify the year of yet another one (in *Almagest* 3.1, 3.4, and 7.2). Namely, those in Table 5.2.

Table 5.2: Summer solstice observations reported in *The Almagest*

observer	summer solstice date	time of day	*Almagest* section	Toomer reference	JD of actual occurrence
Meton	−431-06-27	at dawn	3.1	1984:138	1563813.86844249
Callippus	−329	−	3.1	1984:138	1601068.55544626
Aristarchus	−279	−	3.1	1984:138	1619330.64052040
Hipparchus	−134	−	3.1	1984:139	1672290.72812040
Hipparchus	−127	−	7.2	1984:328	1674847.41251610
Ptolemy	140-06-25	about 2 hours after midnight	3.4	1984:154	1772366.99878601

The purpose of these observations was to establish the length of the tropical (or solar) year by comparing observations made over long periods of time. For example, Meton's and Hipparchus' solstices differ by $432 - 135 = 297$ years, or about 108477 days (as results from subtracting the corresponding Julian days in Table 5.2). Dividing these two numbers gives an average yearlength of

$$\begin{aligned} Y_{MH_0} &= \frac{108477}{297} \text{ days} \\ &= 365.\dot{2}\dot{4} \text{ days} \\ &= 365;14,32,43,38,10... \text{ days} \\ &\approx 365 \text{ days, 5 hours, 49 minutes, 5 seconds,} \end{aligned} \qquad (5.46)$$

which is only 9 seconds over the modern estimate for the length of the tropical year back in 135 BC. Namely,

$$\begin{aligned} T_y &= 365.2423190561... \text{ days} \\ &= 365;14,32,20,54,58... \text{ days} \\ &\approx 365 \text{ days, 5 hours, 48 minutes, 56 seconds,} \end{aligned} \qquad (5.47)$$

as derived from the VSOP2000 theory (Moisson and Bretagnon, 2001; Rocher, 2000:10), which gives the length of the tropical year (in days) by the following algorithm.

$$T_y = 365.2421905166 - \frac{(615607 + (684 - (2630 + 32\,T)\,T)\,T)\,T}{10^{10}}, \qquad (5.48)$$

where

$$T = \frac{JD - 2451545}{365250} \qquad (5.49)$$

is the time in Julian millennia (of 365250 days) from the epoch J2000.0 (or $JD = 2451545.0$), and JD is the Julian day for which the calculation is required (as taken from Table 5.2).

The problem of finding the length of the year, as addressed by the ancient Greeks, appears to have a simple solution. One only needs to find an event that repeats regularly every year and can be reliably measured. It turns out that the length of the shadow cast by a vertical rod (called a **gnomon**) provides a handsome way to achieve this goal. This shadow is easier to measure the shorter it gets, and it gets shortest at the time of summer solstice. Counting the number of days between two such events gives the length of the year between them. Counting the days between many such events and dividing the result by the number of years involved (as in Equation 5.46) gives a value that is the average of all those years and is, therefore, more reliable than a single measurement.

However, errors may creep in the calculations. For example, if someone fails to correctly count the number of days that span the long periods of time involved, the result will be less accurate than it should. This is exactly what happened to the yearlength that follows from Hipparchus' first solstice observation. Presumably, it was Hipparchus himself who counted one day too many and instead of getting the nearly perfect value in Equation 5.46, he got

$$Y_{MH_1} = \frac{108478}{297} \text{ days}$$
$$= 365.\dot{2}4579\dot{1} \text{ days}$$
$$= 365;14,44,50,54,32... \text{ days}$$
$$\approx 365 \text{ days, 5 hours, 53 minutes, 56 seconds,} \tag{5.50}$$

which (rounded to the nearest third-order sexagesimal part of a day) is exactly the same as the yearlength recorded on Babylonian tablet BM 55555, as Rawlins (1991:51) first noticed.[34]

Indeed, line 11 of this tablet gives the length of 18 years as 1,49,34;25,27,18 days (or 6574.42425 days) (Neugebauer, 1975:528; 1995:272). Hence, the length of one year is

$$Y_{Bab} = \frac{1,49,34;25,27,18}{18} \text{ days}$$
$$= 365.2457916 \text{ days}$$
$$= 365;14,44,51 \text{ days}$$
$$\approx 365 \text{ days, 5 hours, 53 minutes, 56 seconds.} \tag{5.51}$$

The connection between Equations 5.50 and 5.51 is clear, and the most probable explanation for it (as Rawlins suggests) is that, at some point in history, some Babylonian scribe wrote it using Hipparchan data. It is to be noticed that line 12 of this tablet describes this yearlength as 'sidereal' (in Neugebauer's translation), when in fact, it is 5 minutes longer than the *tropical* year back then (and would have been only 9 seconds longer, had days been counted correctly), adding to the likelihood that the Babylonian scribe who wrote it misunderstood thy original Greek source.

---

[34] This tablet can be viewed at the British Museum's Collection Database, with registration number 1882-0704-143.

Another thing the ancients noticed is that the time of **summer solstice** (that is, when the Sun gets closest to the north celestial pole, or, technically, the time when the Sun's apparent geocentric longitude is exactly 90 degrees) does not necessarily occur at local noon (since the gnomon's shadow may get shorter at locations further east or west, yet as far north as the observer's). So, one way of refining our estimates is to interpolate the time when the shortest shadow occurs from a number of readings several days before and after the event. Thus, using a rudimentary form of interpolation, such as described by Thurston (2001:154), the ancient Greeks were able to round the time of a solstice to the nearest quarter of a day, typically recording them as occurring either at dawn, noon, dusk, or midnight.

According to Thurston, measurements were made at noon during several days around the solstice. Then, four possibilities arose depending on the lengths of the gnomon's shadow. These are as follows.

1. If the shadow gets shortest on a given day and the shadows an equal number of days before and after this event are equally long, then the solstice occurs at noon.
2. If the shadow gets equally short on two consecutive days and equally long an equal number of days from these two, then the solstice occurs at midnight.
3. If the shadow gets shortest on a given day and is not as long on the previous day as it is on the next, then the solstice occurs at about dawn.
4. If the shadow gets shortest on a given day and is longer on the day before than it is on the day after, then the solstice occurs at about dusk.

The interested reader may like to know that it is possible to improve upon this accuracy by using modern interpolation formulas, such as those given by Meeus (1998:25), the simplest of which use three equally spanned readings. Namely,

$$y = y_2 - \frac{(y_3 - y_1)^2}{8(y_1 + y_3 - 2y_2)}, \qquad (5.52)$$

where the interpolated value y (which, in this case, is the length of the gnomon's shadow) follows from the shadow lengths y_i read at noon an equal number of days either side of the event, with y_2 being the shortest of all those readings. This formula tells us how short the shortest shadow can possibly get.

It is even possible to obtain the time t (in decimal days) when the shortest shadow occurs by means of any of these other formulas,

$$t = t_2 + \frac{(t_2 - t_1)(y_1 - y_3)}{2(y_1 + y_3 - 2y_2)} = t_2 + \frac{(t_3 - t_2)(y_1 - y_3)}{2(y_1 + y_3 - 2y_2)}, \quad (5.53)$$

where t_2 is the day (of the month, or otherwise) when our shortest reading was taken, t_1 and t_3 are, respectively, the days when the first and last readings were taken, and y_1, y_2, and y_3 are as before.

Using equations such as these, or even more complex ones using five (rather than three) readings, Hartner (1977:7) found that a span of 10 days before and after the solstice usually yields an accuracy of about two hours, and an optimal span of 45 days before and after the solstice usually improves the accuracy to about one hour. The ancient Greeks, however, did not have access to the above formulas, so they relied on a simpler procedure, such as that described by Thurston.

As Rawlins (1991:52) points out, it is significant that Ptolemy gives the year and time (usually rounded to the nearest quarter of a day) of all the solar observations in *The Almagest* (including 24 equinoxes and 6 summer solstices) except the summer solstices of Aristarchus and Hipparchus (of which only the year is given) and that of Callippus (of which not even the year is explicitly given), yet these particular solstices were of the utmost importance for the purpose of establishing the length of the year, and therefore, constructing reliable calendars. According to Rawlins, there is a reason why Ptolemy should omit this vital information. Namely, that the times of these solstices were embarrassingly different from those predicted by the algorithms in *The Almagest*. Ptolemy describes those readings as 'conducted rather crudely' (Toomer, 1984:137). His own, however, he describes as 'made with the greatest accuracy', pinpointing the occurrence of his solstice on Mesore 11, 887 of the Nabonassar Era, about 2 hours after midnight (or 140-06-24 at 23:42:26 UT).

Indeed, the algorithms in *The Almagest* predict that this solstice fell on Mesore 11, 887 of the Nabonassar Era, 12 hours, 56 minutes, and 19 seconds (of mean solar time) after local noon in Alexandria (or 140-06-24 at 22:38:45 UT). So, Ptolemy's reading of this solstice, which he says was 'determined as accurately as possible', agrees closely with the *Almagest* prediction, differing from it by just over an hour. But this is all spurious accuracy. In reality (that is, as given by Meeus' 1998:177 algorithms), the actual solstice occurred on 140-06-23 at 11:58:15 UT, so both Ptolemy and his *Almagest* were off the mark by nearly 36 and 35 hours, each. So, in this case, not even an accuracy of a whole day was reached.

Now, since we are assuming (with Rawlins) that Ptolemy omitted the time of Callippus', Aristarchus', and Hipparchus' solstices because they did not agree with the *Almagest* predictions (and therefore was bending facts to suit theories, rather than theories to suit facts), it is logical to think that these solstices occurred at times other than those predicted by *The Almagest*. Otherwise, Ptolemy would have brandished them as proofs of the reliability of his method. The *Almagest* predictions of these solstices are as given in Table 5.3.

The information above makes it possible to attempt a reconstruction of the time of Hipparchus' first solstice. As it happens, if taken to occur exactly 108478 days after Meton's, Hipparchus' solstice matches the yearlength on BM 55555 (to the nearest third-order sexagesimal part of a day), so it may be that the time of day of Hipparchus' solstice is the same as Meton's, as Rawlins (1991:51) proposes.

Ptolemy reports that Meton's solstice fell 'on Phamenoth 21 in the Egyptian calendar, at dawn, in the year when Apseudes was archon at Athens' (Toomer, 1984:138). Meton himself would have recorded the date of this solstice in terms of the Athenian calendar, which happened to be lunisolar, with months starting every New Moon and years starting on the first New Moon after the summer solstice (Hannah, 2008:36).

Meton's solstice was special in that it happened near a Full Moon. This is something we can deduce from a short inscription found at Miletus (Diels and Rehm, 1904:96, 266), which explicitly connects the solstice of Skirophorion 13 under Apseudes to Phamenoth 21 Egyptian (Fotheringham, 1924:383),[35] to which the 1st-century BC Greek historian Diodorus of Sicily adds (in *Library* 12.36.2; Oldfather, 1946:446) that Meton, the son of Pausanias, introduced the 19-year cycle in Athens, starting it on Skirophorion 13. So the solstice fell on Skirophorion 13, which, by definition, corresponds to a 13-day old Moon (that is, a nearly Full Moon). Modern computation shows that the Full Moon in question occurred on June 30, 432 BC, at about 4:46:03 UT, or Skirophorion 15, 6 hours, 20 minutes, and 58 seconds after midnight, local mean solar time in Athens (where days officially started at sunset).

Modern computation also shows that Meton's solstice fell on June 28, 432 BC, at about 8:50:33 UT (as shown in Table 5.3), or Skirophorion 13, 10 hours, 25 minutes, and 28 seconds after local midnight. In other words, the solstice

---

35 This inscription, known as *Parapegma* 84, dates to a little after 109 BC (since it mentions too the solstice of Skirophorion 14 under Polycletus, who was archon that year).

Table 5.3 Summer solstice predictions based on the algorithms in *The Almagest*

observer	predicted Egyptian date and equinoctial time after dawn in Alexandria		corresponding Proleptic Julian date and UT		actual date and UT		time offset
Meton	316 Phamenoth 20	22:37:07	−431-6-27	02:19:33	−431-6-28	08:50:33	−30:31:00
Callippus	418 Pharmouthi 16	02:27:31	−329-6-27	06:09:57	−329-6-28	01:19:51	−19:09:54
Aristarchus	468 Pharmouthi 28	10:27:31	−279-6-26	14:09:57	−279-6-27	03:22:21	−13:12:24
Hipparchus	613 Payni 4	04:51:31	−134-6-26	08:33:58	−134-6-26	05:28:30	03:05:27
Hipparchus	620 Payni 5	22:17:56	−127-6-26	02:00:22	−127-6-25	21:54:01	04:06:21
Ptolemy	887 Mesore 11	18:56:19	140-6-24	22:38:45	140-6-23	11:58:15	34:40:30

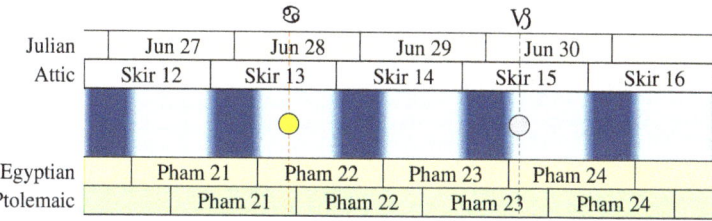

Figure 5.2 The summer solstice of 432 BC fell 1¾ days before a Full Moon. Several calendars are shown for comparison: the Proleptic Julian calendar is solar, with days starting at midnight; the Attic calendar is lunisolar, with days starting at sunset; the Egyptian calendar is solar, with days starting at dawn; and Ptolemy's version of the latter calendar has days starting at noon. Modern computation shows that the summer solstice fell on June 28, at 8:50:33 UT (or Skirophorion 13, 10 hours, 25 minutes, and 28 seconds after local midnight), and the Full Moon that followed fell on June 30, at 4:46:03 UT (or Skirophorion 15, 6 hours, 20 minutes, and 58 seconds after local midnight). Attic times are given in local mean solar time (that is, UT + 1:34:55).

occurred an hour and 35 minutes before noon in Athens (which is noon, to the nearest quarter day). (See Figure 5.2.)

Thurston (2001:156) estimates that, with the primitive technology and method he describes, equinoxes can be read to an accuracy of about three hours, while Hartner (1977:7) estimates that solstices can be read to an accuracy of about two hours. However, the ancients need not always have been this accurate. In particular, this solstice fell about an hour and a half short of noon, yet Ptolemy reports it was dated at dawn by 'the school of Meton and Euctemon' (Toomer, 1984:138). This school included at least one other astronomer called Apollinarius, as we are told (in *The Treatise of the Bright Fixed Stars*) by an unknown writer usually referred to as the Anonymous of the Year 379 (Cumont and Boll, 1904:205; Brennan, 2017:115).

Together, Meton, Euctemon, Apollinarius, and the rest of their school had an important announcement to make. Namely, that a summer solstice would soon be followed by a Full Moon. That is, the highest Sun in the year would be followed by the brightest Moon in the month. This was no ordinary coincidence: it was the brightest of them all, a rare wonder to behold, and a most convenient way to end the year and start the newly introduced cycle, which basically says that the Moon shows the same phase against the same background of stars every 19 years, or 6940 days. Meton's yearlength was, therefore,

$$Y_M = \frac{6940}{19} \text{ days}$$
$$= 365.2631578947... \text{ days}$$
$$= 365;15,47,22,6,18... \text{ days}$$
$$\approx 365 \text{ days, 6 hours, 18 minutes, 57 seconds.} \tag{5.54}$$

This must have made quite an impression at the time, firmly establishing Meton's reputation as an astronomer. The remains of the sundial he set up on the grounds of the Pnyx (probably to observe solstices) are still there (Kourouniotes and Thompson, 1932:207), testifying to the high regard he was held in, as reported by Philochorus (*Atthis* 99; Müller, 1885:400), and somehow casting doubts on the accusation of 'carelessness' that Ptolemy issued against the whole team of astronomers who were responsible for introducing a cycle that is still used today in most lunisolar calendars.

Ptolemy claimed that his own observations were carried out with the utmost care (giving his results to the nearest hour, rather than the customary quarter of a day), but the truth is that 'each of his observations is approximately one day later than the actual moment of equinox or solstice' (Jones, 2005:21). Those observed by other astronomers fare differently when compared with both the truth and the predictions of *The Almagest*. For example, the first three solstices in Table 5.3 fall roughly a day later (rather than earlier) than predicted, while Hipparchus' solstices are the most accurately predicted of all the solar observations recorded in *The Almagest*, falling earlier than predicted by only three hours (the first solstice) and four (the second). Yet, Ptolemy does not give Hipparchus' times either, suggesting that they did not agree with those predicted by *The Almagest*.

Apart from his own, the only solstice Ptolemy dates fully is Meton's, which (as we know) he reports as falling on Phamenoth 21, at dawn. Indeed, *The Almagest* predicts this solstice fell on Phamenoth 20, 16 hours, 37 minutes, and 7 seconds after noon (or 1 hour, 22 minutes, and 53 seconds before sunrise), local mean solar time, at Alexandria, which is the same as 16 hours, 12 minutes, and 10 seconds after noon (or 1 hour, 47 minutes, and 50 seconds before sunrise), local mean solar time, at Athens. So, in both cases, *The Almagest* predicts the date of this solstice as falling shortly before the end of Phamenoth 20, or shortly before the start of Phamenoth 21, supporting Ptolemy's report that Meton's solstice fell on the stated day and time (in the civil Egyptian calendar). Oddly enough, despite having successfully predicted it, Ptolemy warns his readers that this solstice was 'somewhat crudely recorded' (Toomer, 1984:138), unwittingly revealing that there is something wrong with this prediction that does not transpire from the data he gives, but from the data he omits. Namely, that the day predicted by

The Almagest corresponds to Skirophorion 12, not Skirophorion 13 (as required by the Milesian inscription and Diodorus' report). In other words, the fact that Meton's, Euctemon's, and Apollinarius' solstice is described as 'crudely recorded', when (according to the *Almagest* predictions) it should have been described as 'accurately recorded', gives Ptolemy's game away.

The important thing to note is that the Full Moon that followed this solstice fell on Skirophorion 15 (as befits a 15-day old Moon in a hollow, or 29-day long, lunar month). This day corresponds to June 30 in the Proleptic Julian calendar (as shown in Figure 5.2). It immediately follows that Skirophorion 13, the day of the solstice, corresponds to June 28 (not June 27, as *The Almagest* predicts). This is Phamenoth 22, at noon, local mean solar time in Alexandria, where days officially started at dawn, unlike Ptolemy's, which started at noon (Toomer, 1984:170).[36] This means that by the time the Milesian inscription was carved, the Greeks had already adopted the Egyptian way of counting days from dawn. The most practical way of implementing such a change is by shifting the whole Attic system of counting days back by half a day, to bring it in unison with the Egyptian system. Doing so also means shifting Meton's Skirophorion 13 back by half a day, making it correspond exactly to Phamenoth 21. This made it look as if the ancient solstice had been dated one day earlier than it really had.

The fact that Hipparchus put one extra day into Equation 5.50 (thus missing a golden opportunity to get the nearly perfect value in Equation 5.46), suggests that the shift had already been effected by his time. On this basis, Rawlins (1991:51) suggests that the date of Hipparchus' solstice can be reconstructed by adding 108478 days to June 28, 432 BC, and subtracting the extra day Hipparchus mistakenly added, giving June 26, 135 BC. He also assumes that Hipparchus' time of day is the same as Meton's, which Ptolemy reports at dawn. If so, Hipparchus timed his solstice fairly accurately, being less than two hours too early.

So Rawlins assumes that Hipparchus dated his solstice accurately but counted days incorrectly. It is also possible that he counted days correctly but dated his solstice one day too late. However, seeing that Hipparchus is hardly ever wrong by more than 10 hours in his equinox predictions (Jones, 2005:21), the latter possibility is less likely than Rawlins's. As for the word 'dawn' that Ptolemy reports, it must be noted that, in reality, Meton's solstice fell at near noon, while Hipparchus'

---

36 Ptolemy counts days from noon, rather than dawn, for reasons of astronomical convenience. In fact, using Ptolemy's starting point, Meton's solstice fell on Phamenoth 21, 10 hours and 25 minutes after midnight, local mean solar time in Alexandria, only one hour and 35 minutes before the start of Phamenoth 22. Officially, however, Phamenoth 22 had already started (as shown in Figure 5.2).

fell roughly halfway between dawn and noon. This time, *The Almagest* makes a rather good prediction (as mentioned above) and puts Hipparchus' solstice at near noon (as shown in Table 5.3). Since Ptolemy omits giving Hipparchus' day and time (and since we are assuming that Ptolemy omits offending data), we may infer that Hipparchus did not have his solstice falling at noon. Hence, he most likely had it falling at dawn (just as Ptolemy had Meton's). If so, the number of days between the solstices observed by these men is a whole number, which agrees with Equation 5.50 (though, as we saw earlier, it counts one day too many).

Now, let us turn our attention to Callippus. The only way he could have compared his solstice readings with Meton's (if indeed he did) to obtain his famous yearlength (of 365¼ days) is by assuming he dated his own solstice over half a day too late, on June 28, 330 BC, at dusk. Only then does the count of days between his solstice and Meton's add up to the right number. Namely, $365.25 \times (431 - 329) = 37255.5$ days. Thus, Callippus' yearlength is

$$\begin{aligned} Y_{MK} &= \frac{37255.5}{102} \text{ days} \\ &= 365.25 \text{ days} \\ &= 365;15 \text{ days} \\ &= 365 \text{ days, 6 hours.} \end{aligned} \quad (5.55)$$

Incredible though this may sound, it fits the data. For example, had Meton dated his solstice at noon (rather than dawn, as attested), then Callippus would have dated his at midnight, which is even worse (and more incredible) than before; and should Callippus have dated his solstice at either dawn or noon, then Ptolemy would have reported it, because both are equally close to the prediction of *The Almagest*, which puts Callippus' solstice roughly halfway between dawn and noon. The fact that he mentions neither, leaves dusk as the most likely candidate. Furthermore, had Callippus been more accurate, then he would have obtained a shorter yearlength than that in Equation 5.55. This is not historically attested, so the evidence presented so far points to 'dusk' as his most likely final word.

Let us now turn to the other solstices in Table 5.2. In his (lost) work *On the length of the year*, Hipparchus compared his solstice with that of Aristarchus, concluding that the 145 years between them were actually shorter by half a day than 145 Callippic years (of exactly 365¼ days each). In his own words, as quoted by Ptolemy (in *Almagest* 3.1),

Passage 61 It is clear that, over 145 years, the solstice occurs sooner than it would have with a [365]¼-day year by half the sum of the length of day and night (Toomer, 1984:139; see also Heiberg, 1898:206).

So Hipparchus says that the number of days between his first solstice and that observed by Aristarchus is

$$D_{AH_1} = 145 \times 365.25 - 0.5 = 52960.75. \tag{5.56}$$

Dividing this by the number of years elapsed gives

$$\begin{aligned} Y_{AH_1} &= \frac{52960.75}{145} \text{ days} \\ &= 365.2465517241... \text{ days} \\ &= 365;14,47,35,10,20... \text{ days} \\ &\approx 365 \text{ days, 5 hours, 55 minutes, 2 seconds.} \end{aligned} \tag{5.57}$$

From this information, it is possible to reconstruct the time of Hipparchus' second solstice by counting the number of days between his two solstices (which can be found by assuming a Callippic length for the seven years between them, or, alternatively, by subtracting the corresponding Julian days in Table 5.2, rounded to the nearest quarter), adding this count to the number of days in Equation 5.56, and dividing this sum by the number of years elapsed between Hipparchus' second solstice and Aristarchus'. The only way to do this and obtain a result similar to that in Equation 5.57 is by assuming that Hipparchus' second solstice was dated at midnight, on June 26, 128 BC (which, by the way, is only 11 minutes and 35 seconds too soon, making it the most accurate solar observation of all those in *The Almagest*). Thus, the number of days between Hipparchus' recorded solstices is

$$D_{HH} = 365.25 \times (134 - 127) = 2556.75, \tag{5.58}$$

and the average yearlength between Hipparchus' second solstice and Aristarchus' is

$$\begin{aligned} Y_{AH_2} &= \frac{52960.75 + 2556.75}{145 + 7} \text{ days} \\ &= \frac{55517.5}{152} \text{ days} \\ &= 365.2467105263... \text{ days} \\ &= 365;14,48,9,28,25... \text{ days} \\ &\approx 365 \text{ days, 5 hours, 55 minutes, 16 seconds.} \end{aligned} \tag{5.59}$$

The results in Equations 5.57 and 5.59 agree (to the nearest minute of time) with the yearlength used as standard throughout *The Almagest* (Toomer, 1984:140), which is perhaps why the length of the year is given only to the nearest second-order sexagesimal part of a day in that book, where Ptolemy quotes Hipparchus' conclusion on the matter as follows.

> **Passage 62** I have composed a work on the length of the year in one book, in which I show that the solar year (by which I mean the time in which the Sun goes from a solstice back to the same solstice, or from an equinox back to the same equinox) contains 365 days plus a fraction which is less than one quarter by about the three-hundredth part of the sum of one day and night, and not, as the mathematicians suppose, exactly one quarter of a day beyond the above-mentioned number of days (Toomer, 1984:139; see also Heiberg, 1898:208).

So, Hipparchus concludes that the number of days in a tropical year is, on average,

$$Y_{AH} = 365 + \frac{1}{4} - \frac{1}{300} \text{ days}$$
$$= 365.24\dot{6} \text{ days}$$
$$= 365;14,48 \text{ days}$$
$$= 365 \text{ days, 5 hours, 55 minutes, 12 seconds.} \quad (5.60)$$

Note that Equations 5.57 and 5.59 assume that Hipparchus' first and second solstices were dated at dawn and midnight, each, while *The Almagest* predicts them to have fallen at noon and dawn, respectively. This may explain Ptolemy's omission, since he could not ask his readers to believe that Hipparchus had been wrong twice on such important matters, especially after telling them that some of his equinoxes had been observed 'most accurately' and his first solstice had been observed 'again with accuracy' (Toomer, 1984:138, 139, 168).

Giving Hipparchus' discordant solstice times might have backfired, which is something Ptolemy would simply not risk. He might have chosen to specify—as he did with three of Hipparchus' autumnal equinoxes (namely, those of 159 BC, 147 BC, and 143 BC)—that they were off the expected mark by about a quarter of a day, but he chose not to. Fortunately, he gives a clue as to the fate of the missing solstice times when he quotes Hipparchus as saying (in his lost work *On the displacement of the solsticial and equinoctial points*) that

> **Passage 63** In the case of the solstices, I have to admit that both I and Archimedes may have committed errors of up to a quarter of a day in our observations and calculations [of the time] (Toomer, 1984:133; see also Heiberg, 1898:195).

Another remark by Ptolemy, saying that 'it is possible for an error of up to a quarter of a day to occur not only in observations of solstices, but even in equinox observations' (Toomer, 1984:134), gives a further clue as to what may have happened to the omitted solstice times in Table 5.2. As worded, these remarks imply that such errors were found to be more frequent in solstices than they were in equinoxes, and indeed Hipparchus' equinoxes are generally well predicted by, and fully dated in *The Almagest*, though, in fact, virtually all are off the mark by about a quarter of a day (always early in the case of vernal equinoxes, and always late in the case of autumnal equinoxes, as can be checked in Tables 5.5 and 5.6 (on pages 218 and 219), where all the solar observations in *The Almagest* are compared with both prediction and reality). This is likely due to a slight misalignment of the *equinoctial* (or *equatorial*) *armillary* used by Hipparchus (Price, 1957:588), or 'bronze ring', as Ptolemy calls it (Toomer, 1984:133), whose estimate for the latitude of Alexandria, 31 degrees (as assessed by Toomer, 1974a:133), is short of the truth by about 12 minutes of arc.

Note also that the result in Equation 5.60 is less accurate than that in Equations 5.50 and 5.51, meaning that Equation 5.60 may not have been Hipparchus' last (or only) word on the matter (or that of whoever took the trouble to make the calculations leading to the result baked on BM 55555). It also means that the extra day in Equations 5.50 and 5.51 was added to the span of time between Aristarchus' solstice and Hipparchus' first solstice. This is because the error introduced by such a mistake shows more in this shorter span of time than in the longer span of time between Meton's solstice and Hipparchus'. This would not have happened if the extra day had been added between Meton's solstice and Aristarchus'. This is perhaps easier to see if the mentioned yearlengths are compared to each other and reality. Thus, using Equation 5.48, the length of the tropical year back in 280 BC was

$$T_y = 365.2423274015\ldots \text{ days}$$
$$= 365;14,32,22,43,7\ldots \text{ days}$$
$$\approx 365 \text{ days, 5 hours, 48 minutes, 57 seconds,} \qquad (5.61)$$

and those given by Equations 5.50 and 5.60 (which are repeated here for convenience) are, respectively,

$$Y_{MH_1} \approx 365 \text{ days, 5 hours, 53 minutes, 56 seconds,} \qquad (5.50)$$

and

$$Y_{AH} = 365 \text{ days, 5 hours, 55 minutes, 12 seconds.} \qquad (5.60)$$

As we can see, Equation 5.60 is less accurate than Equation 5.50 (when compared with Equation 5.61). This would not have happened had the extra day been added between the solstices of Meton and Aristarchus. Indeed, had the latter been the case, the resulting yearlength would have been

$$Y'_{AH} = \frac{52960.75 - 1}{145} \text{ days}$$
$$= 365.2396551724... \text{ days}$$
$$= 365;14,22,45,31,2... \text{ days}$$
$$\approx 365 \text{ days, 5 hours, 45 minutes, 6 seconds,} \tag{5.62}$$

which is more accurate than Equation 5.50. But the use of this yearlength is not historically attested, which is why we can be sure that the extra day was added between Aristarchus' solstice and the first of Hipparchus' solstices.

This makes historical sense. Aristarchus lived at a time when the Macedonian rule over Egypt had barely started, so it is reasonable to assume that the Attic calendar was still in use among them, starting days at dusk (rather than dawn). So Aristarchus used the same calendar as Meton (or at least knew it very well).

Now, Aristarchus' and Meton's solstices are 152 years (or exactly $152/19 = 8$ Metonic cycles, or $152/76 = 2$ Callippic cycles) apart. According to Meton's rule, these solstices should be $8 \times 19 \times (365 + 5/19) = 55520$ days apart, and according to Callippus' rule, they should be $8 \times 19 \times (365 + 1/4) = 55518$ days apart. Aristarchus was checking which of these men was more correct. Should Meton be right, then the Sun and Moon would be in close opposition two days after the solstice, just as they were 152 years earlier. (Two days is the span of time separating a solstice reportedly dated at dawn, Skirophorion 13, and a lunar phase presumably dated at dawn, Skirophorion 15, as should be clear from inspection of Figure 5.2.) Should Callippus be right, the same thing would repeat on the same days of the lunar month (namely, Skirophorion 13 and 15), but two days earlier in the year.

None of this happened quite as expected. This time, the solstice fell about a whole day of the month earlier than Meton's, on Skirophorion 12, 5 hours, 22 minutes, and 1 second after local midnight (or -279-6-27 at 3:22:21 UT), and the Full Moon that followed was totally eclipsed (that is, perfectly aligned with the Earth and Sun) on Skirophorion 15, 1 hour, 21 minutes, and 56 seconds after local midnight (or -279-6-29 at 23:22:16 UT). The fact that the Sun, Earth, and Moon were perfectly aligned on this occasion allowed Aristarchus to know not only when and where (in the sky) his Full Moon had happened, but also Meton's. Namely, under the Goat's head. (See Figure 5.3.)

The length of the year 211

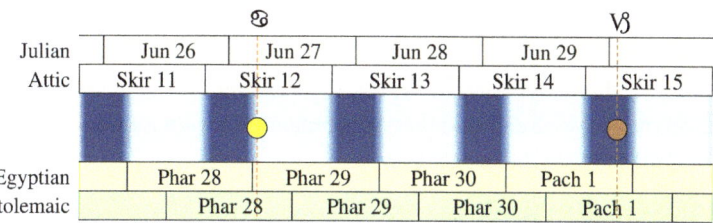

Figure 5.3 The summer solstice of 280 BC fell 2¾ days before an eclipsed Full Moon. The Proleptic Julian, Attic, Egyptian, and Ptolemaic calendars are as in Figure 5.2. Modern computation shows that the summer solstice fell on June 27, at 3:22:21 UT (or Skirophorion 12, 5 hours, 22 minutes, and 1 second after local midnight), and the Full Moon that followed fell on June 29, at 23:22:16 UT (or Skirophorion 15, 1 hour, 21 minutes, and 56 seconds after local midnight). Dates and times are shown in local mean solar time, at Alexandria (that is, UT + 1:59:40).

Now, Ptolemy omitted the time of Aristarchus' solstice presumably because it clashed with the *Almagest* prediction that it should fall roughly halfway between noon and dusk. This means that Aristarchus did not time his solstice either at noon or dusk (otherwise, Ptolemy would have thought this safe and convenient to report), leaving dawn as the most likely candidate. Putting Aristarchus' solstice time at dawn (and therefore only 20 minutes late) makes this the second best timed solar observation (against reality) of all those mentioned in *The Almagest*—the first one being Hipparchus' second solstice, which (as we saw) was only 11 minutes and 35 seconds too early (as shown in Table 5.6).

Also, having dated his peculiar solstice at dawn, on Skirophorion 12, Aristarchus would then go on to count the number of days between his solstice and Meton's. The latter was reportedly timed at dawn too, so there is a whole number of days between these solstices. Namely, 55517. Dividing this number by 152 gives the length of Aristarchus' tropical year as assessed on this momentous occasion. Namely,

$$
\begin{aligned}
Y_{MA} &= \frac{55517}{152} \text{ days} \\
&= 365.2434210526\ldots \text{ days} \\
&= 365;14,36,18,56,50\ldots \text{ days} \\
&\approx 365 \text{ days, 5 hours, 50 minutes, 32 seconds,}
\end{aligned} \qquad (5.63)
$$

which is 1 minute and 35 seconds longer than the modern estimate (in Equation 5.61).

There are alternative ways to express this yearlength. Some, for example, are based on the above assumption that Aristarchus' day count is three days less than the Metonic count,

$$Y_{MA} = 365 + \frac{5}{19} - \frac{3}{152}, \tag{5.64}$$

or one day less than the Callippic count,

$$Y_{MA} = 365 + \frac{1}{4} - \frac{1}{152}. \tag{5.65}$$

But the most interesting (and indeed the concisest) way of putting it is as a *finite continued fraction*. Thus,

$$Y_{MA} = 365 + \cfrac{1}{4 + \cfrac{1}{9 + \cfrac{1}{4}}}$$

$$= 365'4'9'4. \tag{5.66}$$

Remarkably, four centuries later, a contemporary of Ptolemy called Vettius Valens wrote a work on astrology known as *Anthology*, a fragment of which is found in a 13th or 14th century codex (as dated by Cohn et al., 1896:xix) preserved in the Vatican Library. Namely, *Codex Vaticanus Graecus* 381. Folio 163v of it contains a short list of ancient estimates for the length of the year according to various κανονογράφοι (or 'rule writers'). Among them, Aristarchus' is given as either 365'4'104 or 365'4'100'4 (depending on how we wish to read the line over the last two digits). This is identical to the result in Equation 5.66, but for one digit.

Ernst Maass (1892:140), the man who found this list, published this particular item as τξε'δ'ιδ, meaning 365'4'14, but the fact is that the second last digit is faint in the manuscript, and the shape Maass read as an ι (which is Greek for 10), can also be read as a ρ (which is Greek for 100), as he himself points out in a footnote to his text. Pingree (1986:455) reads it as a ρ. Jones (2009:21) accepts both readings as possible. Luckily, inspection of the manuscript itself leaves no room for doubt: the digit in question is a ρ.

Let us have a look at the possible mathematical interpretations that arise. Namely,

$$Y_{VA_1} = 365 + \frac{4}{104} \approx 365 \text{ days, 0 hours, 55 minutes, 23 seconds,} \quad (5.67)$$

$$Y_{VA_2} = 365 + \cfrac{1}{4 + \cfrac{1}{104}} \approx 365 \text{ days, 5 hours, 59 minutes, 8 seconds,} \quad (5.68)$$

$$Y_{VA_3} = 365 + \cfrac{1}{4 + \cfrac{1}{100 + \cfrac{1}{4}}} \approx 365 \text{ days, 5 hours, 59 minutes, 6 seconds.} \quad (5.69)$$

None of these appear to make sense. The first one is too short, as anyone can tell, and the last two are virtually identical to Callippus' yearlength, so one gains little but an increase in cumbersomeness by using them. Besides, no neat fractional part of a day can be added to the number of days between Meton's, Callippus', and Aristarchus' solstices that yields any of the results in Equations 5.67 to 5.69.

However, as is often the case in science, the simplest explanation is usually the best. Thus, in view of the similarity between the 13th (or 14th) century copy of this particular item and Equation 5.66, an emendation suggests itself that the copied ρ be read as a θ (which is Greek for 9). The reasons encouraging this emendation are both mathematical and palaeographical. The former include the above analysis, and the latter requires us to look closely at the manuscript in question and note that the medieval scribe who wrote it used two different kinds of rho. Namely, a tailless and a tailed one, denoted here, respectively, by ρ and ϱ. The former is systematically used in words to denote the letter 'r' and the latter is systematically used in numbers to denote 'a hundred'. The only exception to this rule is the very letter that concerns us here, which ends with a gentle bend in the manuscript, where a tail is expected.

Had this been a mistake, the medieval scribe who made it could simply have added the missing tail anytime afterwards. The tail is used to indicate that the letter is to be read as a number. Its absence in this particular item is eloquent. Whatever the scribe was trying to tell us (if anything) was not worth a marginal note (or so it was deemed). A myriad explanations are possible, but, again, Occam's razor provides a smooth way forward. So, assuming the scribe was too honest to invent anything thou was not seeing, thou refrained from writing a tailed (numerical) rho because that was not what thy eyes were seeing in the text from which thou was copying. So writing a tailless (textual) rho was thy way of conveying thy inability to decide what it was that thou was seeing.

Whatever it was, it was something that looked (to thee) like a textual (non-numerical) rho. Fortunately, the Greek numerals that best fit this description are not many. Namely, ϵ and θ. In a damaged, faint, or unclear original, any of these letters may have looked like the textual rho that the medieval scribe was honest enough to render without the customary tail. If so, this particular item in Vettius Valens' list is the oldest surviving trace of what might well be the most accurate estimate for the length of the tropical year ever achieved in classical antiquity. Namely, that in Equation 5.63.

A revised version of the whole list of 'rule writers' in the mentioned manuscript is given in Table 5.4 below, along with a second extant list that can be found in Vettius Valens' *Anthology* 9.12 (Kroll, 1908:353). The latter is based on a 16th century manuscript preserved in the Bodleian Library (namely, MS. *Archivum Seldenianum* B.19, folio 171r), itself a copy of a 13th century codex in the Vatican Library (namely, *Codex Vaticanus Graecus* 191). The latter, however,

Table 5.4: Vettius Valens' lists of yearlength estimates, as ascribed to several 'rule writers'. (Restorations are shown in grey.)

(a) Valens' list of yearlengths in Vat. Gr. 381 (folio 163v)

canonographer	original	translation
Euctemon, Philippus, Apollinarius	τξεʹθιʹεʹ	365ʹ19ʹ5ʹ
Aristarchus of Samos (Σαβῖνος)	τξεʹδʹρδʹ	365ʹ4ʹ*4ʹ
Babylonian	τξεʹδʹεξʹ	365ʹ4ʹ65ʹ
Sudines	τξεʹδʹγʹεʹ	365ʹ4ʹ3ʹ5ʹ
Garbled (Ζῶν ουξ)	τξεʹδʹϱʹϛʹ	365ʹ4ʹ100ʹ6ʹ

(b) Valens' list of yearlengths in Selden B.19 (folio 171r)

canonographer	original	translation
Meton, Euctemon, Philippus	τξεʹθιʹεʹ	365ʹ19ʹ5ʹ
Aristarchus of Samos	τξεʹδʹκʹξʹβʹ	365ʹ4ʹ20ʹ60ʹ2ʹ
Chaldeans	τξεʹδʹεʹζʹ	365ʹ4ʹ5ʹ7ʹ
Babylonians	τξεʹδʹϱμδʹ	365ʹ4ʹ144ʹ

no longer contains the precious list, as it was mutilated some time after the copy was made.[37]

Note that Vettius Valens confirms that both Euctemon and Apollinarius agreed upon the yearlength that resulted from Meton's solstice observation, adding that Philippus (possibly of Opus, who observed a century later from the Peloponnese) also agreed upon this yearlength, as is also confirmed by Geminus, who attributes the 19-year cycle to 'the astronomers around Euctemon, Philippus, and Callippus' (*Isagoge* 8.50; Evans and Berggren, 2006:183), and Censorinus, who attributes it to Meton of Athens (*Birthday Book* 18.8; Evans and Berggren, 2006:88).

So, apart from Ptolemy's omissions, we seem to have found some evidence—if we are ready to accept the above-proposed emendation—that the number of days between Meton's and Aristarchus' solstices is an integer. This is because we also assumed that both solstices were timed at dawn. The same was assumed for Hipparchus' first solstice, since this agrees with Equation 5.50. So we should expect the count of days between Aristarchus' solstice and Hipparchus' first solstice to be a whole number too. However, Hipparchus himself (as quoted by Ptolemy in Passage 61) says it is not a whole number. Instead, his day count is 52960.75 (as given by Equation 5.56). We may come up with a myriad explanations for this discrepancy, but perhaps the simplest one is as follows.

Unlike Meton and Aristarchus, who presumably used the same calendar, Hipparchus lived at a time when the Attic calendar had been reformed to count days from dawn (after the Egyptian fashion). Hipparchus knew that half a day had been subtracted from the Attic count. So, he calculated the number of days separating his solstice from Aristarchus' by using the Callippic rule (as Passage 61 suggests), then recalled Aristarchus' statement (in Equation 5.65) that the true count is 'one day less than the Callippic count' (which was justified because Hipparchus' first solstice was only 7 years short of 2 Callippic cycles from Aristarchus' solstice, which, in turn, was 2 Callippic cycles from Meton's), and then added the half day that had been subtracted earlier because of calendaric reforms. The result was Equation 5.56, which is reworked below to show the presumed reasoning behind Hipparchus' reckoning of the number of days between his first solstice and Aristarchus'.

$$D_{AH_1} = 145 \times 365.25 - 1 + 0.5 = 52960.75. \tag{5.70}$$

---

37 Vat. Gr. 191 was copied by sixteen hands and assembled between about 1296 and 1298 (Turyn, 1964:89; Acerbi, 2016:192). Significantly enough, this codex also contains the only known version of Ptolemy's *Geography* that is not influenced by Byzantine revisions.

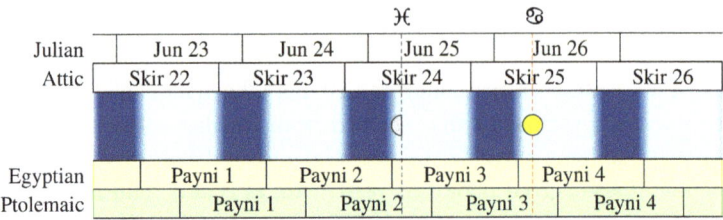

Figure 5.4 The summer solstice of 135 BC fell about a day after a Last Quarter Moon, but this is of no essence, because this time, the Moon plays no role. The calendars are as in Figure 5.2. Modern computation shows that the summer solstice fell on June 26, at 5:28:30 UT (or Payni 4 Egyptian, 7 hours, 28 minutes, and 10 seconds after local midnight). Dates and times are shown in Alexandria mean solar time (that is, UT + 1:59:40).

It is to be noted that the invention of the *scaphe*, or hollowed hemisphere, which Vitruvius (*On Architecture* 9.8.1; Morgan, 1914:273) attributes to Aristarchus along with a flat sundial, may have contributed to the increased accuracy with which solstices were timed by both Aristarchus and Hipparchus (as compared with those of Meton and Callippus). The situation during Hipparchus' first attested solstice is as shown in Figure 5.4.

This observation took place 7 years before the really important one of 128 BC. This is the solstice that gave rise to Equation 5.57 when it was compared to Aristarchus'. This is also the solstice that gave rise to Equations 5.50 and 5.51 when it was compared to Meton's, being the second most successful measurement of the tropical yearlength achieved in classical antiquity and should have been the most accurate one had the day count between both solstices been correct. As it happened, an extra day was added when someone—perhaps Hipparchus—reckoned that Phamenoth 21 Egyptian corresponds to Skirophorion 12 Athenian (rather than the traditionally reported Skirophorion 13)—a discrepancy conveniently omitted in *The Almagest*.

The solstice in 128 BC was 8 Metonic cycles (or 2 Callippic cycles) after Aristarchus', which, in turn, was 8 Metonic cycles (or 2 Callippic cycles) after Meton's. The results Hipparchus obtained on this much awaited occasion were as anticipated in 135 BC. The new situation is as shown in Figure 5.5.

This is the solstice that gave rise to Equation 5.59, which rounds down to Equation 5.60 (which eventually became the standard yearlength throughout *The Almagest*). It showed Hipparchus that Aristarchus had been nearly right but not

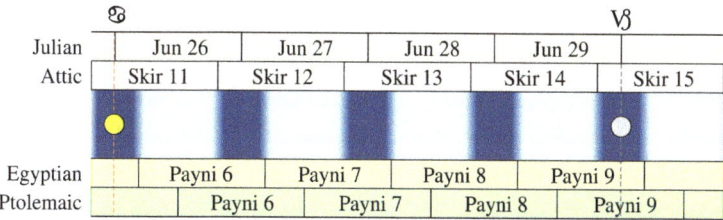

Figure 5.5 The summer solstice of 128 BC fell 4 days before a Full Moon. The calendars displayed are as in Figure 5.2. Modern computation shows that the summer solstice fell on June 25, at 21:54:01 UT (or Payni 5, 23 hours, 53 minutes, and 41 seconds after midnight), and the Full Moon that followed fell on June 29, at 21:55:48 UT (or Payni 9, 23 hours, 55 minutes, and 28 seconds after midnight). Dates and times are in Alexandria mean solar time (that is, UT + 1:59:40).

quite when he said that 8 Metonic cycles are '3 days less than Meton thought', or '1 day less than Callippus thought'. On this occasion, Hipparchus found they were '1 day and a quarter less than Callippus thought' (and so should Aristarchus have found had Meton timed his solstice correctly, at noon). However, instead of using this information to obtain a new (and more accurate) yearlength, Hipparchus added half a day to compensate for intervening calendaric reforms. This, together with his reasoning that the extra day and a quarter over 152 years roughly translates as one day over 145 years, justified his choice of day count in Equation 5.70.

To recap, the days and times of Meton's, Callippus', Aristarchus', and Hipparchus' solstices have been reconstructed based on a web of assumptions whose only tenability is the fact that they smoothly fit all the available data, including Ptolemy's telltale omissions, the *Almagest* predictions, modern predictions, and documentary evidence. Occam's principle that the simplest explanation should be favoured has been followed at all times, and always choosing the most likely one. Under this guidance, the *reported* and *reconstructed* solar observations in *The Almagest* are as given in Tables 5.5 and 5.6 below.

It may be interesting to know what yearlength Aristarchus and Hipparchus should both have found had Meton's solstice been correctly timed (in the former case) and had days been correctly counted (in the latter case). It turns out that both are

Table 5.5: Solstice and equinox predictions based on the algorithms in *The Almagest*

observer	predicted Egyptian date and equinoctial time after dawn in Alexandria		corresponding Proleptic Julian date and UT		actual date and UT		time offset
Meton	316 Phamenoth 20	22:37:07	−431-6-27	02:19:33	−431-6-28	08:50:33	−30:31:00
Callippus	418 Pharmouthi 16	02:27:31	−329-6-27	06:09:57	−329-6-28	01:19:51	−19:09:54
Aristarchus	468 Pharmouthi 28	10:27:31	−279-6-26	14:09:57	−279-6-27	03:22:21	−13:12:24
Hipparchus	613 Payni 4	04:51:31	−134-6-26	08:33:58	−134-6-26	05:28:30	03:05:27
Hipparchus	620 Payni 5	22:17:56	−127-6-26	02:00:22	−127-6-25	21:54:01	04:06:21
Ptolemy	887 Mesore 11	18:56:19	140-6-24	22:38:45	140-6-23	11:58:15	34:40:30
Hipparchus	586 Mesore 30	01:18:01	−161-9-27	05:00:27	−161-9-27	00:28:51	04:31:36
Hipparchus	589 Mesore 30	19:03:37	−158-9-26	22:46:03	−158-9-26	18:00:00	04:46:03
Hipparchus	590 Epagomenae 1	00:58:49	−157-9-27	04:41:15	−157-9-26	23:45:59	04:55:16
Hipparchus	601 Epagomenae 3	18:06:01	−146-9-26	21:48:27	−146-9-26	15:40:45	06:07:42
Hipparchus	602 Epagomenae 4	00:01:13	−145-9-27	03:43:39	−145-9-26	21:34:55	06:08:44
Hipparchus	605 Epagomenae 4	17:46:49	−142-9-26	21:29:15	−142-9-26	15:09:11	06:20:04
Ptolemy	880 Athyr 7	07:51:37	132-9-25	11:34:03	132-9-24	02:29:18	33:04:45
Ptolemy	887 Athyr 9	01:18:01	139-9-26	05:00:27	139-9-24	19:27:57	33:32:30
Hipparchus	602 Mechir 26	23:14:30	−145-3-24	02:56:56	−145-3-24	13:07:19	−10:10:23
Hipparchus	603 Mechir 27	05:09:42	−144-3-23	08:52:08	−144-3-23	18:48:44	−09:56:36
Hipparchus	604 Mechir 27	05:04:54	−143-3-23	14:47:20	−143-3-23	00:36:54	−09:49:34
Hipparchus	605 Mechir 27	17:00:06	−142-3-23	20:42:32	−142-3-24	06:37:12	−09:54:40
Hipparchus	606 Mechir 27	22:55:18	−141-3-23	02:37:44	−141-3-24	12:22:09	−09:44:25
Hipparchus	607 Mechir 28	04:50:30	−140-3-23	08:32:56	−140-3-23	18:09:48	−09:36:52
Hipparchus	613 Mechir 29	16:21:42	−134-3-23	20:04:08	−134-3-24	05:08:18	−09:04:10
Hipparchus	614 Mechir 29	22:16:54	−133-3-24	01:59:20	−133-3-23	10:50:55	−08:51:35
Hipparchus	615 Mechir 30	04:12:06	−132-3-23	07:54:32	−132-3-23	16:33:47	−08:39:15
Hipparchus	616 Mechir 30	10:07:18	−131-3-23	13:49:44	−131-3-23	22:22:42	−08:32:58
Hipparchus	617 Mechir 30	16:02:30	−130-3-23	19:44:56	−130-3-24	04:10:36	−08:25:40
Hipparchus	618 Mechir 30	21:57:42	−129-3-24	01:40:08	−129-3-24	09:58:48	−08:18:40
Hipparchus	619 Phamenoth 1	03:52:54	−128-3-23	07:35:20	−128-3-23	15:41:05	−08:05:45
Hipparchus	620 Phamenoth 1	09:48:06	−127-3-23	13:30:32	−127-3-23	21:28:11	−07:57:39
Ptolemy	887 Pachon 7	06:26:30	140-3-22	10:08:56	140-3-21	14:14:18	19:54:38

The length of the year 219

Table 5.6: Solstices and equinoxes recorded in *The Almagest*

observer	recorded Egyptian date and equinoctial time after dawn in Alexandria		corresponding Proleptic Julian date and UT	actual date and UT	time offset
Meton	316 Phamenoth 22	00:00:00	−431-6-28 03:42:26	−431-6-28 08:50:33	−05:08:07
Callippus	418 Pharmouthi 17	12:00:00	−329-6-28 15:42:26	−329-6-28 01:19:51	14:22:35
Aristarchus	468 Pharmouthi 29	00:00:00	−279-6-27 03:42:26	−279-6-27 03:22:21	00:20:05
Hipparchus	613 Payni 4	00:00:00	−134-6-26 03:42:26	−134-6-26 05:28:30	−01:46:04
Hipparchus	620 Payni 5	18:00:00	−127-6-25 21:42:26	−127-6-25 21:54:01	−00:11:35
Ptolemy	887 Mesore 11	20:00:00	140-6-24 23:42:26	140-6-23 11:58:15	35:44:11
Hipparchus	586 Mesore 30	12:00:00	−161-9-27 15:42:26	−161-9-27 00:28:51	15:13:36
Hipparchus	589 Epagomenae 1	00:00:00	−158-9-27 03:42:26	−158-9-26 18:00:00	09:42:26
Hipparchus	590 Epagomenae 1	06:00:00	−157-9-27 09:42:26	−157-9-26 23:45:59	09:56:27
Hipparchus	601 Epagomenae 3	18:00:00	−146-9-26 21:42:26	−146-9-26 15:40:45	06:01:41
Hipparchus	602 Epagomenae 4	00:00:00	−145-9-27 03:42:26	−145-9-26 21:34:55	06:07:31
Hipparchus	605 Epagomenae 4	12:00:00	−142-9-26 15:42:26	−142-9-26 15:09:11	00:33:15
Ptolemy	880 Athyr 7	08:00:00	132-9-25 11:42:26	132-9-24 02:29:18	33:13:08
Ptolemy	887 Athyr 9	01:00:00	139-9-26 04:42:26	139-9-24 19:27:57	33:14:29
Hipparchus	602 Mechir 27	00:00:00	−145-3-24 03:42:26	−145-3-24 13:07:19	−09:24:53
Hipparchus	603 Mechir 27	06:00:00	−144-3-23 09:42:26	−144-3-23 18:48:44	−09:06:18
Hipparchus	604 Mechir 27	12:00:00	−143-3-23 15:42:26	−143-3-24 00:36:54	−08:54:28
Hipparchus	605 Mechir 27	18:00:00	−142-3-23 21:42:26	−142-3-24 06:37:12	−08:54:46
Hipparchus	606 Mechir 28	00:00:00	−141-3-24 03:42:26	−141-3-24 12:22:09	−08:39:43
Hipparchus	607 Mechir 28	06:00:00	−140-3-23 09:42:26	−140-3-23 18:09:48	−08:27:22
Hipparchus	613 Mechir 29	18:00:00	−134-3-23 21:42:26	−134-3-24 05:08:18	−07:25:52
Hipparchus	614 Mechir 30	00:00:00	−133-3-24 03:42:26	−133-3-24 10:50:55	−07:08:29
Hipparchus	615 Mechir 30	06:00:00	−132-3-23 09:42:26	−132-3-23 16:33:47	−06:51:21
Hipparchus	616 Mechir 30	12:00:00	−131-3-23 15:42:26	−131-3-23 22:22:42	−06:40:16
Hipparchus	617 Mechir 30	18:00:00	−130-3-23 21:42:26	−130-3-24 04:10:36	−06:28:10
Hipparchus	618 Phamenoth 1	00:00:00	−129-3-24 03:42:26	−129-3-24 09:58:48	−06:16:22
Hipparchus	619 Phamenoth 1	06:00:00	−128-3-23 09:42:26	−128-3-23 15:41:05	−05:58:39
Hipparchus	620 Phamenoth 1	12:00:00	−127-3-23 15:42:26	−127-3-23 21:28:11	−05:45:45
Ptolemy	887 Pachon 7	07:00:00	140-3-22 10:42:26	140-3-21 14:14:18	20:28:08

$$Y_{\text{MAH}} = \frac{55516.75}{152} \text{ days}$$
$$= 365.2417763157... \text{ days}$$
$$= 365;14,30,23,41,3... \text{ days}$$
$$\approx 365 \text{ days, 5 hours, 48 minutes, 9 seconds,} \tag{5.71}$$

which is only 48 seconds shorter than the modern estimate in Equation 5.61.

It may also be surprising to learn that Callippus, the man who timed the seasons, should have been wrong by as much as half a day on this occasion (as we have assumed), but Callippus' decision to time his solstice the way he did may have been influenced by an extra factor. Namely, the Moon. Both Meton and Callippus tried to find a cycle that equated a whole number of years to a whole number of months and a whole number of days. The cycles they came up with, namely,

$$19 \text{ years} \approx 235 \text{ synodic months} \approx 6940 \text{ days} \tag{5.72}$$

and

$$76 \text{ years} \approx 940 \text{ synodic months} \approx 27759 \text{ days,} \tag{5.73}$$

respectively, were both designed to be lunisolar, not just solar. They never thought of ruling the Moon out of the equation. Neither did Aristarchus, who was among the first astronomers to benefit from Alexander's conquest of Babylon.

Soon after the conquest of this city, Callisthenes of Olynthus (Aristotle's nephew and Alexander's scientific adviser) ordered a translation to be made into Greek of the whole batch of Babylonian astronomical records (as we saw in Section 5.1). So, suddenly, the Greeks had access to a vast amount of accumulated knowledge. Curiously, eight months after the conquest of Babylon, Callippus' cycle was launched. This does not mean that the cycle bearing his name originated in Babylon. However, it is not so clear even today whether the cycle bearing Meton's name is really by him or originated in Babylon, with scholars like Webb (1921:70) on one side of the argument and Fotheringham (1919:183, 1925:78) on the other. Unfortunately, the present author has found no firm documentary evidence to support either claim. What is not in dispute is that the Babylonians knew about another useful lunisolar cycle, called the *saros*, from at least 575 BC, as can be learned, for example, from tablet BM 38462 (Brack-Bernsen and Steele, 2005:182).[38]

A **saros** is a very interesting cycle, because it basically says that eclipses repeat themselves after 223 lunations (or synodic months). So, if an eclipse is known to have occurred on a certain day, then a similar one will occur 223 months later, and so on during the complete length of a *saros series*, usually comprising a range of 69 to 89 eclipses over periods of 1226 to 1587 years, with a median of 72 eclipses over a period of 1280 years (Espenak and Meeus, 2009b:55). Furthermore, a saros neatly relates all major lunisolar events, namely,

$$\begin{aligned}
1 \text{ saros} &= 223 \text{ synodic months} \\
&\approx 242 \text{ draconic months} \\
&\approx 239 \text{ anomalistic months} \\
&\approx 32539/135 \text{ (or } 241 + 4/135) \text{ sidereal months} \\
&\approx 2434/135 \text{ (or } 18 + 4/135) \text{ Callippic years} \\
&= 6585 \text{ days, 7 hours, 44 minutes} \\
&= 6585.3\dot{2} \text{ days,}
\end{aligned}$$
(5.74)

as Ptolemy explains in the following passage (taken from *Almagest* 4.2).

> **Passage 64** The even more ancient [astronomers] used the somewhat crude estimate that such a period could be found in 6585⅓ days. For they saw that in that interval occurred approximately 223 lunations, 239 returns in anomaly, 242 returns in latitude, and 241 revolutions in longitude plus 10⅔ degrees. which is the amount the Sun travels beyond the 18 revolutions which it performs in the above time (that is, when the motion of Sun and Moon is measured with respect to the fixed stars). They called this interval the 'Periodic', since it is the smallest single period which contains (approximately) an integer number of returns of the various motions. In order to obtain a period with an integer number of days, they tripled the 6585⅓ days, obtaining 19756 days, which they called 'exeligmos'. Similarly, by tripling the other numbers, they obtained 669 lunations, 717 returns in anomaly, 726 returns in latitude, and 723 revolutions in longitude plus 32 degrees, which is the amount the Sun travels beyond its 54 revolutions (Toomer, 1984:175; see also Heiberg, 1898:270).

So, according to Ptolemy, a saros is about 6585⅓ days long. He calls this a 'crude estimate', but in fact, it is a fairly good one, and it goes to the credit of

---

38 The oldest known instance of the word *saros* can be found in *The Suda* (Sigma 148), a large 10th century Byzantine encyclopedia of the ancient Mediterranean world. Ptolemy himself referred to this cycle as 'the period'.

those 'even more ancient [ones]', as he calls them, that they found an even more accurate estimate. Namely, the one Ptolemy reports as 18 Callippic years plus 10 degrees and two-thirds, or

$$1 \text{ saros} \approx \left(18 + \frac{10.\dot{6}°}{360°}\right) \text{ Callippic years}$$
$$= \left(18 + \frac{4}{135}\right) \text{ Callippic years}$$
$$= 6585.3\dot{2} \text{ days}.$$

So Ptolemy has rescued from oblivion two different estimates for the duration of a saros, one slightly more accurate than the other, and both excellent approximations. Namely,

$$1 \text{ saros} \approx 6585.3\dot{2} \text{ days}$$
$$= 6585 \text{ days, 7 hours, 44 minutes} \qquad (5.75)$$

and

$$1 \text{ saros} \approx 6585.\dot{3} \text{ days}$$
$$= 6585 \text{ days, 8 hours}. \qquad (5.76)$$

Now, the French mathematician Paul Tannery (1888:79) showed that the latter of these approximations was used to construct the yearlength ascribed to Aristarchus by the 3rd century Roman grammarian Censorinus in *Birthday Book* 19.2 (Hultsch, 1867:40; Maude, 1900:25). Namely,

$$Y_{AC} = 365 + \frac{1}{4} + \frac{1}{1623} \text{ days} \qquad (5.77)$$
$$\approx 365 \text{ days, 6 hours, 0 minutes, 53 seconds}.$$

Indeed, dividing the less accurate of the above approximations to the number of days in a saros by the number of Callippic years in a saros, we obtain

$$Y_{AT} \approx \frac{6585.\dot{3}}{2434/135} = \frac{889020}{2434} \tag{5.78}$$

$$= 365 + \frac{1}{4} + \frac{1}{1622.\dot{6}} \tag{5.79}$$

$$= 365 + \frac{1}{4} + \frac{3}{4868} \tag{5.80}$$

$$= 365 + \frac{1}{4} + \cfrac{1}{1622 + \cfrac{1}{1 + \cfrac{1}{2}}}, \tag{5.81}$$

which rounds down to Equation 5.77.

This yearlength is only 53 seconds longer than a Callippic year, and it is as useless a yearlength as that in Equation 5.68 (which is 52 seconds less than a Callippic year) but for one little detail: the tiny fraction by which it exceeds the length of a Callippic year is just the difference (in days per year) between Approximations 5.75 and 5.76. That is,

$$\frac{6585.\dot{3} - 6585.3\dot{2}}{2434/135} = \frac{3}{4868}.$$

This means that the yearlength in Equation 5.77 (which along with Censorinus, we will assume is by Aristarchus) is an artificial one. That is, the tiny fraction by which it exceeds the length of a Callippic year is just the result of rounding Approximation 5.75 up to Approximation 5.76. In fact, had Aristarchus used the more accurate of these approximations (rather than the less accurate one), he would have come up with Callippus' yearlength all over again (rather than that reported by Censorinus). The fact that he came up with Equation 5.77 instead means that Aristarchus knew about the saros (in particular, he knew both of the above approximations), and therefore, had access to some Babylonian knowledge. It also means that he knew about the **exeligmos** (or triple saros), since it is necessary to use the rougher of the above approximations in order to obtain a cycle with a whole number of days. This cycle is obtained by tripling Approximation 5.74 (as explained in Passage 64) and is as follows.

1 exeligmos = 669 synodic months
  ≈ 726 draconic months
  ≈ 717 anomalistic months
  ≈ 32539/45 (or 723 4/45) sidereal months
  ≈ 2434/45 (or 54 4/45) Callippic years
  = 19755 days, 23 hours, 12 minutes
  = 19755.9̇6̇ days. (5.82)

The exeligmos says that 669 months after a particular eclipse, a similar one will be seen at about the same time of the day.

Aristarchus then noticed that 45 exeligmos (or 135 saros) span 2434 Callippic years exactly. Hence, he came up with the following new cycle.

45 exeligmos = 30105 synodic months
  ≈ 32670 draconic months
  ≈ 32265 anomalistic months
  ≈ 32539 sidereal months
  ≈ 2434 Callippic years
  = 889018.5 days. (5.83)

Here, everything is made up of whole numbers, but for the last value. Doubling it all would have solved this problem, but Approximation 5.83 is already too long to be a useful cycle, so doubling it was not a practical solution. However, using Approximation 5.76 (rather than its more accurate version) gave him what he wanted. Namely, a cycle where everything is expressed in whole numbers. In order to achieve this goal, he dropped the use of the Callippic yearlength and used a slightly longer one. Namely, the one in Equation 5.78 (as derived by Tannery from Censorinus' report in Equation 5.77). Note that this is not the length of a *tropical* year, but the length of what we may call a *saronic* year. Using this, he corrected Approximation 5.83 to construct the following new cycle.

45 exeligmos = 30105 synodic months
    ≈ 32670 draconic months
    ≈ 32265 anomalistic months
    ≈ 32539 sidereal months
    ≈ 2434 saronic years
    = 889020 days. (5.84)

For the first time in history, a cycle had been found that related all major lunisolar events using only whole numbers (or rather, good approximations to whole numbers). Aristarchus called this new cycle a 'Great Year', as reported by Censorinus in *Birthday Book* 18.11 (Hultsch, 1867:39; Maude, 1900:23). Tannery noticed that the number of years reported by Censorinus (namely, MMCCCCLXXXIIII) was exactly like that in Approximation 5.84, but for an extra L in the Latin rendering of this number, so he reasoned that this extra L was a scribal mistake and that the original number intended by Aristarchus was 2434.

Aristarchus had his Great Year, but he knew it was not as good a cycle as that in Approximation 5.83 (which uses the more accurate of Approximations 5.75 and 5.76). He also knew that everything is fairly accurate in Approximations 5.74 and 5.83, in particular, the number of synodic months and the number of days in either cycle (which had been counted by generations of careful Babylonian astronomers, rather than calculated using Callippus' yearlength). This allowed him (or anyone) to obtain a fairly accurate value for the length of the synodic month. Namely,

$$M_{BA} = \frac{6585.3\dot{2}}{223} \text{ days} \tag{5.85}$$

$$= \frac{889018.5}{30105} \text{ days} \tag{5.86}$$

$$= 29.53059292476... \text{ days} \tag{5.87}$$

$$= 29;31,50,8,4,18,17... \text{ days} \tag{5.88}$$

$$= 29 \text{ days, } 12 \text{ hours, } 44 \text{ minutes, } 3 \text{ seconds, } 13 \text{ jiffies, ...} \tag{5.89}$$

This value differs by only four-fifths of a second from that obtained using the modern expression for the length of the mean synodic month, as derived by Doggett (1992:576) from the lunar ephemeris ELP 2000-85 of Chapront-Touzé and Chapront (1988:342). Namely,

$$M_{syn} = 29.5305888531 + 0.00000021621\,T - 3.64 \times 10^{-10}\,T^2, \tag{5.90}$$

where $T = (JD - 2451545)/36525$ is the time in Julian centuries (of 36525 days each) from the epoch J2000.0 (or $JD = 2451545.0$), and JD is the Julian day for which the calculation is required. Using the Julian day corresponding to the summer solstice observed by Aristarchus (in Table 5.2), the modern estimate for the length of the mean synodic month back in 280 BC is

$$M_{syn} = 29 \text{ days, 12 hours, 44 minutes, 2 seconds, 26 jiffies, ...} \tag{5.91}$$

Whether the value in Equation 5.89 was found by Aristarchus (using Equation 5.86) or someone else (using Equation 5.85) is something we cannot tell from the extant evidence at our disposal. Anyone with knowledge of the saros could have found it, had thou tried. Significantly, among all the Babylonian tablets read so far, only one contains a value for the mean synodic month. Namely, BM 55555, which was shown in Section 5.1 to be post-Hipparchan and contain some mathematics of Greek origin. In particular, the length of the tropical year (in Equation 5.50). Whether the estimate for the length of the mean synodic month in it is also of Greek origin or was conceived in Babylon is something we are now going to consider.

In Babylon, the smallest time unit, the *še* (or 'barleycorn'), was equal to 3 seconds and 20 jiffies (or sixtieths of a second). This unit is exactly the same as the Hebrew ḥeleq (or 'part'), of which there are 1080 in an hour (or 25920 in a day).[39] So, when the value for the length of the mean synodic month (in Equation 5.87) was expressed in the latter units, it was rounded to the nearest barleycorn (or ḥeleq), becoming

$$M_{BJ} = 29;31,50,8,20 \text{ days} \tag{5.92}$$
$$= 29 \text{ days, 12 hours, 44 minutes, 3 seconds, 20 jiffies.} \tag{5.93}$$

This value (which is called the *molad* in Hebrew) differs by about nine-tenths of a second from the modern estimate (in Equation 5.91), so it is only a trifle less accurate than that in Equation 5.89 (assuming the modern estimate is the more accurate of the lot). As such, it was baked on line 6 of BM 55555 and hallowed in the words of the first century doctor of Jewish Law Rabban Gamaliel, which are quoted here below from Maurice Simon's translation (in Epstein, 1935:110).

> **Passage 65** I have it on the authority of the house of my father's father that the renewal of the Moon takes place after not less than twenty-nine days and a half and two-thirds of an hour and seventy-three parts (*Seder Mo'ed*, Tract *Rosh Hashanah* 25a:24).

---

39 Note that the Hebrew word for 'barley' is *śeġora*.

The last word in this sentence (ḥalaqim, in Hebrew) inspired the author to find the link between Equations 5.89 and 5.93. He also found inspiration in Rawlins's (2002:7) theory on the origin of the *molad*, though both reconstructions (Rawlins's and the author's) differ significantly on many points and outcomes, and are not the only ones attempted so far. Ptolemy, for one, ascribes Equation 5.92 to Hipparchus, saying that it was the result of dividing the number of days and months in a cycle found by the latter astronomer after studying observations made by the Chaldeans and in his own time. Ptolemy defines this cycle as follows (in *Almagest* 4.2; Heiberg, 1898:271; Toomer, 1984:175).

1 Hipparchic cycle = 4267 synodic months

$\approx$ 4573 anomalistic months

$\approx (4612 - 7.5/360)$ sidereal months

$\approx (345 - 7.5/360)$ sidereal years

≈ 126007 days and 1 hour. (5.94)

Thus, the length of the mean synodic month, as derived from Hipparchus' cycle, is

$$M_H = \frac{126007 + 1/24}{4267}$$

= 29.53059331302... days (5.95)

= 29;31,50,8,9,20,12... days (5.96)

= 29 days, 12 hours, 44 minutes, 3 seconds, 15 jiffies, ... (5.97)

Ptolemy equated this to Equation 5.92, but the 9th century Arabic translator al-Hajjaj ibn Yusuf ibn Matar (c. 786 – c. 830) noticed that this is not quite so. In fact, when rendered sexagesimally, Equation 5.95 becomes Equation 5.96 (rather than 5.92). In all likelihood, this is just a rounding discrepancy that can be explained as follows. The saronic value for the mean synodic month (as we may call Equation 5.89) and the one derived by Hipparchus (in Equation 5.97) both reduce to the *molad* (Equation 5.93) when rounded up to the nearest barleycorn. So, Hipparchus' mean synodic month may have been expressed in time units first (as in Equation 5.97), then rounded up to the nearest barleycorn (as in Equation 5.93), and finally expressed sexagesimally (as in Equation 5.92).

So Ptolemy's report may be numerically correct after all, but this does not mean that his ascription to Hipparchus is also correct. In fact, the synodic month baked on BM 55555 and hallowed by Gamaliel in the *Babylonian Talmud* may have originated with either the Babylonians or Aristarchus, since they had the mathematical means to obtain (via Equations 5.85 and 5.86) an even better estimate than Hipparchus', and this estimate also rounds up to the *molad*. Censorinus provided the clue that allows us to know that Aristarchus handled both of Approximations 5.75 and 5.76. Using the latter, he obtained his Great Year, which is the closest we may ever get to having a cycle where everything is expressed in whole numbers. But even he knew that this is too good to be true, since any such cycles exist only in the imagination of those who seek them (Depuydt, 1996:31). So, he would not have used his Great Year (Approximation 5.84) to obtain the more accurate of the above estimates for the length of the synodic month (namely, Equation 5.87), but a cycle he knew was slightly more accurate than (though not as neat as) his Great Year. Namely, the saros or a multiple of it (as in Approximations 5.74 and 5.83).

In fact, the saros is a very reliable cycle relating a whole number of synodic months to a fairly stable number of days. Namely,

$$1 \text{ saros} = 223 \text{ synodic months}$$
$$\approx 6585 \text{ days, 7 hours, 43 minutes, 39 seconds,} \qquad (5.98)$$

which is the median of all the lunar eclipse saroses in NASA's current catalogue (Espenak and Meeus, 2009*b*:53).[40] So the Chaldeans got it right in both of their approximations—the first to the nearest minute, the second to the nearest hour.

Now, independently of who it was that first ever thought of using the saros equation to work out the length of the mean synodic month, we may now ask the question of whether anyone ever thought of working out the length of the year using that very same equation. We may find a clue in Table 5.6, where, apart from Ptolemy's outlandish mismeasurements (which are always described in *The Almagest* as 'very accurate'), the only solstice that appears to have been timed beyond the quarter-day margin of error (in Passage 63) is Callippus', of them all. We also saw above that Callippus launched his Callippic cycle eight months after the conquest of Babylon. So, he might have been among the first Greek scientists to benefit from the intellectual booty brought to Athens by Aristotle's nephew. Should Callippus have learned about the saros, it might have occurred to him that dividing the number of days by the number of years in a saros will give the average length of a year.

---

40 Of all the 13132 lunar eclipse saroses in NASA's current catalogue, that starting on May 4, 2090 BC, is the shortest (with a duration of 6585 days, 6 hours, 14 minutes, 31 seconds) and that starting on Nov 8, 1524 BC, is the longest (with a duration of 6585 days, 8 hours, 54 minutes, 24 seconds), as calculated using Meeus's algorithms, which are good to about a few minutes of time.

In fact, we are working here the opposite way to what Ptolemy suggests (in Passage 64). That is, rather than multiplying Callippus' yearlength by the number of years in a saros to obtain the number of days in a saros, Callippus would have divided the number of days in a saros by the number of years in a saros to obtain the average length of a *saronic* (rather than *tropical*) year. Namely,

$$Y_K = \frac{6585.3\dot{2}}{2434/135} = 365.25 \text{ days.} \tag{5.99}$$

This is why Callippus' yearlength is virtually identical to Aristarchus' Equation 5.78 (as derived by Tannery from Censorinus' report). They are both *saronic* yearlengths, the only difference being that Callippus would have used Approximation 5.75, while Aristarchus is known to have used Approximation 5.76. It was always suspicious that a yearlength (like Callippus') that is slightly inaccurate (if taken to be tropical) should have produced the very accurate estimate for the number of days in a saros in Approximation 5.75. It is rather the other way round, and Aristarchus would have known at once that the number of *tropical* years and the number of Callippic (or *saronic*) years in a saros are not quite the same thing, since he had estimated the length of both (the former in Equation 5.63 and the latter in Equation 5.78). Callippus too knew what he was after when he timed his solstice the way he did: he wanted a lunisolar cycle, not a mere solar year.

This is what follows from just one piece of evidence. Namely, Censorinus' report of Aristarchus' (saronic) yearlength. When expressed as a mixture of mixed and continued fractions (as in Equation 5.81), the numbers there vaguely resemble those in Vettius Valens' second list of 'canonographers' (in Table 5.4). The number 1622 in this equation appears to have been split, stripped, and scrambled into 20, 60, and 2, while the first digit appears to have been deleted, thinking it was just the top of a continued fraction, and therefore, redundant. In any case, the list in the Bodleian manuscript appears to be very corrupt, as Jones (2009:21) says, and therefore, it is impossible for the present author to decode its original meaning beyond a mere suggestion. (The possibility that it is correct is left for future work.)

5.3 The irradiation illusion

Aristarchus' heliocentric theory shook the world like no other before or ever since. Apart from demanding a thorough revision of long held and firmly established traditions, there were formidable scientific arguments that threatened to crash it. One of these held that if the stars were so far away that they showed no yearly shift, then they would either not be seen at all or be ridiculously bigger than the Sun. According to what we have made out of Aristarchus' theory so far, he argued that the stars in an infinite universe (like that of Archelaus in Passage 56) must be no less than about a myriad times as far away as the Sun if they are to show no annual parallax. However, should they be fixed to the glassy dome of a hypothetical firmament, then a much, much shorter distance would do.

The fact that Archimedes uses the word 'myriad' to refer to the number of *cosmoses* (as defined in Passage 49) spanning the large universe associated with Aristarchus' theory (as described in Passage 59), along with the possibility of arriving at this value by means of measurements made with primitive technology, such as would have been available to Aristarchus (as discussed in Chapter 3), very strongly suggests that Aristarchus did indeed hold (with Archelaus) that the universe is boundless. Otherwise, as said above, a ridiculously small universe would have been enough to account for the lack of observable stellar parallaxes.

Indeed, according to the American doctor Eugene Ackerman (1962:48), 'experiments have shown that most people cannot resolve two points of light if their separation is as small as 5×10^{-4} radians' (or about $1'43''$), and 'persons with the most acute vision can resolve an angular separation of about 2×10^{-4} radians' (or about $41''$). Assuming this is true and assuming also that all stars are equally far from the Sun, studding a hypothetical glassy sphere, the radius F of such a sphere need not be greater than about 3 astronomical units (AU) for two close line-of-sight stars to be distinguishable to most people when the Earth is closest to those stars, and distinguishable only to the sharpest human eyes when the Earth is furthest from those stars, as can be found by the following equation and approximation (whose derivation is left to the reader as an exercise).

$$\frac{F}{S} = \sqrt{\left(\frac{\sin(\alpha_{max} + \alpha_{min})}{\sin(\alpha_{max} - \alpha_{min})}\right)^2 + \left(\frac{\sin(\alpha_{max} + \alpha_{min})}{\sin(\alpha_{max} - \alpha_{min})} - 1\right)^2 \tan^2 \alpha_{max}} \quad (5.100)$$

$$\approx \frac{D}{S} = \frac{\sin(\alpha_{max} + \alpha_{min})}{\sin(\alpha_{max} - \alpha_{min})}, \quad (5.101)$$

where D is the distance (in astronomical units S) between the Sun and the point halfway between the two stars, and α is half the angular size of the gap between those stars (which is greatest when the Earth is closest to them, and smallest when it is furthest from them). (Both Equation 5.100 and Approximation 5.101 are purely geometric and assume the pair of close line-of-sight stars to lie in the plane of the ecliptic, where the effect is maximized.)[41]

Substituting $\alpha_{max} = 1'43''/2$ and $\alpha_{min} = 41''/2$ (as taken from Ackerman's experiments) into Approximation 5.101, the length (in AU) of the radius of a firmament where no stellar parallax can be detected (with the naked eye) is at least

$$\frac{F}{S} \approx \frac{\sin\left(1'43''/2 + 41''/2\right)}{\sin\left(1'43''/2 - 41''/2\right)} = 2.322... \tag{5.102}$$

As a more realistic example, we may consider, as Rawlins (2008:25) does, the case of Giedi I and Giedi II, on the horn of the Goat, which lie close to the plane of the ecliptic. These stars are about $6'36''$ apart—a gap barely discernible with the naked eye. Taking this to be what most people can discern, and half of this what only the sharpest human eyes can discern, then the length (in AU) of the radius of a firmament that shows some stellar parallax between these stars needs not exceed

$$\frac{F}{S} \approx \frac{\sin\left(3'18'' + 1'39''\right)}{\sin\left(3'18'' - 1'39''\right)} = 2.999... \tag{5.103}$$

If we prefer, as Sufi (964:43) and Bohigian (2008:537) do, to use Mizar and Alcor, which are about $12'$ apart, then, taking this to be what most people can discern, and taking half of this to be what only those with the sharpest eyes can see, and supposing these stars lay close to the plane of the ecliptic, then the length (in AU) of the radius of the firmament that should allow humans to detect some stellar parallax between them (with the naked eye) is at most

$$\frac{F}{S} \approx \frac{\sin\left(6' + 3'\right)}{\sin\left(6' - 3'\right)} = 8.219... \tag{5.104}$$

This is certainly not what Aristarchus had in mind when he spoke of a universe that is about a 'myriad' times as wide as the cosmos, and a cosmos that is about a 'myriad' times as wide as the Earth (as in Passages 58 and 59). In fact, this kind of distances makes sense only if we discard the idea of a fixed starry sphere and consider instead an infinite void sprinkled with stars at random distances from

---

41 Two interactive illustrations of this point can be found online by googling the words 'Distance to the Sphere of the Fixed Stars' and 'Geometric distance to two close line-of-sight stars'.

each other. Measuring stellar parallaxes would then require a completely different approach to that employed in Equation 5.100 or Approximation 5.101. Namely, that employed in modern astronomy, according to which, taking the breadth of the Earth's orbit to be the shorter leg of a right-angled triangle, and the longer leg to be the distance to the closer of a pair of stars, and assuming the further of these stars to be infinitely far away, then, by Pythagoras' theorem and simple trigonometry, the hypotenuse of this triangle, which (geometrically speaking) can be taken to be the greatest possible distance between the Earth and the closer of these stars, is given by

$$\frac{F}{S} = \frac{2}{\sin \theta}, \tag{5.105}$$

where S is equal to one astronomical unit and θ is the angle between the two stars when seen from both ends of the Earth's orbit.

Using this simple approach, along with Ackerman's sharpest discernible angle, we have

$$\frac{F}{S} = \frac{2}{\sin(2 \times 10^{-4} \text{ radians})} \approx 10000, \tag{5.106}$$

which is almost exactly a 'myriad' astronomical units, just like those in *The Sand Reckoner*. This is the best that the best human eye can theoretically do.

So we may safely conclude that Aristarchus' universe was not bounded by any glassy sphere on which the stars were fixed. His stars were at random distances from the Sun, and this explains the vast distances spoken of in *The Sand Reckoner*, even though Archimedes bounded Aristarchus' universe in order to make it fillable. In other words, were all the stars equidistant from the Sun, then a small universe would suffice to account for the lack of stellar parallax; but should they be at random distances from the Sun, then even the closest of them must be placed hugely far away in order to account for the lack of observable parallax. (Note that in both cases, the shift is maximized for stars near the ecliptic.)

Aristarchus' brilliant answer was unassailable, but the 'enemies of the new philosophy', as Voltaire (1775:230) called them, had another ace up their sleeves. Namely, the one mentioned at the beginning of this section: that 'the stars must be unbelievably bigger than the Sun if they are to be visible at all and yet as far away as Aristarchus said they were'. In other words, should the stars be about as big as the Sun and about as far away as he proposed, they should not be seen at all. 'Hence', they thought, 'Aristarchus' theory makes no geometric sense'.

Hipparchus, for one, believed that the angular size of the stars could be measured (with the naked eye), in the same way as that of the Sun and Moon could. He said that the angular size of the Sun is thirty times greater than that of the faintest stars (as reported by Ptolemy in *Planetary Hypotheses* 1.2.8; Heiberg, 1907:70; Goldstein, 1967:8; Hamm, 2011:69), which means that, if the Sun is 30' wide (as Aristarchus estimated), then the faintest stars are 1' wide, which in turn means that, at least as far back as Hipparchus' time, the smallest angle the human eye was deemed to resolve was then, as is now, typically around 1'. Hipparchus also estimated the apparent size of Aphrodite, the brightest planet, to be a tenth that of the Sun. That is, the brightest planet was thought to be 3' wide. He also assessed the angular size of the brightest stars to be the same as Ares', namely, 1'30", and that of Zeus, Hermes, and Chronos, he estimated to be 2'30", 2', and 1'40", each.

Ptolemy also claims having repeated Hipparchus' measurements only to find the same results (although he does not say how these were obtained). Tycho, too, thought these results were trustworthy and estimated the angular sizes of first to sixth magnitude stars to be 2', 1'30", 1'5", 45", 30", and 20", each (Brahe, 1602:482, 1915:431; Helden, 1985:50). So, several of the most prominent astronomers of all times agreed that the angular size of the stars could be measured with the naked eye and had found similar results. All seemed to indicate that heliocentrism had been debunked for good.

Galileo thought otherwise. He devised an ingenious experiment that yielded different results (Hughes, 2001a:268). This experiment was an adaptation of that described in *Sand Reckoner* 1.12 (and on page 144 of the present book) with both of Archimedes' cylinders replaced by a single hanging rope or cord, and the Sun replaced by a first magnitude star, such as Vega. Galileo (1632:209, 1967:362) found that the angular size of such a star is 'no more than five seconds'. He even advanced an explanation of why his predecessors had agreed on a larger size. Namely, they were misled by what he called 'irradiation' (Galilei, 1632:197, 208, 1967:338, 360; Sheehan, 2018:19), a phenomenon that is now fully understood thanks to the work of the German physiologist Hermann von Helmholtz (1867:321-7) and the German neuroscientists Jens Kremkow et al. (2014:3173), who identified an effect termed *neuronal blurring* (or *brain magnification*), which 'causes bright planets and stars to appear enlarged', as Sheehan (2018:22) explains. This is why in practice we see stars that in theory we should not (since the brightest of them are no wider than four jiffies of arc), and this is also why we see them at night, not in daylight.

The **irradiation illusion** makes bright objects appear enlarged when seen against a dark background. This effect disappears when the object is not bright or when the background is not dark. The opposite is also true. That is why sunspots are usually engulfed by their brighter surroundings and can only be seen (with the naked eye) when the Sun's brightness is dimmed by either fog or smoke, or by being low on the horizon, as Chinese and Korean astronomers first observed in the tenth century (Hayakawa et al., 2015:3).[42] This illusion can be used to produce spectacular effects like that in Figure 5.6 below.

Galileo (1632:196, 1967:338) also noticed that Venus appeared to be larger than Jupiter when seen with the naked eye, but not when seen through a spyglass; then, it is actually Jupiter that has the greater angular size. The reason is that Venus is closer to the Sun, and hence, brighter, and, as we have seen, our brain magnifies bright objects in darkness, and the brighter, the greater the magnification.

Galileo did not say whether his experiment was inspired by that of Archimedes, who, in turn, did not say whether his own experiment was inspired by that (or those) of Aristarchus. However, in regard to his own experiment with

Figure 5.6 Light objects are enlarged by the brain, as in the *bulging draughtboard illusion*. This is why stars, which are too small to be seen in daylight, can be seen at night.

---

42 For safety reasons, the reader is strongly advised never to look into the Sun, not even with filters.

cylinders, Archimedes said that 'observations of this type have often been reported' (*Sand Reckoner* 1.11; Heiberg, 1881:250; Vardi, 1997:3). This suggests that 'observations of this type', as he calls them, originated with Aristarchus, since it was he who pioneered the high-precision measurement of the Sun's angular size, and such experiments 'had often been repeated' only to confirm him into astronomical stardom.

Whether it is along these (or other) lines that heliocentrism was said by Plutarch (Goodwin, 1878*b*:438) to have been 'proved' by the Babylonian (or Erythraean) astronomer Seleucus of Seleukia (c. 150 BC) is something we cannot tell, because time has deprived each and every single one of us of our right to know. All we can do is wonder at Plutarch's tantalizing words in the passage below.

> **Passage 66** Was it necessary to conceive that the Earth, 'rolling about the axis stretched through the universe' [as Plato says], was not represented as being held together and at rest, but as turning and revolving, as Aristarchus and Seleucus afterwards proved, the former stating it only as a hypothesis, the latter showing it by reasoning? (Waerden, 1987:528; see also Heath, 1913:305; Plato, 1888:132).

Waerden (1987:529) has hypothesized that Aristarchus did not develop his theory into a mathematical model capable of predicting the positions of the planets, but Seleucus, having been born at a time when trigonometry was more developed, just might. This is, of course, one of a myriad possible hypotheses that might explain Plutarch's words (in Passage 66).

Plutarch lived in the first century AD. Cleanthes of Assos, however, did not think the upheaval his contemporary Aristarchus had caused was so hypothetical, but of course, there is no accounting for tastes.

Another possibility is that Aristarchus, being a mathematician, would have known it is impossible to formally prove anything astronomical and would have used the more rigorous word 'hypothesis', as Heath (1913:306) points out. Under this light, Seleucus could not have found any mathematical 'proof' either, but perhaps a series of effective refutations of arguments raised against the new theory. Thus, Seleucus may have been to Aristarchus what Galileo was to Copernicus, or Huxley to Darwin: a staunch, articulate defender of a theory destined to shed light on the true workings of the world.

5.4 Conclusion

In this chapter, we have seen how Aristarchus may have measured the Moon's distance to the precision conveyed both in *The Sand Reckoner* (Equation 4.35) and by the *lunar eclipse method* (Equation 4.95), and how he may have measured the length of the year, both *tropical* (Equation 5.63) and *saronic* (Equation 5.78). We have also seen how some of the most formidable arguments ever raised against heliocentrism were eventually shown to be flawed. This did not prevent his theory from being swept under the carpet for nearly two thousand years, only to be revived by a man who did not acknowledge him, and who, as Osiander said (in Copernicus, 1543:8), introduced the revived idea as if it was a hypothesis, just as Plutarch said Aristarchus did. Clever move! Perhaps we would have less than we do had he done otherwise.

As we know, none of Aristarchus' works has survived apart from his first book *On Sizes*, but it is likely (from what we have gathered so far) that he wrote a second book *On Sizes* (explaining his novel heliocentric theory of the universe), a book *On the length of the year* (explaining the difference between tropical, saronic, and great years), a book *On Optics* (explaining his views on human vision), and perhaps on other subjects as well, as Vitruvius suggests in Passage 1 and in the passage (*On Architecture* 9.2.3) where he describes him as 'a highly gifted mathematician who left numerous demonstrations in his teachings about the Moon and its phases' (author's translation; see also Granger, 1934:228; Rowland and Howe, 2001:112).

Whether he was 'highly gifted' or not is something on which scholars are still divided. The present book hopes to have made it easier for the reader to fairly judge on this and other issues, such as whether the importance of the man who moved the Earth without using any levers has been correctly appraised through history, or whether he, the giant on whose shoulders all other giants have stood, should have been indicted, and if so, why? Is it because he shattered the foundations of geocentrism, or is it because he smashed our *egocentrism* to smithereens, revealing our true, unprivileged place in the cosmos? Whatever the answer, this magnificent scheme of things of which we are but a small part may well forever remain too grand to fathom or fill with sand, except, of course, for the likes of Aristarchus and Archimedes.

A The length of the Moon's shadow

Let s denote the radius of the Sun, and m that of the Moon. Also, let the distance between the centres A and C of the Sun and Moon be denoted by H, and let the Moon's shadow extend from the point C to the point T and be denoted by L (as in Figure A.1).

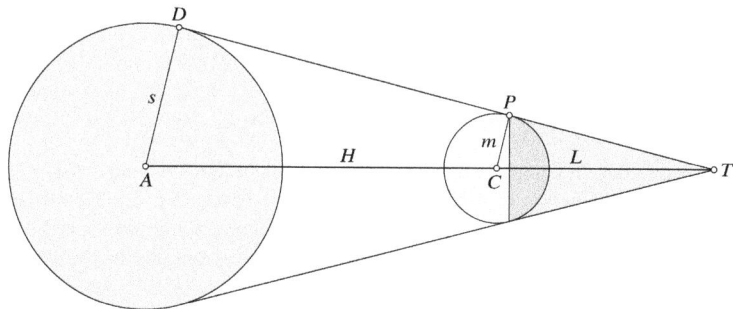

Figure A.1 The angle FAD is invisible to the human eye because it is much smaller than the limit of what the latter can see.

Then, we can see that

$$\frac{H+L}{s} = \frac{H}{s-m}.$$

Making L the subject of this equation (and simplifying), the length of the Moon's shadow is given by

$$L = \frac{Hm}{s-m}. \tag{A.1}$$

Also, let the point C be at right angles with A and the eye of an earthy observer positioned at point B (not shown in Figure A.1), and let the Sun's distance from B be denoted by S, and that of the Moon, by M.

Since ACB is a right-angled triangle, it is easy to see (by Pythagoras' theorem and the sine law) that the distance between the Sun and Moon can be expressed variously as

$$H = \sqrt{S^2 - M^2} \tag{A.2}$$

$$= \frac{s-m}{\sin \zeta_{\mathbb{C}}}, \tag{A.3}$$

where $\zeta_{\mathbb{C}}$ is half the *geometric* angular size of the Moon, which is not to be confused with $\rho_{\mathbb{C}}$ (or half the *apparent* size of the Moon). An expression for the former (as derived from

Equations A.2 and A.3) is

$$\zeta_{\mathbb{C}} = \sin^{-1}\left(\frac{s-m}{\sqrt{S^2 - M^2}}\right). \tag{A.4}$$

Substituting Equations A.2 and A.3 into Equation A.1, we obtain the following equivalent expressions for the length of the Moon's shadow.

$$L = \frac{m\sqrt{S^2 - M^2}}{s - m} \tag{A.5}$$

$$= \frac{m}{\sin \zeta_{\mathbb{C}}}. \tag{A.6}$$

We can also find a relation between M and L. That is, between the Moon's distance from the point B, which happens to be the same as the *apparent* length of the Moon's shadow (when its tip is at B), and the *geometric* length of the Moon's shadow (which is the length of this shadow whenever it is not going through the Earth's atmosphere). Thus, using the sine law, we have

$$\frac{M}{\sin 90°} = \frac{m}{\sin \rho_{\mathbb{C}}}$$

and

$$\frac{L}{\sin 90°} = \frac{m}{\sin \zeta_{\mathbb{C}}}$$

Combining these equations, we have

$$M \sin \rho_{\mathbb{C}} = L \sin \zeta_{\mathbb{C}}.$$

Rearranging and substituting for $\sin \zeta_{\mathbb{C}}$, as derived from Equations A.2 and A.3, we obtain a curious formula which uses all relevant parameters, including $\rho_{\mathbb{C}}$, which is an *apparent* one, yet gives a *geometric* outcome. Namely, the length of the geometric shadow of the Moon.

$$L = \frac{M \sin \rho_{\mathbb{C}}}{\sin \zeta_{\mathbb{C}}} = \left(\frac{\sqrt{S^2 - M^2}}{s - m}\right) M \sin \rho_{\mathbb{C}}. \tag{A.7}$$

When seen from space, during a total eclipse of the Sun (like the one filmed by the geostationary satellite JMA Himawari-8 on March 9, 2016), the shadow of the Moon on the Earth looks rather blurred because almost all of it is penumbra and only a tiny spot in the centre is umbra proper.

B Trigonometry for Proposition 13

Ratios associated with the width of the Earth's shadow

In Proposition 13 of *On Sizes*, Aristarchus found upper and lower bounds for the ratio of the line ON in Figure 2.8 (on page 66) to the widths ($2s$ and $2m$) of the Sun and Moon. These bounds are given by Inequalities 2.24 and 2.25 (on page 66). We can now use trigonometry to find the exact values of these ratios. This is done as follows.

The line BC, joining the centres of the Earth and Moon (in Figure 2.8), is equal to the Moon's distance. The Moon's radius is then given by

$$m = M \sin \rho_{\mathbb{C}}, \tag{B.1}$$

where M is the Moon's distance and $\rho_{\mathbb{C}}$ is half the angular size of the Moon.

Similarly, the Sun's radius is given by

$$s = S \sin \rho_{\odot}, \tag{B.2}$$

where S is the Sun's distance and $\rho_{\odot}$ is half the angular size of the Sun, which Proposition 8 equates to that of the Moon, so we shall denote them both by ρ from now on.

Also, the line BL, which is as long as the line OB (in Figure 2.8), has length

$$BL = M \cos \rho. \tag{B.3}$$

Now, the angular size of the Earth's shadow at the distance of the point L is 2φ, so half the line ON is given by $M \cos \rho \sin 2\varphi$, and therefore,

$$ON = 2M \cos \rho \sin 2\rho. \tag{B.4}$$

Making M the subject of Equation B.1 and using it in Equation B.4, we have

$$ON = \frac{2m \cos \rho \sin 2\rho}{\sin \rho} = \frac{2m \sin 2\rho}{\tan \rho}. \tag{B.5}$$

Dividing through by $2m$ gives the exact relation

$$\frac{ON}{2m} = \frac{\sin 2\rho}{\tan \rho}. \tag{B.6}$$

Also, dividing through Equation B.5 by $2s$ and simplifying, we have

$$\frac{ON}{2s} = \frac{m \sin 2\rho}{s \tan \rho}. \tag{B.7}$$

Now, substituting for s (as given by Equation 2.18 on page 62), we have

$$\frac{ON}{2s} = \frac{m \sin 2\rho \sin 3°}{m \tan \rho} = \frac{\sin 2\rho \sin 3°}{\tan \rho}. \tag{B.8}$$

Thus, Equations B.6 and B.8 are the exact, modern equivalents of Aristarchus' bounds (as given by Inequalities 2.24 and 2.25 on page 66).

Finding expressions for M and m

It is to be noted that, given the condition (of Hypothesis 5) that the Earth's shadow is twice as wide as the Moon at the point where they meet, it is possible to find mathematical expressions for M and m in terms of ρ. (That is, we can know both M and m if we know ρ.) These are given by the Cartesian coordinates of the point O, which is the intersection of the lines through OB and OZ (in Figure 2.8). The equations of these lines are, respectively,

$$y = x \tan 2\rho \quad \text{and} \quad y = \frac{e}{\cos \zeta} - x \tan \zeta, \tag{B.9}$$

where x and y are the Cartesian coordinates, e is the Earth's radius, ρ is as before, and ζ is half the angular size of the Earth as seen from the tip of its own shadow (that is, the angle the points B and G make with the tip Z of the Earth's shadow).

Equating the right-hand sides of the above equations gives

$$x \tan 2\rho = \frac{e}{\cos \zeta} - x \tan \zeta,$$

Making x the subject of this equation, we have

$$x = \frac{e}{(\tan 2\rho + \tan \zeta) \cos \zeta}. \tag{B.10}$$

Combining Equations B.9 and B.10 gives

$$y = \frac{e \tan 2\rho}{(\tan 2\rho + \tan \zeta) \cos \zeta}. \tag{B.11}$$

Furthermore, the length of the line OB is given by $M \cos \rho$ (by Equation B.3), so the horizontal and vertical components of the point O are given by

$$x = M \cos \rho \cos 2\rho \tag{B.12}$$

and

$$y = M \cos \rho \sin 2\rho. \tag{B.13}$$

Combining Equations B.10 and B.12 and rearranging, we have

$$\frac{M}{e} = \frac{1}{(\tan 2\rho + \tan \zeta) \cos \zeta \cos \rho \cos 2\rho}. \tag{B.14}$$

Substituting this into Equation B.1 gives

$$\frac{m}{e} = \frac{\sin \rho}{(\tan 2\rho + \tan \zeta) \cos \zeta \cos \rho \cos 2\rho}$$
$$= \frac{\tan \rho}{(\tan 2\rho + \tan \zeta) \cos \zeta \cos 2\rho}. \tag{B.15}$$

Alternatively, we can use the triangle BZG (in Figure 2.8) to easily find the length of the Earth's (geometric) shadow as

$$\frac{W}{e} = \frac{1}{\sin \zeta}, \tag{B.16}$$

where e and ζ are as before. This is also the length of the Earth's shadow in the triangle OBZ. But we know (from Equation B.12) that the x-component of the line OB is $M \cos \rho \cos 2\rho$, so the x-component of the line OZ is the difference between this and Equation B.16. We also know (from Equation B.13) that the y-component of the line OB is $M \cos \rho \sin 2\rho$, but this is also the y-component of the line OZ. Hence, by Pythagoras' theorem, the length of the line OZ is

$$\sqrt{(M \cos \rho \sin 2\rho)^2 + (e/\sin \zeta - M \cos \rho \cos 2\rho)^2}.$$

Also, by the sine law, we have

$$\frac{\sin \zeta}{M \cos \rho \sin 2\rho} = \frac{\sin 90°}{\sqrt{(M \cos \rho \sin 2\rho)^2 + (e/\sin \zeta - M \cos \rho \cos 2\rho)^2}}.$$

Squaring both sides and rearranging gives

$$\sin^2 \zeta = \frac{(M \cos \rho \sin 2\rho)^2}{(M \cos \rho \sin 2\rho)^2 + (e/\sin \zeta - M \cos \rho \cos 2\rho)^2},$$

or, equivalently,

$$\frac{1}{\sin^2 \zeta} = \frac{(M \cos \rho \sin 2\rho)^2 + (e/\sin \zeta - M \cos \rho \cos 2\rho)^2}{(M \cos \rho \sin 2\rho)^2}$$

$$= 1 + \left(\frac{e/\sin \zeta - M \cos \rho \cos 2\rho}{M \cos \rho \sin 2\rho}\right)^2.$$

Subtracting 1 from both sides gives

$$\frac{1}{\sin^2 \zeta} - 1 = \left(\frac{e/\sin \zeta - M \cos \rho \cos 2\rho}{M \cos \rho \sin 2\rho}\right)^2.$$

That is,

$$\frac{1 - \sin^2 \zeta}{\sin^2 \zeta} = \left(\frac{e/\sin \zeta - M \cos \rho \cos 2\rho}{M \cos \rho \sin 2\rho}\right)^2.$$

Using the trigonometric identity $\cos^2 \zeta + \sin^2 \zeta = 1$, we have

$$\frac{\cos^2 \zeta}{\sin^2 \zeta} = \left(\frac{e/\sin \zeta - M \cos \rho \cos 2\rho}{M \cos \rho \sin 2\rho}\right)^2.$$

Taking roots and multiplying through by $\sin \zeta$ gives

$$\cos \zeta = \frac{e - M \cos \rho \cos 2\rho \sin \zeta}{M \cos \rho \sin 2\rho}.$$

Multiplying through by $M \cos \rho \sin 2\rho$ gives

$$M \cos \rho \sin 2\rho \cos \zeta = e - M \cos \rho \cos 2\rho \sin \zeta,$$

which can be rewritten as

$$M \cos \rho \sin 2\rho \cos \zeta + M \cos \rho \cos 2\rho \sin \zeta = e,$$

Dividing through by $M \cos \rho$ gives

$$\sin 2\rho \cos \zeta + \cos 2\rho \sin \zeta = \frac{e}{M \cos \rho}.$$

Applying a suitable trigonometric addition formula, we have

$$\sin(2\rho + \zeta) = \frac{e}{M \cos \rho}.$$

Rearranging gives

$$\frac{M}{e} = \frac{1}{\sin(2\rho + \zeta) \cos \rho} = \frac{\sec \rho}{\sin(2\rho + \zeta)}, \tag{B.17}$$

which is a neater alternative to Equation B.14. Substituting this into Equation B.1 gives

$$\frac{m}{e} = \frac{\sin \rho}{\sin(2\rho + \zeta) \cos \rho} = \frac{\tan \rho}{\sin(2\rho + \zeta)}, \tag{B.18}$$

which again is a neater alternative to Equation B.15. (It is to be noted that these formulas are purely geometric. That is, they do not take into account the distorting effects of the Earth's atmosphere.)

Now, by analogy with Equation A.1 (on page 237), the length of the Earth's geometric shadow is

$$\frac{W}{e} = \frac{S}{S - e}, \tag{B.19}$$

where S is the Sun's distance from Earth, and e and s are as before.

Trigonometry for Proposition 13 243

Combining Equations B.16 and B.19, we have

$$\frac{W}{e} = \frac{1}{\sin \zeta} = \frac{S}{s-e}. \tag{B.20}$$

So

$$\sin \zeta = \frac{s-e}{S}. \tag{B.21}$$

This equation can be used in Equations B.17 and B.18 to find expressions for M and m in terms of ρ, S, s, and e, should they be needed.

Finding expressions for ζ

It is even possible to find an expression for ζ as a function of ρ alone. This can be done by rewriting Equation B.21 as follows.

$$\sin \zeta = \frac{s}{S} - \frac{e}{S}.$$

Using Equations B.2 and 2.13 gives

$$\sin \zeta = \sin \rho - \frac{e \sin 3°}{M}.$$

Making M the subject of this equation, we obtain

$$\frac{M}{e} = \frac{\sin 3°}{\sin \rho - \sin \zeta}. \tag{B.22}$$

Combining this with Equation B.17, we have

$$\frac{M}{e} = \frac{\sin 3°}{\sin \rho - \sin \zeta} = \frac{\sec \rho}{\sin (2\rho + \zeta)},$$

which can be rearranged and simplified to give

$$\sin (2\rho + \zeta) = \frac{\sin \rho - \sin \zeta}{\sin 3° \cos \rho}, \tag{B.23}$$

as promised. This is an implicit equation in ζ which is valid only when $\rho = 1°$. It can be solved as follows. Expanding the left-hand side gives

$$\sin 2\rho \cos \zeta + \cos 2\rho \sin \zeta = \frac{\sin \rho - \sin \zeta}{\sin 3° \cos \rho}.$$

Multiplying through by $\sin 3° \cos \rho$ gives

$$\sin 3° \cos \rho \sin 2\rho \cos \zeta + \sin 3° \cos \rho \cos 2\rho \sin \zeta = \sin \rho - \sin \zeta.$$

Dividing through by $\sin \zeta$ gives

$$\frac{\sin 3° \cos \rho \sin 2\rho \cos \zeta}{\sin \zeta} + \sin 3° \cos \rho \cos 2\rho = \frac{\sin \rho}{\sin \zeta} - 1,$$

which can be rearranged as

$$\frac{\sin \rho - \sin 3° \cos \rho \sin 2\rho \cos \zeta}{\sin \zeta} = \sin 3° \cos \rho \cos 2\rho + 1.$$

Squaring both sides, we have

$$\frac{(\sin \rho - \sin 3° \cos \rho \sin 2\rho \cos \zeta)^2}{\sin^2 \zeta} = (\sin 3° \cos \rho \cos 2\rho + 1)^2.$$

Using the trigonometric identity $\cos^2 \zeta + \sin^2 \zeta = 1$, we have

$$\frac{(\sin \rho - \sin 3° \cos \rho \sin 2\rho \cos \zeta)^2}{1 - \cos^2 \zeta} = (\sin 3° \cos \rho \cos 2\rho + 1)^2.$$

Multiplying through by $1 - \cos^2 \zeta$ gives

$$(\sin \rho - \sin 3° \cos \rho \sin 2\rho \cos \zeta)^2 = (\sin 3° \cos \rho \cos 2\rho + 1)^2 (1 - \cos^2 \zeta).$$

Expanding the brackets gives

$$\sin^2 \rho - 2 \sin \rho \sin 3° \cos \rho \sin 2\rho \cos \zeta + (\sin 3° \cos \rho \sin 2\rho)^2 \cos^2 \zeta$$
$$= (\sin 3° \cos \rho \cos 2\rho + 1)^2 - (\sin 3° \cos \rho \cos 2\rho + 1)^2 \cos^2 \zeta.$$

Rearranging and collecting like terms gives

$$\left(\sin^2 3° \cos^2 \rho \sin^2 2\rho + \sin^2 3° \cos^2 \rho \cos^2 2\rho + 2 \sin 3° \cos \rho \cos 2\rho + 1\right) \cos^2 \zeta$$
$$+ (-2 \sin 3° \cos \rho \sin \rho \sin 2\rho) \cos \zeta$$
$$+ \sin^2 \rho - (\sin 3° \cos \rho \cos 2\rho + 1)^2 = 0,$$

which is a quadratic in $\cos \zeta$. Using the identities $\cos^2 \rho + \sin^2 \rho = 1$ and $\sin 2\rho = 2 \sin \rho \cos \rho$, this quadratic simplifies as

$$\left(\sin^2 3° \cos^2 \rho + 2 \sin 3° \cos \rho \cos 2\rho + 1\right) \cos^2 \zeta$$
$$+ \left(-4 \sin 3° \cos^2 \rho \sin^2 \rho\right) \cos \zeta$$
$$+ \sin^2 \rho - (\sin 3° \cos \rho \cos 2\rho + 1)^2 = 0. \tag{B.24}$$

The solution to this equation can be found by the quadratic formula. Taking the positive root (which is the one making sense in this context) and applying some trigonometric manipulations, the author obtained the following solution for ζ.

$$\zeta = \cos^{-1}\left(\frac{2dp^2q^2 + \left(d\left(2q^3 - q\right) + 1\right)\sqrt{q^2\left(d^2 + 1\right) + 2dq\left(q^2 - p^2\right)}}{dq\left(dq + 4q^2 - 2\right) + 1}\right)$$
$$= 0.850862178523857...°, \tag{B.25}$$

where $d = \sin 3°$, $p = \sin \rho$, and $q = \cos \rho$. This solution is rather unwieldy and valid only when $\rho = 1°$. Yet, it is both exact and good enough for our purposes.

C The eclipse method

The **eclipse method** is a way of using the distance (from Earth) of one of the two luminaries to find that of the other. There are two versions of it. Namely, the *lunar eclipse method* and the *solar eclipse method*. Both methods complement each other and were used by Hipparchus in his two books bearing the same title as Aristarchus' *On the Sizes and Distances of the Sun and Moon*. In all, four books were written bearing this title: two by Aristarchus and two by Hipparchus, but only the first book in this series is extant. We know about the other three books from several sources, and thanks to the outstanding work of Swerdlow (1969), Toomer (1974a), and Carman (2020:180), we know a great deal about the contents of Hipparchus' two books *On Sizes*. An explanation of the eclipse method (along the lines of these scholars plus additions by the author) is now to be given.

C.1 The lunar eclipse method

The lunar eclipse method is an offshoot of Proposition 13 in Aristarchus' first book *On Sizes*. It is based on a diagram like that in Figure C.1, which is a variant of that in Figure 2.8 and that used by Carman (2020:179).

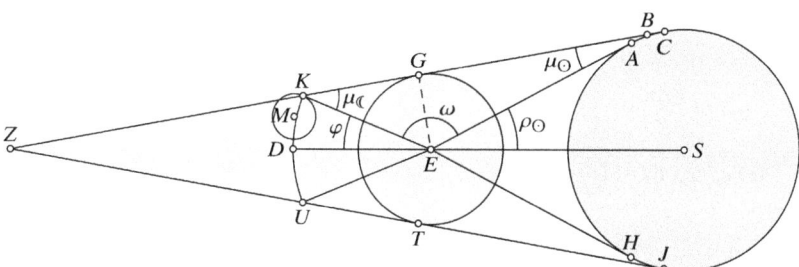

Figure C.1 Lunar eclipse diagram

We start by noting that the distance EK (between the centre E of the Earth and the last point on the Moon's disc K to enter the Earth's shadow) is the same as the distance ED (between E and the centre D of the Earth's shadow). The distance ED is almost the same as the distance EM (between E and the centre M of the Moon). Similarly, the distance EA (between E and the outer end of the Sun's disc A) is almost the same as the distance ES (between E and the centre S of the Sun).

Note also that the lines EA and GC (in Figure C.1) meet at the point B, which lies slightly above the Sun's surface, somewhere between the points A and C. The angles GBE and EKG are the horizontal parallax of the Sun and Moon, respectively, and will be denoted by $\mu_\odot$ and $\mu_\mathbb{C}$. The angles SEA and KEM are half the angular size of the Sun and Moon, respectively, and will be denoted by $\rho_\odot$ and $\rho_\mathbb{C}$. The angle KED is half the angular size of the Earth's shadow and will be denoted by φ. Finally, the supplementary angle KEA will be denoted by ω.

By inspection of the diagram, we see that

$$\mu_\odot + \mu_\mathbb{C} + \omega = 180° \tag{C.1}$$

and

$$\varphi + \rho_\odot + \omega = 180°. \tag{C.2}$$

Subtracting these two equations and rearranging, we obtain

$$\mu_\odot + \mu_\mathbb{C} = \varphi + \rho_\odot, \tag{C.3}$$

and hence,

$$\mu_\mathbb{C} = \varphi + \rho_\odot - \mu_\odot \quad \text{and} \quad \mu_\odot = \varphi + \rho_\odot - \mu_\mathbb{C}. \tag{C.4}$$

Also, by the Sine Rule, we have

$$EK = \frac{EG}{\sin \mu_\mathbb{C}} \quad \text{and} \quad EB = \frac{EG}{\sin \mu_\odot}, \tag{C.5}$$

which can be rearranged as

$$\mu_\mathbb{C} = \sin^{-1}\left(\frac{EG}{EK}\right) \quad \text{and} \quad \mu_\odot = \sin^{-1}\left(\frac{EG}{EB}\right). \tag{C.6}$$

Putting together Equations C.4 and C.5 gives

$$EK = \frac{EG}{\sin(\varphi + \rho_\odot - \mu_\odot)} \quad \text{and} \quad EB = \frac{EG}{\sin(\varphi + \rho_\odot - \mu_\mathbb{C})}. \tag{C.7}$$

Substituting for $\mu_\odot$ and $\mu_\mathbb{C}$ from Equation C.6 into Equation C.7, we have

$$EK = \frac{EG}{\sin\left(\varphi + \rho_\odot - \sin^{-1}\left(\frac{EG}{EB}\right)\right)} \quad \text{and} \quad EB = \frac{EG}{\sin\left(\varphi + \rho_\odot - \sin^{-1}\left(\frac{EG}{EK}\right)\right)}. \tag{C.8}$$

As written, these equations are true equalities.

Since the difference between EB and ES and the difference between EK and EM are both negligible, we can use the above equations as approximations to the distances of the Sun and Moon. Thus, using the letters S and M to denote the latter distances, as well as the letter e to denote the Earth's radius, we have

$$\frac{M}{e} \approx \frac{1}{\sin\left(\varphi + \rho_\odot - \sin^{-1}\left(\frac{e}{S}\right)\right)} \quad \text{and} \quad \frac{S}{e} \approx \frac{1}{\sin\left(\varphi + \rho_\odot - \sin^{-1}\left(\frac{e}{M}\right)\right)}. \tag{C.9}$$

Thus, if we know the distance S between the Earth and the Sun, we automatically know the distance M between the Earth and the Moon, and the other way round, provided we assume (like Aristarchus, Hipparchus, and Ptolemy) that the apparent size of both luminaries is the same (that is, $\rho_\odot = \rho_\mathbb{C}$).

It is a problem of mathematical interest whether we can find equations, rather than approximations, relating these distances, and it turns out that, in fact, we can. To do this, we start by noting that, using simple trigonometry, the distance between E and A in the right-angled triangle EAS is

$$EA = ES \cos \rho_\odot, \tag{C.10}$$

and the radius of the Sun is

$$AS = ES \sin \rho_\odot. \tag{C.11}$$

Alternatively, using the right-angled triangle SAB, where the point B is as defined above, and noting that the angle BSA is half of $\mu_\odot$, the radius of the Sun is

$$AS = BS \cos\left(\frac{\mu_\odot}{2}\right) \tag{C.12}$$

and the distance from A to B is

$$AB = BS \sin\left(\frac{\mu_\odot}{2}\right). \tag{C.13}$$

Making BS the subject of the latter equation gives

$$BS = \frac{AB}{\sin\left(\frac{\mu_\odot}{2}\right)}. \tag{C.14}$$

Also, applying Pythagoras' Theorem to the right-angled triangle SAB, we have

$$BS = \sqrt{AB^2 + AS^2}. \tag{C.15}$$

Hence, combining Equations C.14 and C.15, we have

$$\sqrt{AB^2 + AS^2} = \frac{AB}{\sin\left(\frac{\mu_\odot}{2}\right)}. \tag{C.16}$$

Squaring both sides gives

$$AB^2 + AS^2 = \frac{AB^2}{\sin^2\left(\frac{\mu_\odot}{2}\right)}. \tag{C.17}$$

Dividing through by AB^2 gives

$$1 + \frac{AS^2}{AB^2} = \frac{1}{\sin^2\left(\frac{\mu_\odot}{2}\right)}. \tag{C.18}$$

Subtracting 1 from both sides (and simplifying) gives

$$\frac{AS^2}{AB^2} = \frac{1}{\sin^2\left(\frac{\mu_\odot}{2}\right)} - 1 = \frac{1 - \sin^2\left(\frac{\mu_\odot}{2}\right)}{\sin^2\left(\frac{\mu_\odot}{2}\right)} = \frac{\cos^2\left(\frac{\mu_\odot}{2}\right)}{\sin^2\left(\frac{\mu_\odot}{2}\right)} = \frac{1}{\tan^2\left(\frac{\mu_\odot}{2}\right)}. \tag{C.19}$$

Taking the positive root of both sides and rearranging, we have

$$AB = AS \tan\left(\frac{\mu_\odot}{2}\right). \tag{C.20}$$

We also know that the triangle EGB is right-angled, and therefore, the equation for the radius of the Earth is

$$EG = EB \sin \mu_\odot, \tag{C.21}$$

which can be rearranged as

$$EB = \frac{EG}{\sin \mu_\odot}. \tag{C.22}$$

But the distance from E to B is

$$EB = EA + AB. \tag{C.23}$$

So, combining Equations C.22 and C.23, we have

$$\begin{aligned}
\frac{EG}{\sin \mu_\odot} &= EA + AB \\
&= ES \cos \rho_\odot + AS \tan\left(\frac{\mu_\odot}{2}\right) &&\text{(by Equations C.10 and C.20)} \\
&= ES \cos \rho_\odot + ES \sin \rho_\odot \tan\left(\frac{\mu_\odot}{2}\right) &&\text{(by Equation C.11)} \\
&= ES \left(\cos \rho_\odot + \sin \rho_\odot \tan\left(\frac{\mu_\odot}{2}\right)\right),
\end{aligned} \tag{C.24}$$

which can be rearranged as

$$\frac{EG}{ES} = \sin\mu_\odot \left(\cos\rho_\odot + \sin\rho_\odot \tan\left(\frac{\mu_\odot}{2}\right)\right) \tag{C.25}$$

$$= \sin\left(2 \times \frac{\mu_\odot}{2}\right)\left(\cos\rho_\odot + \tan\left(\frac{\mu_\odot}{2}\right)\sin\rho_\odot\right) \tag{C.26}$$

$$= 2\sin\left(\frac{\mu_\odot}{2}\right)\cos\left(\frac{\mu_\odot}{2}\right)\left(\cos\rho_\odot + \tan\left(\frac{\mu_\odot}{2}\right)\sin\rho_\odot\right) \tag{C.27}$$

$$= 2\sin\left(\frac{\mu_\odot}{2}\right)\left(\cos\left(\frac{\mu_\odot}{2}\right)\cos\rho_\odot + \sin\left(\frac{\mu_\odot}{2}\right)\sin\rho_\odot\right) \tag{C.28}$$

$$= 2\sin\left(\frac{\mu_\odot}{2}\right)\cos\left(\frac{\mu_\odot}{2}\right)\cos\rho_\odot + 2\sin^2\left(\frac{\mu_\odot}{2}\right)\sin\rho_\odot \tag{C.29}$$

$$= \sin\mu_\odot \cos\rho_\odot + 2\sin^2\left(\frac{\mu_\odot}{2}\right)\sin\rho_\odot \tag{C.30}$$

$$= \sin\mu_\odot \cos\rho_\odot + 2 \times \frac{1}{2}(1 - \cos\mu_\odot)\sin\rho_\odot \tag{C.31}$$

$$= \sin\mu_\odot \cos\rho_\odot + \sin\rho_\odot - \cos\mu_\odot \sin\rho_\odot \tag{C.32}$$

$$= \sin\rho_\odot + \sin\mu_\odot \cos\rho_\odot - \cos\mu_\odot \sin\rho_\odot \tag{C.33}$$

$$= \sin\rho_\odot + \sin(\mu_\odot - \rho_\odot), \tag{C.34}$$

where the double-angle formula $\sin 2\alpha = 2\sin\alpha\cos\alpha$ has been used in the third and sixth steps, the half-angle formula $\sin^2\alpha = \frac{1}{2}(1 - \cos 2\alpha)$ has been used in the seventh step, and the addition formula $\sin(\alpha - \beta) = \sin\alpha\cos\beta - \cos\alpha\sin\beta$ has been used in the last step.

Now, turning our attention to the triangle EKM, we see that it is right-angled, and therefore, the distance from E to K is

$$EK = EM \cos\rho_\mathrm{C} \tag{C.35}$$

and the radius of the Moon is

$$MK = EM \sin\rho_\mathrm{C}. \tag{C.36}$$

Using Equations C.4, C.6 and C.35, we have

$$\mu_\odot = \varphi + \rho_\odot - \mu_\mathrm{C}$$

$$= \varphi + \rho_\odot - \sin^{-1}\left(\frac{EG}{EK}\right)$$

$$= \varphi + \rho_\odot - \sin^{-1}\left(\frac{EG}{EM \cos\rho_\mathrm{C}}\right). \tag{C.37}$$

Substituting for $\mu_\odot$ from Equation C.37 into Equation C.34 gives

$$\frac{EG}{ES} = \sin\rho_\odot + \sin\left(\varphi + \rho_\odot - \sin^{-1}\left(\frac{EG}{EM \cos\rho_\mathrm{C}}\right) - \rho_\odot\right) \tag{C.38}$$

$$= \sin\rho_\odot + \sin\left(\varphi - \sin^{-1}\left(\frac{EG}{EM \cos\rho_\mathrm{C}}\right)\right). \tag{C.39}$$

The reciprocal of this equation is the exact solution to the lunar eclipse method and can be made to look slightly neater by using the letters S for the distance to the Sun, M for the distance to the Moon, and e for the Earth's radius (as is customary in this book). Thus,

$$\frac{S}{e} = \frac{1}{\sin\rho_\odot + \sin\left(\varphi - \sin^{-1}\left(\frac{e}{M\cos\rho_\mathbb{C}}\right)\right)}. \tag{C.40}$$

This equation can be reversed to obtain the distance of the Moon from that of the Sun. Thus, subtracting $\sin\rho_\odot$ from both sides of Equation C.39 gives

$$\frac{EG}{ES} - \sin\rho_\odot = \sin\left(\varphi - \sin^{-1}\left(\frac{EG}{EM\cos\rho_\mathbb{C}}\right)\right). \tag{C.41}$$

Taking the inverse sine of both sides gives

$$\sin^{-1}\left(\frac{EG}{ES} - \sin\rho_\odot\right) = \varphi - \sin^{-1}\left(\frac{EG}{EM\cos\rho_\mathbb{C}}\right), \tag{C.42}$$

which can be rearranged as

$$\sin^{-1}\left(\frac{EG}{EM\cos\rho_\mathbb{C}}\right) = \varphi - \sin^{-1}\left(\frac{EG}{ES} - \sin\rho_\odot\right). \tag{C.43}$$

Taking the sine of both sides, we have

$$\frac{EG}{EM\cos\rho_\mathbb{C}} = \sin\left(\varphi - \sin^{-1}\left(\frac{EG}{ES} - \sin\rho_\odot\right)\right), \tag{C.44}$$

which can be rearranged as

$$\frac{EG}{EM} = \cos\rho_\mathbb{C} \sin\left(\varphi - \sin^{-1}\left(\frac{EG}{ES} - \sin\rho_\odot\right)\right). \tag{C.45}$$

Again, taking reciprocals and replacing the letters ES, EM and EG with the letters S, M and e, the exact solution of the lunar eclipse method for the Moon's distance is

$$\frac{M}{e} = \frac{1}{\cos\rho_\mathbb{C} \sin\left(\varphi - \sin^{-1}\left(\frac{e}{S} - \sin\rho_\odot\right)\right)}. \tag{C.46}$$

In summary, two equations have been found that give the distance to the Sun and Moon by the lunar eclipse method. The ancients, however, did not have the mathematical tools needed to find these equations, but they managed to find two approximations that are only very slightly short of the truth. The equations found above and the approximations found in antiquity are as follows.

$$\frac{S}{e} = \left(\sin\rho_\odot + \sin\left(\varphi - \sin^{-1}\left(\frac{e}{M\cos\rho_\mathbb{C}}\right)\right)\right)^{-1} \tag{C.40}$$

$$\approx \left(\sin\left(\varphi + \rho_\odot - \sin^{-1}\left(\frac{e}{M}\right)\right)\right)^{-1}, \tag{C.9}$$

$$\frac{M}{e} = \left(\cos\rho_{\mathbb{C}}\sin\left(\varphi - \sin^{-1}\left(\frac{e}{S} - \sin\rho_{\odot}\right)\right)\right)^{-1} \tag{C.46}$$

$$\approx \left(\sin\left(\varphi + \rho_{\odot} - \sin^{-1}\left(\frac{e}{S}\right)\right)\right)^{-1}. \tag{C.9}$$

C.2 The solar eclipse method

In *Almagest* 5.11 (Toomer, 1984:243), Ptolemy reports that Hipparchus estimated the distance to the Moon by making assumptions about that of the Sun and putting them into a formula relating these distances. In *Almagest* 5.15 (Toomer, 1984:255), Ptolemy gives a derivation of this formula, and in *Almagest* 5.14 (Toomer, 1984:254), he says that this derivation is like that 'followed' by Hipparchus. This derivation is repeated in the works of Swerdlow (1969:294) and Toomer (1974a:130), as well as here below, where modern mathematical notation is used (for the reader's convenience). The original diagram for the solar eclipse method is simpler, but similar to the one in Figure C.2, which is the author's choice for the present work.

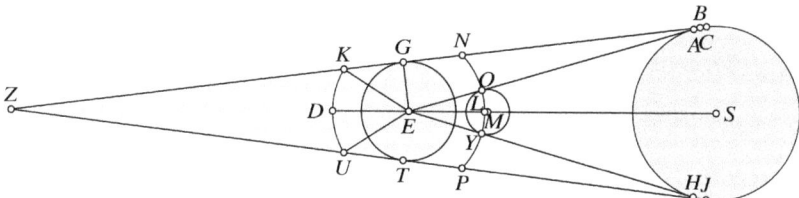

Figure C.2 Solar eclipse diagram

We start by noting that the distance EM (between the centre E of the Earth and the centre M of the Moon) is almost the same as the distance EL (between E and the centre L of the Moon's disc when it exactly covers the Sun). We note also that the distance LO (between L and O, the foremost point on the Moon's disc) is given, to a good approximation, by

$$LO \approx EO\sin\rho_{\mathbb{C}} = EL\sin\rho_{\mathbb{C}}, \tag{C.47}$$

where $\rho_{\mathbb{C}}$ is half the angular size of the Moon.

Similarly, we note that the length of the chord DK (between D, the centre of the Earth's shadow where it cuts the orbit of the Moon's disc, and K, the point where this orbit first cuts this shadow) is given, very nearly, by

$$DK \approx \frac{\varphi}{\rho_{\mathbb{C}}} \times LO, \tag{C.48}$$

where φ is half the angular size of the Earth's shadow where it meets the orbit of the Moon's disc.

Combining Approximations C.47 and C.48, we have

$$DK \approx \frac{\varphi}{\rho_{\mathbb{C}}} EL \sin \rho_{\mathbb{C}}. \tag{C.49}$$

We may also note that the ratio of the arc LN (between L and the point N, where the orbit of the Moon's disc meets the tangent to the Sun and Earth) to the arc LO (between L and O) is exactly the same as the ratio between the angles φ and $\rho_{\mathbb{C}}$. Hence, the ratio of the chord LN to the chord LO is very closely approximated by the ratio between the angles φ and $\rho_{\mathbb{C}}$. That is,

$$\frac{LN}{LO} \approx \frac{\varphi}{\rho_{\mathbb{C}}}. \tag{C.50}$$

Note also that the length of the chord LN minus the length of the Earth's radius is about the same as the length of the Earth's radius minus the length of the chord DK. That is,

$$LN - EG \approx EG - DK. \tag{C.51}$$

This can be rearranged as

$$LN \approx 2EG - DK. \tag{C.52}$$

Substituting for DK (from Approximation C.49), we have

$$LN \approx 2EG - \frac{\varphi}{\rho_{\mathbb{C}}} EL \sin \rho_{\mathbb{C}}. \tag{C.53}$$

We may also note that the length of the arc ON is exactly the same as the length of the arc LN minus the length of the arch LO, and that the length of the chord ON is approximately the same as the difference between the lengths of the chords LN and LO. That is,

$$ON \approx LN - LO. \tag{C.54}$$

Substituting for LN (from Approximation C.53) and LO (from Approximation C.47) into Approximation C.54, we have

$$ON \approx 2EG - \frac{\varphi}{\rho_{\mathbb{C}}} EL \sin \rho_{\mathbb{C}} - EL \sin \rho_{\mathbb{C}}, \tag{C.55}$$

which reduces to

$$ON \approx 2EG - \left(\frac{\varphi}{\rho_{\mathbb{C}}} + 1\right) EL \sin \rho_{\mathbb{C}}. \tag{C.56}$$

Also, we note that the distance ES (from E to the centre S of the Sun) is the sum of the distances EL and LS, and therefore,

$$LS = ES - EL. \tag{C.57}$$

Finally, the chord ON bears to the Earth's radius EG a ratio similar to the ratio between the line segments LS and ES. That is,

$$\frac{ON}{EG} \approx \frac{LS}{ES}. \tag{C.58}$$

Substituting for LS (from Approximation C.57) into this, we have

$$\frac{ON}{EG} \approx \frac{ES - EL}{ES} = 1 - \frac{EL}{ES}, \tag{C.59}$$

which can be rearranged as

$$\begin{aligned}\frac{EL}{ES} &\approx 1 - \frac{ON}{EG} \\ &\approx 1 - \frac{1}{EG}\left(2EG - \left(\frac{\varphi}{\rho_{\mathbb{C}}} + 1\right)EL \sin \rho_{\mathbb{C}}\right) \quad \text{(by Approximation C.56)} \\ &= 1 - \left(2 - \left(\frac{\varphi}{\rho_{\mathbb{C}}} + 1\right)\frac{EL}{EG} \sin \rho_{\mathbb{C}}\right) \\ &= \left(\frac{\varphi}{\rho_{\mathbb{C}}} + 1\right)\frac{EL}{EG} \sin \rho_{\mathbb{C}} - 1. \end{aligned} \tag{C.60}$$

Multiplying through by EG/EL gives

$$\frac{EG}{ES} \approx \left(\frac{\varphi}{\rho_{\mathbb{C}}} + 1\right)\sin \rho_{\mathbb{C}} - \frac{EG}{EL}. \tag{C.61}$$

Taking the reciprocal of both sides and noting that the distance EL is almost the same as the distance EM, we have

$$\frac{ES}{EG} \approx \frac{1}{\left(\dfrac{\varphi}{\rho_{\mathbb{C}}} + 1\right)\sin \rho_{\mathbb{C}} - \dfrac{EG}{EM}}. \tag{C.62}$$

Finally, using the letters S and M to denote the distance to the Sun and Moon, respectively, and the letter e to denote the Earth's radius, we obtain the expression

$$\frac{S}{e} \approx \left(\left(\frac{\varphi}{\rho_{\mathbb{C}}} + 1\right)\sin \rho_{\mathbb{C}} - \frac{e}{M}\right)^{-1}, \tag{C.63}$$

where, as before, $\rho_{\mathbb{C}}$ is half the angular size of the Moon (when it is equal to that of the Sun), and φ is half the angular size of the Earth's shadow (as cast on the Moon's disc). This is the formula referred to by Ptolemy, and it is reversible, so that it can be used to approximate the distance to the Moon when that of the Sun is known. Thus,

$$\frac{M}{e} \approx \left(\left(\frac{\varphi}{\rho_{\mathbb{C}}} + 1\right)\sin \rho_{\mathbb{C}} - \frac{e}{S}\right)^{-1}. \tag{C.64}$$

It is to be noted that using the small angle approximation allows us to write alternative approximations that look nicer than those above. Namely,

$$\frac{S}{e} \approx \left(\sin\varphi + \sin\rho_{\mathbb{C}} - \frac{e}{M}\right)^{-1} \tag{C.65}$$

and

$$\frac{M}{e} \approx \left(\sin\varphi + \sin\rho_{\mathbb{C}} - \frac{e}{S}\right)^{-1}. \tag{C.66}$$

C.3 Linear approximations

The approximations found in antiquity (namely, Approximations C.9, C.63 and C.64) are **scalar fields** (or functions of many variables), but the number of variables needed can be reduced by choosing an initial set of parameters (such as Ptolemy's or al-Battani's), then creating linear equations for those that are fixed (namely, $\rho_\odot$, $\rho_{\mathbb{C}}$, φ, and e), and finally, putting them into the above scalar fields to turn them into functions of a single variable (namely, M). Thus, using the typical equation for a line through two points adapted to, say, $\rho_\odot$ as a function of M, we have

$$\frac{\rho_\odot - \rho_{\odot\min}}{M - M_{\max}} = \frac{\rho_{\odot\max} - \rho_{\odot\min}}{M_{\min} - M_{\max}}, \tag{C.67}$$

where $\rho_\odot$ is half the angular size of the Sun, $\rho_{\odot\max}$ and $\rho_{\odot\min}$ are the upper and lower bounds assigned to this parameter (by Ptolemy, al-Battani, or otherwise), M is the Moon's distance (in Earth radii), and $M_{\max}$ and $M_{\min}$ are the upper and lower bounds assigned to this parameter (by the cited authors).

Making $\rho_\odot$ the subject of this equation, we have

$$\rho_\odot = \left(\frac{M_{\max} - M}{M_{\max} - M_{\min}}\right)(\rho_{\odot\max} - \rho_{\odot\min}) + \rho_{\odot\min}. \tag{C.68}$$

Working similarly for $\rho_{\mathbb{C}}$ and φ, we have

$$\rho_{\mathbb{C}} = \left(\frac{M_{\max} - M}{M_{\max} - M_{\min}}\right)(\rho_{\mathbb{C}\max} - \rho_{\mathbb{C}\min}) + \rho_{\mathbb{C}\min} \tag{C.69}$$

and

$$\varphi = \left(\frac{M_{\max} - M}{M_{\max} - M_{\min}}\right)(\varphi_{\max} - \varphi_{\min}) + \varphi_{\min}, \tag{C.70}$$

where the upper and lower bounds are to be read off the sets of parameters of either Ptolemy (on page 154), al-Battani (on page 157), or otherwise. Note that the closer the Moon is, the larger it looks, and the same applies to the Earth's shadow. Hence, $\rho_{\mathbb{C}\max}$ and $\varphi_{\max}$ correspond to $M_{\min}$, and $\rho_{\mathbb{C}\min}$ and $\varphi_{\min}$ correspond to $M_{\max}$.

The point of intersection of Equations C.68 and C.69 can be found by equating them. That is, the value of M for which $\rho_\odot = \rho_\mathbb{C}$ can be found by making M the subject of

$$\left(\frac{M_{\max} - M}{M_{\max} - M_{\min}}\right)(\rho_{\odot\max} - \rho_{\odot\min}) + \rho_{\odot\min}$$
$$= \left(\frac{M_{\max} - M}{M_{\max} - M_{\min}}\right)(\rho_{\mathbb{C}\max} - \rho_{\mathbb{C}\min}) + \rho_{\mathbb{C}\min}. \tag{C.71}$$

Thus, according to the solar eclipse method, the lunar distance (in Earth radii e) at which the Sun and Moon both have exactly the same angular size (as seen from Earth) is

$$\frac{M}{e} = \frac{(\rho_{\odot\min} - \rho_{\mathbb{C}\min})(M_{\max} - M_{\min})}{\rho_{\odot\max} - \rho_{\odot\min} - \rho_{\mathbb{C}\max} + \rho_{\mathbb{C}\min}} + M_{\max}, \tag{C.72}$$

and the solar distance at which this happens can be found by putting the lunar distance resulting from Equation C.72 into Approximation C.63.

C.4 The eclipse of March 14, 190 BC

According to Toomer (1974a:132), Hipparchus used a diagram like that in Figure C.3 to investigate the solar eclipse of March 14, 190 BC.

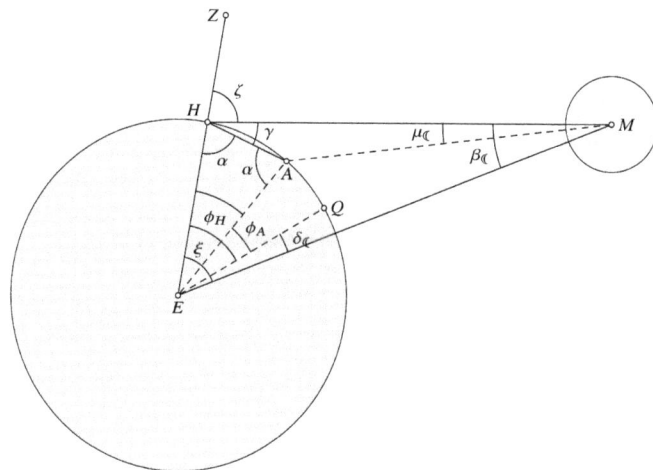

Figure C.3 Eclipse of March 14, 190 BC (not to scale)

He reasoned that the distance EM (between the centres of the Earth and the Moon) is about the same as the Earth's radius EA plus the distance AM between Alexandria and the centre of the Moon. That is,

$$EM \approx EA + AM. \tag{C.73}$$

By the Sine Rule,

$$\frac{AM}{\sin \gamma} = \frac{AH}{\sin \mu_{\mathbb{C}}}, \tag{C.74}$$

where $\mu_{\mathbb{C}}$ is the Moon's parallax, AH is the distance between Alexandria and the Hellespont, and γ is the angle AHM (between Alexandria A, the Hellespont H, and the Moon's centre M) in Figure C.3.

Also, the angle ε (in the same figure) is the difference in latitude between H and A. That is,

$$\varepsilon = \phi_H - \phi_A. \tag{C.75}$$

But the triangle AHE (between A, H, and the Earth's centre E) is isosceles, and therefore $\varepsilon + \alpha + \alpha = 180°$. Hence,

$$\alpha = \frac{180° - \varepsilon}{2} = 90° - \frac{\phi_H - \phi_A}{2}. \tag{C.76}$$

Also, by the Sine Rule,

$$\frac{AH}{\sin \varepsilon} = \frac{EA}{\sin \alpha}. \tag{C.77}$$

At this point, Toomer (1974a:132) makes the assumption that $\zeta \approx \xi$, where ζ is the angular distance between the Moon and the zenith at the Hellespont, and ξ is the same zenith distance as seen from the centre of the Earth. However, Carman (2020:189) saw no need for this simplification, and neither does the author, so from this point on, we shall follow Carman's reconstruction which is closely based on Toomer's.

From inspection of the diagram in Figure C.3, we see that

$$\delta_{\mathbb{C}} = \delta_{\odot} + \beta_{\mathbb{C}}, \tag{C.78}$$

where $\delta_{\odot}$ is the Sun's declination, $\delta_{\mathbb{C}}$ is the Moon's declination, and $\beta_{\mathbb{C}}$ is the Moon's latitude. Also,

$$\begin{aligned}\zeta &= \phi_H - \delta_{\mathbb{C}} + \beta_{\mathbb{C}} \\ &= \phi_H - \delta_{\odot},\end{aligned} \tag{C.79}$$

and

$$\xi = \phi_H - \delta_{\mathbb{C}}. \tag{C.80}$$

The eclipse of March 14, 190 BC   259

It is also easy to see that

$$\zeta + \gamma + \alpha = 180°, \tag{C.81}$$

and therefore (using Equations C.76 and C.79),

$$\begin{aligned}
\gamma &= 180° - \alpha - \zeta \\
&= 180° - \left(90° - \frac{\phi_H - \phi_A}{2}\right) - (\phi_H - \delta_\odot) \\
&= 90° - \frac{\phi_H + \phi_A}{2} + \delta_\odot.
\end{aligned} \tag{C.82}$$

Hence, using Approximation C.73 and Equations C.74, C.77 and C.82, the distance to the Moon is

$$\begin{aligned}
EM &\approx EA + AM \\
&= EA + \frac{AH}{\sin \mu_\mathbb{C}} \times \sin \gamma \\
&= EA + \frac{AH}{\sin \mu_\mathbb{C}} \times \sin\left(90° - \frac{\phi_H + \phi_A}{2} + \delta_\odot\right) \\
&= EA + \frac{EA \sin \varepsilon}{\sin \alpha \sin \mu_\mathbb{C}} \times \sin\left(90° - \frac{\phi_H + \phi_A}{2} + \delta_\odot\right).
\end{aligned} \tag{C.83}$$

Dividing through by EA and substituting for ε (from Equation C.75) and α (from Equation C.76), we have

$$\begin{aligned}
\frac{EM}{EA} &\approx 1 + \frac{\sin(\phi_H - \phi_A)\sin\left(90° - \frac{\phi_H + \phi_A}{2} + \delta_\odot\right)}{\sin \mu_\mathbb{C} \sin\left(90° - \frac{\phi_H - \phi_A}{2}\right)} \\
&= 1 + \frac{\sin(\phi_H - \phi_A)\cos\left(\frac{\phi_H + \phi_A}{2} - \delta_\odot\right)}{\sin \mu_\mathbb{C} \cos\left(\frac{\phi_H - \phi_A}{2}\right)}
\end{aligned} \tag{C.84}$$

Using the letters M and e to denote the Moon's distance and the Earth's radius, and keeping all other variables as before, we can rewrite this as

$$\frac{M}{e} \approx 1 + \frac{\sin(\phi_H - \phi_A)\cos\left(\frac{\phi_H + \phi_A}{2} - \delta_\odot\right)}{\sin \mu_\mathbb{C} \cos\left(\frac{\phi_H - \phi_A}{2}\right)}. \tag{C.85}$$

Using an approximation like this, it is likely that Hipparchus found the lower bound for the Moon's distance in his first book *OnSizes*, as Toomer first showed.

C.5 Summary

In summary, the distance to the Sun and the distance to the Moon are given either by exact scalar fields such as those found above by the author, or by approximations such as those derived in antiquity from either the *lunar eclipse method* or the *solar eclipse method*, or those derived above. In that order, these equations and approximations are as follows.

$$\frac{S}{e} = \left(\sin\rho_\odot + \sin\left(\varphi - \sin^{-1}\left(\frac{e}{M} \times \sec\rho_\mathbb{C}\right)\right)\right)^{-1} \tag{C.40}$$

$$\approx \left(\sin\left(\varphi + \rho_\odot - \sin^{-1}\left(\frac{e}{M}\right)\right)\right)^{-1} \tag{C.9}$$

$$\approx \left(\left(\frac{\varphi}{\rho_\mathbb{C}} + 1\right)\sin\rho_\mathbb{C} - \frac{e}{M}\right)^{-1} \tag{C.63}$$

$$\approx \left(\sin\varphi + \sin\rho_\mathbb{C} - \frac{e}{M}\right)^{-1}, \tag{C.65}$$

$$\frac{M}{e} = \left(\cos\rho_\mathbb{C} \sin\left(\varphi - \sin^{-1}\left(\frac{e}{S} - \sin\rho_\odot\right)\right)\right)^{-1} \tag{C.46}$$

$$\approx \left(\sin\left(\varphi + \rho_\odot - \sin^{-1}\left(\frac{e}{S}\right)\right)\right)^{-1} \tag{C.9}$$

$$\approx \left(\left(\frac{\varphi}{\rho_\mathbb{C}} + 1\right)\sin\rho_\mathbb{C} - \frac{e}{S}\right)^{-1} \tag{C.64}$$

$$\approx \left(\sin\varphi + \sin\rho_\mathbb{C} - \frac{e}{M}\right)^{-1}, \tag{C.66}$$

where all variables are as defined above.

It is to be noted that these are all purely geometric expressions that do not take into account the slight enlargement of the Earth's shadow caused by the distorting effects of the Earth's atmosphere. One way of dealing with this effect is to multiply the value of φ, the angular radius of the Earth's shadow, by a factor such as 50/51, after Chauvenet (1891:542), or 100/101, after Danjon (1951:53).

Glossary

angular resolution of the human eye Smallest angle (typically about a minute of arc) that the human eye can resolve.

anomalistic month Period of time (of about 27⅔ days) between closest approaches of the Moon to Earth.

Anti-Sun Imaginary object equal to the Sun in size and distance, but opposite to it relative to Earth.

antichthon Planet the Pythagoreans placed between the Central Fire and Earth.

antisolar point Point opposite the Sun in the celestial sphere.

aphelion Furthest distance of the Sun (from Earth).

apogee Furthest distance of the Moon (from Earth).

Callippic cycle Period of time equating 76 years to either 940 synodic months or 27759 days.

Callippic year Period of time equal to 365¼ days.

central (or **nearly central**) **eclipse** One in which the centres of the Sun and Moon get very close together at the time of greatest eclipse.

chiliagon Regular polygon of a thousand sides.

congruence Time when the Sun and Moon cut exactly the same figure in the sky.

cosmometer (or **cosmeter**, or **quad box**) Instrument for measuring the *cosmos* (as defined in *The Sand Reckoner*) by Aristarchus' half-Moon method.

cosmos 1. According to Anaximander of Miletus, beautiful order of all there is. 2. According to Anaxagoras of Clazomenae, sphere bounded by the Sun's orbit and containing all there is. 3. According to Philolaus of Croton, sphere bounded by the orbit of the outermost planet. 4. According to Archimedes of Syracuse, sphere bounded by the Sun's orbit.

dichotometer (or **dichometer**) Name preferred by John Westfall to refer to what the author calls a *cosmometer*.

dichotomy Perfect splitting of the Moon's disc into two equal halves (as seen from the observer's position).

draconic (or nodal) month Period of time (of about $27\frac{1}{5}$ days) taken by the Moon to cross the plane of the Earth's orbit at (about) the same point, called a **node**.

eclipse method Method of finding the distance to one of the two luminaries when that of the other is known, based on a geometric study of either lunar or solar eclipses.

eclipse method equations Exact mathematical solution of the eclipse method, as found in Section C.1.

elongation Angle between two celestial objects (as seen from the observer's position).

equatorial dichotomy Time when the lunar terminator crosses the centre of the Moon's disc.

exeligmos Period of time equal to three saros, or $223 \times 3 = 669$ lunar months, after which eclipses repeat themselves under very similar circumstances at about the same time of the day.

firmament *See* Sphere of the Fixed Stars.

geocentrism Theory according to which the Earth is the centre around which all celestial objects move.

gnomon Vertical rod whose shadow gives information about the position of the Sun and the passage of time.

great year Period of time after which solar years and lunar months recur.

half-Moon method Method of measuring the Sun's distance based on the observation of the Moon at *orthogony*, as devised by Aristarchus of Samos.

half-Moon's life Period of time during which the human eye perceives a halved Moon as neither clearly cupped nor gibbous.

heliocentrism Theory of the universe according to which the planets (including the Earth) move around the Sun.

hestiocentrism (or pyrocentrism) Theory according to which the centre of the universe is a central fire called **Hestia** (or 'Hearth'), around which all other objects (including the Earth and the Sun) move.

irradiation illusion Physiological phenomenon according to which bright sources of light are enlarged by the brain when seen against a dark background.

jiffy Unit of time or angle measurement equal to the sixtieth part of a second.

Metonic cycle Period of time equating 19 years to either 235 synodic months or 6940 days.

local minimum Lowest point in a given region in the graph of a function.

lunar terminator Line dividing the bright and dark sides of the Moon.

lunisolar distance field A function relating the distance to the Sun and the distance to the Moon, as derived from the lunar eclipse method.

minicosmos method Method of measuring the Sun's distance by means of a devise called a *cosmometer*.

molad Average length of a lunar (synodic) month as described in the *Babylonian Talmud* (Epstein, 1935:110), equal to 29 days, 12 hours, 44 minutes, 3 seconds, and 20 jiffies.

myriad Greek word for 10000, also meaning a very large number.

neat solilunar distance field Alternative approximation to the *eclipse method* found in Section C.1 that is neater and slightly more accurate than that described in *Almagest* 5.15 (Toomer, 1984:255).

octad A number equal to a myriad myriads (or 10^8).

orthogony Time when the plane in which the lunar terminator lies is parallel to the plane passing through the centre of the Moon and the observer's eye.

orthotomy Time when the plane in which the lunar terminator lies passes through the observer's eye.

perigee Closest distance of the Moon (to Earth).

perihelion Closest distance of the Sun (to Earth).

peripatetic A follower of the Macedonian philosopher Aristotle of Stagira.

pyrocentrism *See* hestiocentrism.

Pythagorean triangle One that is right-angled and whose sides are measured in whole numbers.

quad line (or **quadrature line**) Line drawn at the bottom of a *cosmometer*, perpendicular to both the Sun's beams and the Moon's direction.

quadrature Time when the Moon's elongation from the Sun is exactly 90 degrees.

topocentric As seen from the observer's location.

radian Angle cutting an arc equal in length to the radius of the circle containing this arc.

saronic year Number of days very closely approximated by either Callippus' or Aristarchus' yearlengths (of $365\frac{1}{4}$ and $365\frac{1}{4}\frac{3}{4868}$ days, each).

saros Cycle of 223 synodic months after which eclipses repeat themselves under very similar circumstances.

scalar field A function of many variables.

solilunar distance field A function relating the distance to the Sun and the distance to the Moon, as derived from the solar eclipse method.

solstice Time when the Sun gets closest to one of the celestial poles.

sidereal month Period of time (of about $27\frac{1}{3}$ days) taken by the Moon to cross the zodiac (as seen from Earth).

Sphere of the Fixed Stars (or **firmament**) Imaginary sphere formerly believed to be studded with stars.

synodic month Period of time (of about $29\frac{1}{2}$ days) between New Moons.

tropical year Period of time between summer solstices.

vortex Swirling flow of particles in a fluid.

References

Acerbi, Fabio (2016), 'Byzantine Recensions of Greek Mathematical and Astronomical Texts: A Survey', *Estudios bizantinos* **4**, 133–213.

Ackerman, Eugene (1962), *Biophysical science*, Prentice-Hall, Englewood Cliffs, New Jersey.

Africa, Thomas (1961), 'Copernicus' Relation to Aristarchus and Pythagoras', *Isis: A Journal of the History of Science* **52**, 403–409.

Agarwal, Ravi P. (2020), 'Pythagorean Triples before and after Pythagoras', *Computation* **8**(3), 62.

al-Battani, Abu Abd Allah Muhammad ibn Jabir (1116), *De motu stellarum*, interpretatio latina Platonis Tiburtini, Caudex Vat. lat. 3098, Civitas Vaticana.

al-Battani, Abu Abd Allah Muhammad ibn Jabir (1899), *al-Zij al-Sabi* (c. 910), published in its original Arabic and translated into Latin by Carolo Alphonso Nallino, *al-Battani sive Albatenii opus astronomicum*, Vol. 3, Ulrichum Hoeplium, Milan.

Aquinas, Thomas (1886), *In libros Aristotelis De caelo et mundo expositio*, in *Opera omnia, iussu impensaque Leonis XIII. P.M. edita*, Vol. 3, Romae Typographia Polyglotta, Roma.

Aquinas, Thomas (1964), *Exposition of Aristotle's treatise On the Heavens*, translated by Fabian R. Larcher and Pierre H. Conway, College of St. Mary of the Springs, Columbus, Ohio.

Aristotle (1922), *De Caelo*, translated by John Leofric Stocks, Clarendon Press, Oxford.

Aristotle (1929), *Physics*, with an English translation by Philip H. Wicksteed and Francis M. Cornford, Vol. 1, William Heinemann, London.

Aristotle (1933), *Metaphysics*, with an English translation by Hugh Tredennick, William Heinemann Ltd, London.

Aristotle (1952), *Meteorologica*, with an English translation by Henry Desmond Pritchard Lee, Vol. 397 of *Loeb Classical Library*, Harvard University Press.

Aristotle (1956), *The Nicomachean Ethics*, translated by Harris Rackham, Vol. 73 of *Loeb Classical Library*, Harvard University Press.

Bakker, Arthur (2003), 'The early history of average values and implications for education', *Journal of Statistics Education* **11**(1).

Barnes, Jonathan (2001), *Early Greek Philosophy*, Penguin Books, London.

Bean, Michael (1978), 'Thomas Harriot – astronomer', *Journal of the British Astronomical Association* **88**, 578–590.

Beech, Martin (2008), 'In the shadow of Aristarchus and the lunar eclipse of February 20th, 2008', *Journal of the Royal Astronomical Society of Canada* **102**(3), 98–99.

Berggren, J. Lennart and Nathan Camillo Sidoli (2007), 'Aristarchus's on the sizes and distances of the Sun and the Moon: Greek and Arabic texts', *Archive for history of exact sciences* **61**(3), 213–254.

Björnbo, Axel Anthon, Rasmus Olsen Besthorn and Heinrich Suter (1914), *Die astronomischen tafeln des Muhammed ibn Mūsā al-Khwārizmī in der bearbeitung des Maslama ibn Ahmed al-Madjrītī und der latein*, AF Høst & søn, Kopenhagen.

Bohigian, George M. (2008), 'An Ancient Eye Test – Using the Stars', *Survey of Ophthalmology* **53**(5), 536–539.

Boter, Gerard J. (2007), 'A textual problem in Archimedes, *Arenarius* 218, 14 Heiberg', *Rheinisches Museum für Philologie* **150**(H. 3/4), 424–429.

Bowen, Alan C. and Robert B. Todd (2004), *Cleomedes' Lectures on Astronomy: a Translation of The Heavens*, Vol. 42, University of California Press.

Brack-Bernsen, Lis and John M. Steele (2005), 'Eclipse prediction and the length of the Saros in Babylonian astronomy', *Centaurus* **47**(3), 181–206.

Brahe, Tycho (1602), *Astronomiae instauratae progymnasmata*, Typis Inchoata Uraniburgi Daniae, Pragae Bohemiae.

Brahe, Tycho and Iohannes Ludovicus Aemilius Dreyer (1915), *Tychonis Brahe Dani Opera omnia*, Vol. 2, Hauniae.

Brennan, Chris (2017), *Hellenistic astrology: The study of fate and fortune*, Amor Fati Publications.

Brodrick, James (1961), *Robert Bellarmine: saint and scholar*, Newman Press, Westminster, MD.

Brown, Kevin (2011), 'Platonic Solids and Plato's Theory of Everything', in *Antiquities*, Lulu, Morrisville, North Carolina.

Brown, T. M. and J. Christensen-Dalsgaard (1998), 'Accurate determination of the solar photospheric radius', *The Astrophysical Journal Letters* **500**(2), L195.

Bruins, Evert M. and Marguerite Rutten (1961), 'Textes mathématiques de Suse', *Mémoires de la Mission archéologique en Iran*.

Burkert, Walter (1972), *Lore and science in ancient Pythagoreanism*, Harvard University Press.

Burton, Harry Edwin (1945), 'The Optics of Euclid', *Journal of the Optical Society of America* **35**(5), 357–372.

Calcidius and John Magee (2016), *On Plato's Timaeus*, Vol. 41, Harvard University Press.

Capella, Martianus Minneus Felix (1866), *De nuptiis Philologiae et Mercurii*, Teubner, Lipsiae.

Carman, Christián Carlos (2009), 'Rounding numbers: Ptolemy's calculation of the Earth–Sun distance', *Archive for history of exact sciences* **63**(2), 205–242.

Carman, Christián Carlos (2014), 'Two problems in Aristarchus's treatise on the sizes and distances of the Sun and Moon', *Archive for history of exact sciences* **68**(1), 35–65.

Carman, Christián Carlos (2018), 'The first Copernican was Copernicus: the difference between Pre-Copernican and Copernican heliocentrism', *Archive for History of Exact Sciences* **72**(1), 1–20.

Carman, Christián Carlos (2020), 'On the distances of the Sun and Moon according to Hipparchus', in *Instruments – Observations – Theories: Studies in the History of Astronomy in Honor of James Evans*, The Ancient World Online.

Cassini, Jacques (1740), *Tables astronomiques du soleil, de la lune, des planètes, des étoiles fixes, et des satellites de Jupiter et de Saturne: avec l'explication et l'usage de ces mêmes tables*, De l'Imprimerie royale, Paris.

Chapront-Touzé, Michelle and Jean Chapront (1988), 'ELP 2000-85 A semi-analytical lunar ephemeris adequate for historical times', *Astronomy and Astrophysics* **190**, 342–352.
Chauvenet, William (1891), *A Manual of Spherical and Practical Astronomy*, Vol. 1, Joshua Ballinger Lippincott Company, Philadelphia, Pennsylvania.
Christianidis, Jean, Dimitris Dialetis and Kostas Gavroglu (2002), 'Having a knack for the non-intuitive: Aristarchus's heliocentrism through Archimedes's geocentrism', *History of science* **40**(2), 147–168.
Cicero, Marcus Tullius and Harris Rackham (1933), *De natura deorum, Academica*, William Heinemann, London.
Clagett, Marshall (1964), *Archimedes in the Middle Ages: Volume II. The translations from the Greek by William of Moerbeke*, University of Wisconsin Press.
Cohn, Leopold, Paul Wendland and Siegfried Reiter (1896), *Philonis Alexandrini opera quae supersunt*, Vol. 1, Typis et impensis Georgii Reimerii, Berolini.
Collinder, Per (1964), 'Dicaearchus and the Lysimachian Measurement of the Earth', *Sudhoffs Archiv für Geschichte der Medizin und der Naturwissenschaften* **48**(1), 63–78.
Commandino, Federico (1572), *Aristarchi De magnitudinibus, et distantiis solis, et lunae, liber cum Pappi Alexandrini explicationibus quibusdam. A Federico Commandino in Latinum conuersus, ac commentariis illustratus*, apud Camillum Francischinum, Pisauri.
Copernicus, Nicolaus (1543), *De revolutionibus orbium coelestium*, apud Iohannes Petreium, Norimbergae.
Cumont, Francis and Francis Boll (1904), *Catalogus Codicum Astrologorum Graecorum*, Vol. 5.1, Lamertin, Brussels.
Danjon, André (1951), 'Les éclipses de Lune par la pénombre en 1951', *L'Astronomie* **65**, 51–53.
Datta, Bibhutibhushan (1926), 'Hindu values of π', *Journal of the Asiatic Society of Bengal, New Series* **22**, 25–42.
Delambre, Jean Baptiste Joseph (1817), *Histoire de l'astronomie ancienne*, Vol. 6, Ve Courcier, Paris.
Depuydt, Leo (1996), 'The Egyptian and Athenian dates of Meton's observation of the summer solstice (−431)', *Ancient Society* **27**, 27–45.
Diehl, Ernst (1906), *Procli Diadochi in Platonis Timaeum commentaria*, Vol. 3, in aedibus Benedicti Gotthelf Teubneri, Lipsiae.
Diels, Hermann (1879), *Doxographi Graeci*, in aedibus Georgii Reimeri, Leipzig.
Diels, Hermann (1882), *Simplicii in Aristotelis Physicorum libros quattuor priores commentaria*, Typis et Impensis Georgii Reimeri.
Diels, Hermann (1906), *Die Fragmente der Vorsokratiker, griechisch und deutsch*, Weidmann, Berlin.
Diels, Hermann and Albert Rehm (1904), *Parapegmenfragmente aus Milet*, Akademie der Wissenschaften.
Diodorus Siculus (1957), *Library of History*, Vol. 11, 279 of *Loeb Classical Library*, Harvard University Press.

Doggett, Leroy Elsworth (1992), Calendars, in *Explanatory Supplement to the Astronomical Almanac*, ed. P. Kenneth Seidelmann, University Science Books, California.

Dupuis, Jean (1892), *Theōnos Smyrnaiou Platōnikou tōn kata to mathēmatikon chrēsimōn eis tēn Platōnos anagnōsin: Théon de Smyrne, philosophe platonicien. Exposition des connaissances mathématiques utiles pour la lecture de Platon*, Hachette, Paris.

Eastwood, Bruce Stansfield (1992), 'Heraclides and heliocentrism: Texts, diagrams, and interpretations', *Journal for the History of Astronomy* **23**(4), 233–260.

Ehrle, Franz (1890), *Historia bibliothecae romanorum pontificum: tum Bonifatianae tum Avenionensis*, Vol. 1, typis Vaticanis.

Epstein, Isidore (1935), *The Babylonian Talmud, Seder Mo'ed*, Vol. 13, Soncino Press, London.

Erhardt, Rudolf von and Erika von Erhardt-Siebold (1942), 'Archimedes' Sand-reckoner: Aristarchus and Copernicus', *Isis* **33**(5), 578–602.

Espenak, Fred and Jean Meeus (2006a), *Five Year Millennium Canon of Solar Eclipses: –1999 to +3000 (2000 BCE to 3000 CE)*, National Aeronautics and Space Flight Administration, Goddard Space Flight Center.

Espenak, Fred and Jean Meeus (2006b), *Five Year Millennium Canon of Lunar Eclipses: –1999 to +3000 (2000 BCE to 3000 CE)*, National Aeronautics and Space Flight Administration, Goddard Space Flight Center.

Espenak, Fred and Jean Meeus (2009a), *Five Millennium Catalog of Solar Eclipses: –1999 to +3000 (2000 BCE to 3000 CE) revised*, National Aeronautics and Space Flight Administration, Goddard Space Flight Center.

Espenak, Fred and Jean Meeus (2009b), *Five Millennium Catalog of Lunar Eclipses: –1999 to +3000 (2000 BCE to 3000 CE)*, National Aeronautics and Space Administration, Goddard Space Flight Center.

Evans, James (1998), *The history and practice of ancient astronomy*, Oxford University Press.

Evans, James and J. Lennart Berggren (2006), *Geminos's Introduction to the Phenomena*, Princeton University Press.

Evelyn-White, Hugh Gerard et al. (1920), *Hesiod, the Homeric hymns, and Homerica*, Vol. 57, W. Heinemann.

Fairbanks, Arthur (1898), *The First Philosophers of Greece*, Scribner, London.

Finocchiaro, Maurice A. (1989), *The Galileo Affair: A Documentary History*, University of California Press, Berkeley.

Fortenbaugh, William W. (2017), *Strato of Lampsacus: Text, Translation and Discussion*, Routledge, Abingdon-on-Thames.

Fortenbaugh, William W. and Eckart Schütrumpf (2001), *Dicaearchus of Messana Text, Translation, and Discussion*, Vol. 10 of *Rutgers University Studies in Classical Humanities*, Transaction Publishers, New Brunswick.

Fortenbaugh, William W. and Elizabeth E. Pender (2009), *Heraclides of Pontus: Discussion*, Transaction Publishers, New Brunswick, NJ.

Fotheringham, John Knight (1919), 'Cleostratus', *The Journal of Hellenic Studies* **39**, 164–184.

Fotheringham, John Knight (1924), 'The Metonic and Callippic Cycles', *Monthly Notices of the Royal Astronomical Society* **84**(5), 383–392.

Fotheringham, John Knight (1925), 'Cleostratus (III)', *The Journal of Hellenic Studies* **45**(1), 78–83.

Friedlein, Gottfried (1873), *In primum Euclidis elementorum librum commentarii*, Vol. 263, in aedibus Benedicti Gotthelf Teubneri, Lipsiae.

Gaisford, Thomas et al. (1839), *Theodoreti episcopi Cyrensis Graecarum affectionum curatio*, e typographeo academico, Oxonii.

Galilei, Galileo (1610), *Sidereus Nuncius*, Apud Thomam Baglionum, Venetiis.

Galilei, Galileo (1632), *Dialogo dei Massimi Sistemi Del Mondo, Tolemaico e Copernicano*, Per Gio Batista Landini, Fiorenza.

Galilei, Galileo (1967), *Dialogue concerning the two chief world systems*, University of California press.

Galilei, Galileo (2016), *Sidereus Nuncius, or the sidereal messenger*, University of Chicago Press.

Gassendi, Pierre (1658), *Petri Gassendi Diniensis ecclesiae praepositi et in academia Parisiensi matheseos regii professoris Opera omnia in sex tomos divisa: quorum seriem pagina praefationes proximè sequens continet*, Vol. 6, Sumptibus Laurentii Anisson, & Ioannis Baptistae Devenet.

Geus, Klaus (2014), 'A Day's Journey in Herodotus' *Histories*', in Klaus Geus and Martin Thiering (Eds.), *Features of Common Sense Geography: Implicit Knowledge Structures in Ancient Geographical Texts*, LIT Verlag Münster, Wien.

Gingerich, Owen (1985), 'Did Copernicus owe a debt to Aristarchus?', *Journal for the History of Astronomy* **16**(1), 37–42.

Gingras, Bruno (2003), 'Johannes Kepler's Harmonices mundi: A "Scientific" Version of the Harmony of the Spheres, Part II', *Journal of the Royal Astronomical Society of Canada* **97**(12), 259–265.

Gioè, Angelo (2007), *Aristarque de Samos: sur les dimensions et les distances du soleil et de la lune: édition critique, présentation, traduction et notes*, PhD thesis, Università Paris IV, Sorbonne.

Goldstein, Bernard Raphael (1967), 'The Arabic version of Ptolemy's *Planetary Hypotheses*', *Transactions of the American Philosophical Society* **57**(4), 3–55.

Gomez, Alberto Gomez (2013), *Aristarchos of Samos, the Polymath: A Collection of Interrelated Papers*, AuthorHouse.

Goodwin, William Watson (1878a), *Plutarch's Morals*, Vol. 3, Little, Brown, Boston.

Goodwin, William Watson (1878b), *Plutarch's Morals*, Vol. 5, Little, Brown, Boston.

Gottschalk, H. B. (1980), *Heraclides of Pontus*, Clarendon Press, Oxford, New York.

Graham, Daniel W. (2010), *The texts of early Greek philosophy: the complete fragments and selected testimonies of the major Presocratics*, Vol. 1, Cambridge University Press.

Graham, Daniel W. (2015), 'On Philolaus' Astronomy', *Archive for History of Exact Sciences* **69**(2), 217–230.

Granger, Frank Stephen (1931), *Vitruvius, On Architecture*, Vol. 1, Harvard University Press.

Granger, Frank Stephen (1934), *Vitruvius, On Architecture*, Vol. 2, Harvard University Press.

Groten, Erwin (2004), 'Fundamental parameters and current (2004) best estimates of the parameters of common relevance to astronomy, geodesy, and geodynamics', *Journal of Geodesy* 77, 724–797.

Guterman, Lila (2000), 'Are mathematicians past their prime at 35?', *Chronicle of Higher Education* 47(14), A18–A20.

Gwilt, Joseph (1874), *The Architecture of Marcus Vitruvius Pollio*, Lockwood, London.

Haberreiter, Margit, W. Schmutz and A. G. Kosovichev (2008), 'Solving the discrepancy between the seismic and photospheric solar radius', *The Astrophysical Journal Letters* 675(1), L53.

Hamm, Elizabeth Anne (2011), *Ptolemy's Planetary Theory: An English Translation of Book One, Part A of the* Planetary Hypotheses *with Introduction and Commentary*, University of Toronto PhD Dissertation.

Hannah, Robert (2008), *Time in antiquity*, Routledge.

Hartner, Willy (1977), 'The role of observations in ancient and medieval astronomy', *Journal for the History of Astronomy* 8(1), 1–11.

Hayakawa, Hisashi, Harufumi Tamazawa, Akito Davis Kawamura and Hiroaki Isobe (2015), 'Records of sunspot and aurora during CE 960–1279 in the Chinese chronicle of the Sòng dynasty', *Earth, Planets and Space* 67(1), 1–14.

Heath, Thomas Little (1897), *The works of Archimedes*, Cambridge University Press.

Heath, Thomas Little (1910), *Diophantus of Alexandria: A study in the history of Greek algebra*, Cambridge University Press.

Heath, Thomas Little (1913), *Aristarchus of Samos, the ancient Copernicus: a history of Greek astronomy to Aristarchus, together with Aristarchus's Treatise on the sizes and distances of the Sun and Moon: a new Greek text with translation and notes*, Clarendon Press, Oxford.

Heath, Thomas Little (1921a), *A history of Greek mathematics*, Vol. 1, Clarendon Press, Oxford.

Heath, Thomas Little (1921b), *A history of Greek mathematics*, Vol. 2, Clarendon Press, Oxford.

Heiberg, Johan Ludvig (1881), *Archimedis Opera omnia cum commentariis Eutocii*, Vol. 2, Teubner, Lipsiae.

Heiberg, Johan Ludvig (1888), *Euclidis opera omnia*, Vol. 1, Teubner, Lipsiae.

Heiberg, Johan Ludvig (1894), *Simplicii in Aristotelis de caelo commentaria*, in aedibus Georgii Reimeri, Berolini.

Heiberg, Johan Ludvig (1898), *Claudii Ptolemaei Opera quae exstant omnia: Syntaxis mathematica*, Pars I: Libros I-VI continens, in aedibus Benedicti Gotthelf Teubneri, Lipsiae.

Heiberg, Johan Ludvig (1907), *Claudii Ptolemaei Opera quae exstant omnia: Opera Astronomica Minora*, in aedibus Benedicti Gotthelf Teubneri, Lipsiae.

Heiberg, Johan Ludvig (1972), *Archimedis Opera omnia cum commentariis Eutocii, corrigenda adiecit Evangelos S. Stamatis*, Vol. 2, Teubner, Stutgardiae.

Helden, Albert van (1977), 'The invention of the telescope', *Transactions of the American Philosophical Society* **67**(4), 1-67.
Helden, Albert van (1985), *Measuring the universe: cosmic dimensions from Aristarchus to Halley*, University of Chicago Press.
Helmholtz, Hermann von (1867), *Handbuch der physiologischen Optik, Allgemeine Encyclopdie der Physik*, Leopold Voss, Leipzig.
Hevelius, Johannes and Jeremiah Horrocks (1662), *Mercurius in Soles visus Gedani et Venus in Sole visa*, Simon Reiniger, Gedani.
Hicks, Robert Drew (1907), *Aristotle de Anima: With Translation, Introduction, and Notes*, Cambridge University Press.
Hippolytus (1851), *Origenis Philosophumena, sive, Omnium Haeresium refutatio: e codice Parisino*, E Typographeo Acadademico, Oxonia.
Hippolytus (1921), *Philosophumena, or the Refutation of All Heresies, translated from the text of Cruice by Francis Legge*, Vol. 1, Society for promoting Christian knowledge, London.
Hippolytus (2016), *Refutation of All Heresies. Translated with an Introduction and Notes by M. David Litwa*, Vol. 40 of *Writings from the Greco-Roman World*, Society of Biblical Literature Press, Atlanta.
Hoag, Arthur Allen (1989), 'Aristarchos Revisited', *Publications of the Astronomical Society of the Pacific* **101**(644), 885r.
Hooke, Robert (1705), *The posthumous works of Robert Hooke*, Richard Waller, London.
Howarth, Leonard Jones (1917), *The Geography of Strabo*, Vol. 1 of *Loeb Classical Library*, Heinemann, Putnam's Sons, London, New York.
Huby, Pamela and Dimitri Gutas (1999), *Theophrastus of Eresus, Commentary*, Vol. 4, Brill, Leiden.
Huffman, Carl A. (1993), *Philolaus of Croton: Pythagorean and Presocratic: A Commentary on the Fragments and Testimonia with Interpretive Essays*, Cambridge University Press.
Hughes, David W. (2001*a*), 'Galileo's measurement of the diameter of a star, and of the eye's pupil', *Journal of the British Astronomical Association* **111**, 266-270.
Hughes, David W. (2001*b*), 'Six stages in the history of the astronomical unit', *Journal of Astronomical History and heritage* **4**, 15-28.
Hultsch, Friedrich Otto (1867), *Censorini De die natali liber*, Teubner, Lipsiae.
Hultsch, Friedrich Otto (1877), *Pappi Alexandrini collectionis quae supersunt*, Vol. 2, apud Weidmannos, Berolini.
James of Cremona (1544), *Archimedes ta mechri nyn sozomena hapanta. Archimedis Syracusani philosophi ac geometrae excellentissimi opera, quae quidem extant, omnia, multis iam seculis desiderata, atque a quam paucissimis hactenus visa, nuncque primum et Graece et Latine in lucem edita*, Ioannes Heruagius, Basileae.
John Paul II (2003), *Papal Addresses to the Pontifical Academy of Sciences 1917-2002 and to the Pontifical Academy of Social Sciences 1994-2002*, Vol. 100 of *Scripta Varia*, Pontifical Academy of Sciences, Vatican City.
Jones, Alexander (1986), 'William of Moerbeke, the Papal Greek Manuscripts and the Collection of Pappus of Alexandria in Vat. gr. 218', *Scriptorium* **40**(40), 16-31.

Jones, Alexander (2005), In order that we should not ourselves appear to be adjusting our estimates to make them fit some predetermined amount, *in* 'Wrong for the right reasons', Springer, pp. 17–39.

Jones, Alexander (2009), *Ptolemy in perspective: use and criticism of his work from antiquity to the nineteenth century*, Vol. 23, Springer Science & Business Media.

Kepler, Johannes (1606), *De stella nova in pede Serpentarii*, Ex officina calcographica Pauli Sessii, Pragae.

Kepler, Johannes (1617), *Ephemerides Novae Motuum Coelestium*, excudebat Johannes Plancus, sumptibus Authoris, Lincii Austriae.

Kepler, Johannes (1620), *Epitome astronomiae copernicanae*, Vol. IV, excudebat Johannes Plancus, Lentiis ad Danubium.

Kepler, Johannes (1630), *Tomi primi Ephemeridum Ioannis Kepleri pars prima et secunda ab anno MDCXXI ad MDCXXVIII*, in Typographeio Ducali, sumptibus Authoris, Sagani Silesiorum.

Kepler, Johannes (1859), *Joannis Kepleri astronomi opera omnia*, Vol. 2, Heyder & Zimmer, Francofurti ad Moenum et Erlangae.

Kepler, Johannes (1866), *Joannis Kepleri astronomi opera omnia*, Vol. 6, Heyder & Zimmer, Francofurti ad Moenum et Erlangae.

Kepler, Johannes (1868), *Joannis Kepleri astronomi opera omnia*, Vol. 7, Heyder & Zimmer, Francofurti ad Moenum et Erlangae.

Kidd, Ian Gray (1999), *Posidonius: The Translation of the Fragments*, Vol. 3, Cambridge University Press. Cambridge Classical Texts and Commentaries.

Kirk, Geoffrey Stephen, John Earle Raven, Malcolm Schofield et al. (1957), *The Presocratic Philosophers: A Critical History with a Selection of Texts*, Cambridge University Press.

Knorr, Wilbur R. (1978), 'Archimedes and the *Elements*: Proposal for a revised chronological ordering of the Archimedean corpus', *Archive for history of exact sciences* **19**(3), 211–290.

Kourouniotes, Konstantinos and Homer A Thompson (1932), 'The Pnyx in Athens', *Hesperia: the journal of the American School of Classical Studies at Athens* **1**, 90–217.

Kremkow, Jens, Jianzhong Jin, Stanley J. Komban, Yushi Wang, Reza Lashgari, Xiaobing Li, Michael Jansen, Qasim Zaidi and Jose Manuel Alonso (2014), 'Neuronal nonlinearity explains greater visual spatial resolution for darks than lights', *Proceedings of the National Academy of Sciences* **111**(8), 3170–3175.

Kroll, Wilhelm (1908), *Vettii Valentis anthologiarum libri*, Weidmann, Berlin.

Kugler, Franz Xaver (1907), *Sternkunde und Sterndienst in Babel*, Vol. 1, Aschendorff, Münster in Westfalen.

Kuhn, Thomas S. (1962), *The structure of scientific revolutions*, University of Chicago press, Chicago.

la Hire, Philippe de (1687), *Tabularum astronomicarum pars prior, de motibus solis et lunae*, apud Stephanum Michallet, Paris.

Laertius, Diogenes (1925*a*), *Lives of Eminent Philosophers*, Vol. 1, translated by Robert Drew Hicks, William Heinemann, London and New York.

Laertius, Diogenes (1925*b*), *Lives of Eminent Philosophers*, Vol. 2, translated by Robert Drew Hicks, William Heinemann, London and New York.

Lansberge, Johan Philip van (1631), *Philippi Lansbergii Uranometriae libri tres*, apud Zachariam Romanum.
Lawlor, Robert and Deborah Lawlor (1979), *Theon of Smyrna. Mathematics useful for understanding Plato*, Wizards Bookshelf, San Diego, CA.
Lehn, Waldemar H. and Siebren van der Werf (2005), 'Atmospheric refraction: a history', *Applied optics* **44**(27), 5624–5636.
Leo XIII (1893), *Providentissimus Deus: Encyclical of Pope Leo XIII on the Study of Holy Scripture*, Vatican City, Libreria Editrice Vaticana.
Livius, Titus (1876), *Ab Urbe Condita Libri*, Vol. 2, Teubner, Lipsiae.
Luzum, Brian, Nicole Capitaine, Agnès Fienga, William Folkner, Toshio Fukushima, James Hilton, Catherine Hohenkerk, George Krasinsky, Gérard Petit, Elena Pitjeva et al. (2011), 'The IAU 2009 system of astronomical constants: the report of the IAU working group on numerical standards for Fundamental Astronomy', *Celestial Mechanics and Dynamical Astronomy* **110**(4), 293–304.
Maass, Ernst (1892), *Aratea*, Weidmannsche Buchhandlung, Berlin.
Macrobius, Ambrosius Theodosius (1952), *Commentary on the dream of Scipio, translated with an introduction and notes by William Harris Stahl*, Records of Western Civilization, Columbia University Press, New York.
Macrobius, Ambrosius Theodosius (1963), *Commentarii in somnium Scipionis*, Teubner, Lipsiae.
Maniatis, Yiorgo N. (2009), 'Pythagorean Philolaus' Pyrocentric Universe', *Schole* **3**(2), 401–415.
Manitius, Karl (1894), *Hipparchi in Arati et Eudoxi Phaenomena commentariorum libri tres*, in aedibus Benedicti Gotthelf Teubneri, Lipsiae.
Manitius, Karl (1974), *Proklou Diadochou Hypotypōsis tōn astronomikōn hypotheseōn*, in aedibus Benedicti Gotthelf Teubneri, Lipsiae.
Mansfeld, Jaap and David Runia (2009), *Aetiana II: The Method and Intellectual Context of a Doxographer. The Compendium*, Vol. 114 of *Philosophia Antiqua*, Brill, Leiden.
Mansfeld, Jaap and David Runia (2018), *Aetiana IV: Papers of the Melbourne Colloquium on Ancient Doxography*, Vol. 148 of *Philosophia Antiqua*, Brill, Leiden.
Mansfeld, Jaap and David Runia (2020), *Aetiana V: An Edition of the Reconstructed Text of the* Placita *with a Commentary and a Collection of Related Texts*, Vol. 153 of *Philosophia Antiqua*, Brill, Leiden.
Martin, Thomas Henri (1849), *Theonis Smyrnaei Platonici Liber de Astronomia, cum Sereni fragmento: Textum Primus Edidit, Latine Vertit, Descriptionibus Geometricis, Dissertatione et Notis Illustravit*, e Republicae Typographeo, Paris.
Maude, William (1900), *The Natal Day*, The Cambridge Encyclopedia, New York.
McKirahan, Richard D. (2011), *Philosophy before Socrates: An introduction with texts and commentary*, Hackett Publishing Company, Indianapolis.
Meeus, Jean (1997), *Mathematical astronomy morsels*, Willmann-Bell, Richmond, Virginia.
Meeus, Jean (1998), *Astronomical algorithms, Second Edition*, Willmann-Bell, Richmond, Virginia.

Meeus, Jean (2002), *More mathematical astronomy morsels*, Willmann-Bell, Richmond, Virginia.
Meeus, Jean (2007), *Mathematical astronomy morsels IV*, Willmann-Bell, Richmond, Virginia.
Meeus, Jean (2009), *Mathematical astronomy morsels V*, Willmann-Bell, Richmond, Virginia.
Mercati, Giovanni (1938), *Codici latini Pico Grimani Pio e di altra biblioteca ignota del secolo XVI esistenti nell'Ottoboniana e i codici greci Pio di Modena con una digressione per la storia dei codici di S. Pietro in Vaticano*, Biblioteca Apostolica Vaticana, Civitate Vaticana.
Moisson, Xavier and Pierre Bretagnon (2001), 'Analytical planetary solution VSOP2000', *Celestial Mechanics and Dynamical Astronomy* **80**(3), 205–213.
Morgan, Morris Hicky (1914), *Vitruvius: the ten books on architecture*, Harvard University Press.
Morrow, Glenn R. et al. (1970), *A commentary on the first book of Euclid's Elements*, Princeton University Press, Princeton.
Müller, Karl (1885), *Fragmenta Historicorum Graecorum*, Ambrosio Firmin Didot, Paris.
Napolitani, Pier Daniele (2013), 'Between myth and mathematics: the vicissitudes of Archimedes and his work', *Lettera Matematica* **1**(3), 105–112.
Netz, Reviel (2003), 'The Goal of Archimedes' *Sand Reckoner*', *Apeiron* **36**(4), 251–290.
Netz, Reviel (2004), *The Works of Archimedes: Volume 1, the Two Books on the Sphere and the Cylinder: Translation and Commentary*, Cambridge University Press.
Neugebauer, Otto (1995), *Astronomical cuneiform texts: Babylonian ephemerides of the Seleucid period for the motion of the Sun, the Moon, and the planets*, Lund Humphries, London.
Neugebauer, Otto Eduard (1969), *The exact sciences in antiquity*, Vol. 9, Courier Corporation.
Neugebauer, Otto Eduard (1975), *A History of Ancient Mathematical Astronomy*, Springer-Verlag Berlin Heidelberg New York, Würzburg.
Neugebauer, Otto Eduard and Richard Anthony Parker (1969), *Egyptian astronomical texts, III: Decans, planets, constellations and zodiacs*, Brown University Press, London.
Nicomachus of Gerasa (1866), *Nicomachi Geraseni Pythagorei introductionis arithmeticae libri II*, Teubner, Lipsiae.
North, John (2008), *Cosmos: an illustrated history of astronomy and cosmology*, University of Chicago Press, Chicago.
Offusius, Johannes Francus (1570), *De divina facultate astrorum in larvatam astrologiam*, ex Typographia Iohannis Royerii, Paris.
Oldfather, Charles Henry (1946), *Library of History*, Vol. 4, 375 of *Loeb Classical Library*, Harvard University Press, Cambridge, Massachusetts.
Open University (2009), *Mathematical methods and fluid mechanics, Block 2*, Open University, Milton Keynes.
Osborne, Catherine (1983), 'Archimedes on the Dimensions of the Cosmos', *Isis* **74**(2), 234–242.

Pannekoek, Antoine (1952), The Astronomical System of Herakleides, in 'Circular no. 4 of the Astronomical Institute of the University of Amsterdam', pp. 373–381.

Pappus (1878), Synagogue, Liber VIII, apud Weidmannos, Berolini.

Parker, Richard Anthony (1974), 'Ancient Egyptian Astronomy', Philosophical Transactions of the Royal Society of London. Series A, Mathematical and Physical Sciences 276(1257), 51–65.

Pingree, David (1986), Vettii Valentis Antiocheni anthologiarum libri novem, De Gruyter.

Pitjeva, Elena Vladimirovna and Erland Myles Standish (2009), 'Proposals for the masses of the three largest asteroids, the Moon-Earth mass ratio and the Astronomical Unit', Celestial Mechanics and Dynamical Astronomy 103(4), 365–372.

Plato (1854), The Works of Plato: A New and Literal Version Chiefly from the Text of Stallbaum, Vol. 6, Henry G. Bohn, London.

Plato (1888), Timaeus, with an English translation by Richard Dacre Archer-Hind and others, Macmillan and Company, London.

Plato, Calcidius and Johann Wrobel (1876), Platonis Timaeus Interprete Chalcidio Cum Eiusdem Commentario Ad Fidem Librorum Manu Scriptorum, Teubner, Lipsiae.

Plutarch (1957), Plutarch's Moralia in fifteen volumes, with an English translation by Frank Cole Babbitt and others, Vol. 12, 406 of Loeb Classical Library, Harvard University Press.

Polybius (1966), The Histories, with an English translation by William Roger Paton, Vol. 3, 138 of Loeb Classical Library, Harvard University Press.

Price, Derek John de Solla (1957), Precision instruments: to 1500, in 'A history of technology', Vol. 3, Clarendon Press, Oxford, pp. 582–619.

Raeder, Hans Henning, Bengt Strömgren and Elis Strömgren (1946), Tycho Brahe's description of his instruments and scientific work as given in Astronomiae instauratae mechanica, Munksgaard, Copenhagen.

Raper, Simon (2017), 'The shock of the mean', Significance 14(6), 12–17.

Rawlins, Dennis (1991), 'Hipparchos' ultimate solar orbit & the Babylonian tropical year', Dio 1, 49–66.

Rawlins, Dennis (2002), 'Aristarchos and the "Babylonian" System B Month', DIO 11, 4–9.

Rawlins, Dennis (2008), 'Aristarchos Unbound: Ancient Vision. The Hellenistic Heliocentrists' Colossal Universe-Scale', DIO 14, 13–32.

Rawlins, Dennis (2012), 'Archimedes' hidden measure in degrees-sunsize disguise: his solar diameter; Hellenistic astronomers' high empiricism confirmed by accurate solar brackets; Babylonian degree-measure already Greek-adopted by 3rd century BC', DIO 20, 3–6.

Riccioli, Giovanni Battista (1651), Almagestum novum, ex typographia haeredis Victorii Benatii, Boloniae.

Rigaud, Stephen Peter (1833), Supplement to Dr. Bradley's Miscellaneous Works With an Account of Harriot's Astronomical Papers, Oxford University Press.

Rocher, Patrick (2000), Les calendriers, in 'Temps et Calendriers', pp. 9–20.

Rogers, Eric M. (1960), Physics for the Inquiring Mind, Princeton University Press.

Rome, Adolphe (1931), Commentaires de Pappus et de Théon d'Alexandrie sur l'Almageste. Tome I. Pappus d'Alexandrie, Commentaire sur les livres 5 et

6 *de l'Almageste*, Studi e Testi 54, Biblioteca Apostolica Vaticana, Civitate Vaticana.

Rome, Adolphe (1943), *Commentaires de Pappus et de Théon d'Alexandrie sur l'Almageste. Théon d'Alexandrie. Tome III. Theon d'Alexandrie. Commentaire sur les livres 3 et 4 de l'Almageste*, Studi e Testi 106, Biblioteca Apostolica Vaticana, Civitate Vaticana.

Rosen, Edward (1947), *The Naming of the Telescope*, Henry Schuman, New York.

Rowland, Ingrid Drake and Thomas Noble Howe (2001), *Vitruvius: Ten Books on Architecture*, Cambridge University Press.

Russo, Lucio and Silvio M. Medaglia (1996), 'Sulla presunta accusa di empietà ad Aristarco di Samo', *Quaderni urbinati di cultura classica* 53(2), 113–121.

Said, Said S. and F. Richard Stephenson (1997), 'Solar and lunar eclipse measurements by medieval Muslim astronomers, II: Observations', *Journal for the History of Astronomy* 28(1), 29–48.

Sarton, George (1952), *Ancient science through the golden age of Greece*, Harvard University Press.

Schemmel, Matthias (2008), *The English Galileo: Thomas Harriot's work on motion as an example of preclassical mechanics*, Vol. 268, Springer Science & Business Media.

Schiaparelli, Giovanni Virginio (1873), *I precursori di Copernico nell'antichità*, Ulrico Hoepli, Milano.

Schütrumpf, Eckart, Peter Stork, Jan van Ophuijsen and Susan Prince (2008), *Heraclides of Pontus: Texts and Translation*, Transaction Publishers, New Brunswick, NJ.

Shapiro, Alan E. (1975), 'Archimedes's measurement of the Sun's apparent diameter', *Journal for the History of Astronomy* 6(2), 75–83.

Sheehan, William Patrick (2018), 'Two important cases of the irradiation illusion in astronomy', *The Antiquarian Astronomer* 12, 17–28.

Sheehan, William Patrick and John Edward Westfall (2010), *The transits of Venus*, Prometheus Books, New York.

Sidoli, Nathan Camillo (2007), 'What we can learn from a diagram: The case of Aristarchus' *On the Sizes and Distances of the Sun and the Moon*', *Annals of Science* 64(4), 525–547.

Sidoli, Nathan and Takanori Kusuba (2014), 'Al-Harawi's Version of Menelaus' Spherics', *Suhayl* 13, 149–212.

Smith, William (1849), *Dictionary of Greek and Roman biography and mythology*, Vol. 1, C.C. Little and J. Brown, Boston.

Stahl, William Harris and Richard Johnson (1971), *Martianus Capella and the seven liberal arts*, Vol. 2, Columbia University Press, New York.

Stephenson, F. Richard and Louay J. Fatoohi (1993), 'Lunar eclipse times recorded in Babylonian history', *Journal for the history of astronomy* 24(4), 255–267.

Sufi, Abd al-Rahman al- (964), *kitab suwar al-kawakib [Book of the Shapes of Stars]*, Bodleian Library, MS. Marsh 144.

Swerdlow, Noel Mark (1969), 'Hipparchus on the Distance of the Sun', *Centaurus* 14(1), 287–305.

Swerdlow, Noel Mark (1973), 'Al-Battani's Determination of the Solar Distance', *Centaurus* 17(2), 97–105.

Swerdlow, Noel Mark (2002), Shadow measurement: the *Sciametria* from Kepler's *Hipparchus* – a translation with commentary, *in* P. M. Harman and A. E. Shapiro, eds, 'The investigation of difficult things: Essays on Newton and the history of the exact sciences in honour of D. T. Whiteside', Cambridge University Press, pp. 19–70.

Tannery, Paul (1888), 'La Grande année d'Aristarque de Samos', *Mémoires de la Société des sciences physiques et naturelles de Bordeaux* **4**, 79.

Taylor, Thomas (1820), *The Commentaries of Proclus on the Timaeus of Plato*, Vol. 2, The author, Walworth, London.

Thims, Libb (2016), *Hmolpedia: A-Z Encyclopedia of Human Thermodynamics, Human Chemistry, and Human Physics*, Vol. 1, LuLu.

Thorndike, Lynn (1941), *A History of Magic and Experimental Science*, Vol. 6, Columbia University Press, New York.

Thurston, Hugh (2001), 'Early Greek solstices and equinoxes', *Journal for the History of Astronomy* **32**(2), 154–156.

Toomer, Gerald James (1968), 'The size of the lunar epicycle according to Hipparchus', *Centaurus* **12**(3), 145–150.

Toomer, Gerald James (1974*a*), 'Hipparchus on the Distances of the Sun and Moon', *Archive for History of Exact Sciences* pp. 126–142.

Toomer, Gerald James (1974*b*), 'The chord table of Hipparchus and the early history of Greek trigonometry', *Centaurus* **18**(1), 6–28.

Toomer, Gerald James (1984), *Ptolemy's Almagest*, Gerald Duckworth, London.

Treweek, Athanasius Pryor (1957), 'Pappus of Alexandria, The manuscript tradition of the Collectio Mathematica', *Scriptorium* **11**(2), 195–233.

Turyn, Alexander (1964), *Codices Graeci Vaticani saeculis XIII et XIV scripti annorumque notis instructi*, Codices e Vaticanis selecti phototypice expressi/Series maior, Bybliotheca Apostolica Vaticana.

Uri, John (1787), *Catalogus, Bibliothecae Bodleianae Codicum Manuscriptorum Orientalium*, Typographeo Clarendoniano, Oxonium.

Valdes, Benito Daza (1623), *Uso de los antoios para todo genero de vistas*, Diego Perez, Sevilla.

Valla, Giorgio (1488), *De magnitudinibus et distantiis solis et lunae*, Antonius de Strata Cremonensis, Venetia.

Vardi, Ilan (1997), *Archimedes, The Sand Reckoner*, preprint.

Voltaire (1775), Système, *in* 'Dictionnaire Philosophique', Vol. 6, Cramer, Geneva.

Voltaire (1901), System, *in* 'Philosophical Dictionary, translated by John Morley, Tobias George Smollett, William F Fleming, and Oliver Herbrand Gordon Leigh', Vol. 7.2, Craftsmen of the St. Hubert Guild, New York.

Wachsmuth, Kurt and Otto Hense (1884), *Joannis Stobaei Anthologium*, Weidmann, Berlin.

Waerden, Bartel Leendert van der (1987), 'The heliocentric system in Greek, Persian and Hindu astronomy', *Annals of the New York Academy of Sciences* **500**(1), 525–545.

Webb, Edmund James (1921), 'Cleostratus Redivivus', *The Journal of Hellenic Studies* **41**(1), 70–85.

Webster, Colin (2014), 'Euclid's *Optics* and Geometrical Astronomy', *Apeiron* 47(4), 526–551.
Wendelin, Godefroy (1626), *Loxias seu De obliquitate Solis*, Apud Hieronymum Verdussium, Antuerpiae.
Wendelin, Godefroy (1644), *Eclipses lunares ab anno 1573 ad 1643 observatae*, Apud Hieronymum Verdussium, Antuerpiae.
Westfall, John Edward (2019), 'Reconstructing Aristarchos' Sizes and Distances', *The Strolling Astronomer* **61**(2), 40 – 54.
Westfall, John Edward and William Patrick Sheehan (2014), *Celestial shadows: Eclipses, transits, and occultations*, Vol. 410 of *Astrophysics and Space Science Library*, Springer.
Westman, Robert S. (2011), *The Copernican question: Prognostication, skepticism, and celestial order*, University of California Press.
Whatton, Arundell Blount (1859), *The Transit of Venus across the Sun. A Translation of the Celebrated Discourse thereupon by the Rev. J. Horrox*, William MacIntosh, London.
Zhmud, Leonid (2008), *The origin of the History of Science in Classical Antiquity*, Vol. 19, Walter de Gruyter.
Ziegler, Hermann Rudolf (1891), *Cleomedis De motu circulari corporum caelestium libri duo ad novorum codicum fidem edidit et latina interpretatione instruxit Hermannus Ziegler*, Teubner, Lipsiae.
Zimmermann, Johann Jakob (1679), *Prodromus biceps Cono-Ellipticae*, Sumptibus Johannis Gottfridi Zubrod, Typis Pauli Trew, Stuttgardiae.

List of Figures

Figure 1.1:	Philolaus' cosmos	25
Figure 2.1:	Proposition 4	50
Figure 2.2:	First half-Moon	52
Figure 2.3:	Last half-Moon	52
Figure 2.4:	Lunar orthogony	53
Figure 2.5:	Lunar quadrature	53
Figure 2.6:	Aristarchus' triangle	59
Figure 2.7:	Spherical version of Aristarchus' triangle	61
Figure 2.8:	Proposition 13	66
Figure 2.9:	Sizes in *On Sizes*	69
Figure 2.10:	Proposition 14	70
Figure 2.11:	Proposition 19	72
Figure 3.1:	Aristarchus Procedure 1	96
Figure 3.2:	Aristarchus Procedure 2	98
Figure 3.3:	Aristarchus Procedure 3	99
Figure 3.4:	Aristarchus Procedure 4	100
Figure 3.5:	Aristarchus Procedure 5	100
Figure 3.6:	The *minicosmos*	102
Figure 3.7:	The *dichotometer*	115
Figure 3.8:	Dichotometer mounted on a triaxial tripod	116
Figure 3.9:	The *cosmometer*	123
Figure 3.10:	Aristarchan versus true orthogony (1)	127
Figure 3.11:	Aristarchan versus true orthogony (2)	128
Figure 4.1:	Aristarchus' model of the universe	132
Figure 4.2:	Aristarchus' revised model of the universe	134
Figure 4.3:	Archimedes' solar width-to-distance ratio	147
Figure 4.4:	Ptolemy-based solilunar distance function	162
Figure 5.1:	Simplified lunar eclipse diagram	188
Figure 5.2:	Meton's summer solstice of 432 BC	203
Figure 5.3:	Aristarchus' summer solstice of 280 BC	211
Figure 5.4:	Hipparchus' summer solstice of 135 BC	216
Figure 5.5:	Hipparchus' summer solstice of 128 BC	217
Figure 5.6:	Bulging draughtboard illusion	234
Figure A.1:	The Moon's shadow	237
Figure C.1:	Lunar eclipse diagram	247
Figure C.2:	Solar eclipse diagram	253
Figure C.3:	Eclipse of March 14, 190 BC	257

List of Tables

Table 1.1:	Ancient planetary arrangements	31
Table 1.2:	Early comments on the Earth's motion	36
Table 3.1:	Harriot's 1611-1-21 field notes	81
Table 3.2:	Harriot's 1611-4-19 field notes	84
Table 3.3:	Harriot's 1611-4-19 half-Moon timings	85
Table 3.4:	Hoag's data	89
Table 3.5:	Hoag's lunar phase estimations	90
Table 3.6:	Hoag's lunar elongation estimations	90
Table 3.7:	Gomez's 2016-03-15 data	105
Table 3.8:	Gomez's 2016-05-13 data	108
Table 3.9:	Gomez's 2016-08-10 data	110
Table 3.10:	Gomez's 2016-08-25 data	113
Table 3.11:	Westfall's waxing half-Moons	119
Table 3.12:	Westfall's waning half-Moons	120
Table 3.13:	Half-Moon method results	125
Table 4.1:	Pre-Archimedean approximations to pi	149
Table 4.2:	Sun's distance-to-width ratios	150
Table 4.3:	Archimedes' cosmic ladder	181
Table 5.1:	Earth's extreme and mean shadow lengths	192
Table 5.2:	Summer solstice observations in *The Almagest*	196
Table 5.3:	Summer solstice predictions based on *The Almagest*	202
Table 5.4:	Vettius Valens' lists of 'rule writers'	214
Table 5.5:	Solstice and equinox predictions based on *The Almagest*	218
Table 5.6:	Solstices and equinoxes recorded in *The Almagest*	219

Index

Aetius, 20–28, 31–34, 36, 130, 134
air bending, 110, 112, 118–120, 122, 144, 176
al-Battani, 152, 153, 156–159, 161, 162, 171, 174–176, 182, 256
    parameters, 156, 157, 159, 174, 175
al-Hajjaj ibn Yusuf ibn Matar, 227
Alberto III Pio, 19
Alexander of Aphrodisias, 35
Alexander the Great, 20, 22, 193, 220
Alexandria, 20–23, 48, 49, 130, 166, 168–170, 179, 200, 202, 204, 205, 211, 216–219, 257, 258
    latitude, 169, 209
Anaxagoras of Clazomenae, 27, 31, 32, 44, 135
Anaximander of Miletus, 28, 44, 129, 143
angular resolution of the human eye, 49–51, 233
angular size, 40, 42, 43, 55, 56, 63, 64, 67, 69, 77, 86, 107, 138, 139, 141, 143, 144, 147, 151, 152, 154, 155, 157, 158, 162, 166, 173, 176, 181–185, 187–191, 193, 195, 230, 232–235, 237, 239, 240, 247, 248, 253, 255–257
Anonymous of the Year 379, 203
Antichthon, 23, 24, 27, 31–33, 140
Antikythera mechanism, 60
antisolar point, 102
Aphrodite, *see* Venus
Apollinarius, 203, 205, 214, 215
apparent size, *see* angular size
Apseudes of Athens, 201
Aquinas, Saint Thomas, 36, 37
Aratus of Soli, 163, 164
Archelaus, 135, 230
Archimedes of Syracuse, 13, 15, 16, 18, 19, 28–30, 39, 42–46, 48, 77, 87, 105, 107, 123, 129–135, 138–153, 162, 176–183, 187, 190, 195, 208, 230, 232, 235
    sunwidth measurement, 143–145, 233, 234
    volume of sphere, 62
Ares, *see* Mars
Aristarchus of Samos, 13–23, 34, 36–43, 45, 47–51, 54, 56–69, 71, 73–77, 79, 81, 85, 87, 88, 92–95, 97, 98, 102–107, 112–114, 117, 120, 122, 123, 125–128, 130, 131, 133, 135, 136, 138, 139, 142–146, 148–153, 162–165, 171–174, 176, 177, 181–184, 187, 189–191, 193–196, 200–202, 206, 207, 209–220, 222–236, 239, 247, 249
    empiricist, 22, 23, 50, 51, 61, 93, 143, 191
    model of the universe, 13, 15, 18, 67, 129–136, 138–142
    on air bending, 112
    on light, colour, and vision, 21–23, 51, 236
    on planetary motion, 139–141
    optimization problem, 67, 68
    Procedure
        I, 96, 97
        II, 97–99
        III, 97, 99, 103, 104
        IV, 99, 100
        V, 99–101, 104, 105, 108, 110, 113
    triangle, 59, 80, 101, 126
    spherical, 61
Aristotle of Stagira, 20, 22, 23, 30, 31, 34, 44, 94, 135, 139, 140, 142, 183, 193, 220, 228
    fifth element, 134
arithmetic mean, *see* Pythagorean means
armillary, 209
Aryabhata II, 153
Athens, 20–22, 39, 201, 203, 204, 228
Autolycus of Pitane, 16

averaging, 94, 124, 126, 191, 194, 197, 198, 207, 208, 228, 229

barleycorn, 226, 227
Beech, Martin, 183
Bellarmine, Saint Robert, 16
Berenice I, 20
BM 38462, 221
BM 55555, 198, 201, 209, 226, 227
Bodleian Library, 17, 214, 229
Boter, Gerard, 133, 134
Brahe, Tycho, 38, 96, 152, 163, 171, 233
brain magnification, *see* neuronal blurring
bulging draughtboard illusion, 234

Calcidius, 37, 38
calendar, 210
    Attic, 201, 203, 205, 210, 211, 215
    Egyptian, 200, 201, 203–205, 211, 215
    Proleptic Julian, 203, 205, 211
    Ptolemaic, 203, 211
Callippic cycle, 210, 216, 220, 228
Callippus of Cyzicus, 196, 200–202, 206, 210, 213, 215–220, 223, 225, 228, 229
Callisthenes of Olynthus, 193, 220
canonographer, 212, 214, 229
Capella, Martianus, 38
Carman, Christián Carlos, 65, 139, 164, 165, 169, 247, 258
Cassini, Giovanni Domenico, 54
Censorinus, 26, 215, 222–225, 228, 229
Central Fire, *see* Hestia
Cesi, Federico, 86
chiliagon, 142, 143, 146, 177
Chronos, *see* Saturn
Cicero, Marcus Tullius, 29, 30, 33, 36
Cleanthes of Assos, 39, 40, 235
Cleomedes, 135, 142, 143, 170
Commandino, Federico, 13
cone, 48, 53, 61, 64, 72, 183, 184
congruence, 155, 159, 174, 175
continued fraction, 212, 229
Copernicus, Nicolaus, 33, 37, 139–141, 152, 163, 164, 171, 235

cosmeter, *see* cosmometer
cosmometer, 102, 123, 126
cosmos, 43–45, 129–131, 133, 135–138, 142, 143, 177, 178, 180–182, 196, 231, 236
Counter-Earth, *see* Antichthon
Crabtree, William, 87

deferent, 166
Demisiani, John, 86
Dicaearchus of Messana, 142
dichometer, *see* dichotometer
dichotometer, 95, 115–118, 123
dichotomy, 51, 80, 88, 91, 93–95, 97, 98, 114, 116–118, 121
    equatorial, 51, 57, 61
Diodorus of Sicily, 16, 201, 205
Diogenes Laertius, 20, 23, 27, 40
Diophantus of Alexandria, 179
distance field
    lunisolar, 165, 175
    solilunar, 154, 156, 159, 164, 165, 174

Earth
    circumference, 142, 143, 152, 153, 177
    girth, *see* circumference
    shadow, 54, 64, 65, 67, 68, 73, 75, 154–156, 158, 166, 176, 183–196, 239–242, 247, 248, 253, 255, 256, 260
    enlargement, 54, 176
    motion, 191
eclipse, 31–33, 38, 40, 54, 55, 60–62, 65, 66, 68, 69, 72, 93, 140, 156, 164, 166, 168–170, 173, 185–188, 193, 194, 210, 211, 221
    annular, 53–55, 61, 155
eclipse diagram, 188, 247, 248, 253, 257, 258
eclipse method, 158, 164, 165, 167, 169, 171–176, 182, 190, 247
    lunar, 164, 165, 173, 235, 247, 252, 260

Index

solar, 164, 165, 173, 175, 247, 253, 257, 260
Ecphantus of Syracuse, 33, 34, 36
elongation, 51, 53, 58–61, 77, 80–82, 85, 88–92, 94, 95, 97, 101–108, 110–113, 115–122, 141, 181, 182
    formula, 110
Empedocles of Akragas, 21, 44
Epicurus of Samos, 21, 22
epicycle, 166
equinox, 200, 203–205, 208, 209, 218, 219
Eratosthenes of Cyrene, 48
Erhardt, Rudolf von, and Erika von Erhardt-Siebold, 15, 18, 43–45, 133, 140
Euclid of Alexandria, 16, 48, 49, 51, 98, 133
Euctemon, 203, 205, 214, 215
Eudemus of Rhodes, 28, 30
Eudoxus of Cnidus, 28, 44, 143
Eutocius of Ascalon, 19
Evans, James, 93
exeligmos, 221, 223, 224

firmament, *see* sphere of the fixed stars

Gaia, 25
Galilei, Galileo, 16, 17, 85, 121, 233–235
Gassendi, Pierre, 87
Gelo II of Syracuse, 19, 45, 129, 146
Geminus of Rhodes, 35, 215
geocentrism, 15, 16, 30, 43, 58, 68, 131, 139, 141, 169, 236
geometric mean, *see* Pythagorean means
Giedi, 231
Gioè, Angelo, 17, 41
gnomon, 88, 197, 199
Gomez, Alberto Gomez, 125
great year
    Aristarchus, 225, 228, 236
    Philolaus, 26, 33
Guckelsberger, Kurt, 98, 104

half-Moon method, 88, 95, 113, 120, 122, 123, 125, 126, 174

half-Moon's life, 102, 113, 121, 126
Harriot, Thomas, 80, 81, 83–85, 87, 88, 92, 104, 125
Heath, Thomas Little, 13, 42, 43, 49, 58, 61, 179
heliocentrism, 13, 15–17, 35, 37–41, 43, 76, 94, 130, 131, 133, 141, 229, 233, 235, 236
heliosphere, 138, 182
Hellespont, 166, 168, 169, 258
Heraclides of Pontus, 33–39
Hermes, *see* Mercury
Hestia, 24, 25, 27, 31–33, 140, 141
hestiocentrism, 24, 25, 30, 32, 141
Hevelius, Johannes, 87
Hicetas of Syracuse, 23, 33, 36
Hiero II of Syracuse, 45, 129
Hipparchic cycle, 227
Hipparchus of Nicaea, 139, 153, 165–174, 182, 196–198, 200–202, 204–211, 215–219, 227, 228, 232, 233, 247, 249, 253, 257, 259
    parameters, 166
Hippolytus of Rome, 29, 31
Hoag, Arthur Allen, 88–92, 94, 101, 104, 113, 114, 117, 125
Hooke, Robert, 50, 57
Horrocks, Jeremiah, 87, 88, 109, 125, 174
Hypsicles, 16

Iacopo da San Cassiano, 19, 132
ill-conditioning, 155, 173, 176
irradiation, 121, 233
Isidore of Miletus, 19

James of Cremona, *see* Iacopo da San Cassiano
John of Stobi, *see* Stobaeus, Johannes
Jones, Alexander, 19, 204, 205, 212, 229
JPL Horizons Online Ephemeris System, 118, 121, 122
Julian century, 226
Julian day, 197, 207, 226
Jupiter, 25, 30, 32, 233, 234

Kepler, Johannes, 85–87, 125, 174, 176
Kuiper belt, 138

la Hire, Philippe de, 54, 176
Lansberge, Philip, 125
Laurentian Library, 19, 20
Leon, The Mathematician, 19
Lincean Academy, 86
Little Astronomy, 16
Livius, Titus, 129
Livy, *see* Livius, Titus
lunation, *see* synodic month
Lysimachia, 142

Maass, Ernst, 212
Macrobius Ambrosius Theodosius, 29
Manitius, Karl, 42
Mars, 25, 30, 32, 233
Martin, Thomas Henri, 37, 38
Medici, Lorenzo de, 19
Mercury, 25, 29, 30, 32, 38, 39, 139, 141, 233
Meton of Athens, 196, 197, 201–206, 209–211, 213–220
Metonic cycle, 201, 210, 215–217, 220
Milesian inscription, 205
minicosmos, 95, 101–103
Mizar and Alcor, 50, 231
molad, 226–228
month
    anomalistic, 221, 223, 224, 227
    draconic, 221, 223, 224
    sidereal, 30, 191, 193, 221, 223–225, 227
    synodic, 93, 220, 221, 223–228
        280 BC, 226
        ELP 2000-85, 225
        Hipparchus, 227
Moon, 25, 32
    distance, 53–56, 59, 63, 67–69, 73, 74, 76, 86, 94, 102, 105, 107, 126, 139, 152, 154–162, 166–170, 172, 181, 183, 184, 186, 189, 192, 194, 195, 235, 238, 239, 247, 249, 252–257, 259, 260
    shadow, 53–56, 61, 63, 186, 187, 237, 238
    unevenness, 86–88
myriad, 87, 105, 107, 108, 110, 112, 120, 126, 142, 143, 152, 153, 177–180, 213, 215, 230–232, 235

Nabonassar Era, *see* calendar, Egyptian
Nasir al-Din al-Tusi, 13, 17, 64, 183
Neugebauer, Otto Eduard, 17, 38, 88, 92, 114
neuronal blurring, 233

Offusius, Jofrancus, 163, 164, 171, 173
orthogony, 51–53, 57, 58, 60, 101–104, 106, 108–113, 121, 122, 126–128
orthotomy, 52, 57, 80, 83, 85–87, 91, 92, 97, 99, 101, 107, 121, 126, 127
Osiander, Andreas, 236

Pappus of Alexandria, 16–18, 41, 48, 153, 165, 167–170
parallax, 125, 166, 167, 230–232
Pausanias, 201
Peloponnese, 215
Perdiccas, 22, 23
phase angle, 51, 53, 57, 60
Phidias, 28, 45, 77, 143
Philip of Opus, 29, 31, 214, 215
Philippus of Opus, *see* Philip of Opus
Philochorus of Athens, 204
Philolaus of Croton, 23–28, 30–32, 44, 93, 130, 135, 139–141
pi, 143, 147–151, 153, 162, 173, 181
planetary order, 24, 28–32, 141
Plato of Athens, 28, 34, 37, 44, 48, 139, 141, 153, 163
    aether, 134
Platonic solids, 163
Plutarch, 20, 36, 39, 235, 236
Pnyx, 204
Poliziano, Angelo, 19

Polybius of Megalopolis, 129
Polycletus of Athens, 201
Pope Clement V, 19
Pope John Paul II, 16
Pope Nicholas V, 19
poppy seed, 142, 179–181
Posidonius of Apamea, 35, 36, 79, 81, 135, 153
Proclus of Athens, 34, 36, 38, 48, 153, 156, 162
pseudo-Plutarch, 20, 22–24
Ptolemy I Soter, 20, 48
Ptolemy Philadelphus, 20–23, 45
Ptolemy, Claudius, 16, 24, 29, 30, 48, 140, 141, 152, 153, 155–166, 168, 169, 171, 172, 174, 175, 182, 196, 200–206, 208, 209, 211, 212, 215, 217–219, 221, 222, 227–229, 233, 249, 253, 255, 256
    parameters, 154, 155, 157, 162, 172, 173, 182, 190
pyrocentrism, 24
Pythagoras of Samos, 22, 101, 232, 237, 241, 249
Pythagorean means, 107, 109, 124
    arithmetic mean, 124, 126, 150, 151, 182
    geometric mean, 150, 151, 182

quad box, *see* quad gauge
quad gauge, 95, 101, 102, 106, 122, 123
quad line, 99–101, 107, 109, 110, 123, 124
quadrature, 51, 53, 58, 81, 83, 85, 93, 95–97, 99, 101, 102, 104, 106, 108–110, 112, 116–118, 122, 127
quadrature line, *see* quad line
Qusta ibn Luqa, 13

radian, 146–148, 151, 165, 173, 174
Rawlins, Dennis, 15, 40–42, 145, 198, 200, 201, 205, 227, 231
refraction, *see* air bending
resolution, *see* angular resolution
Riccioli, Giovanni Battista, 79, 87, 88, 121, 125

saros, 220–224, 226, 228, 229
Saturn, 25, 30, 32, 233
scaphe, 216
Schiaparelli, Giovanni Virginio, 37
Seleucus of Seleukia, 235
Sheehan, William Patrick, 79, 92, 93, 114
Simplicius of Cilicia, 28, 34–38, 130
solstice, 196, 197, 199–204, 206, 208, 209, 216, 218, 219
    128 BC, 201, 207, 208, 211, 216–219
    135 BC, 197, 198, 201, 205–210, 215–219
    280 BC, 196, 201, 206, 207, 209–211, 213, 215, 217–219, 226
    330 BC, 201, 206, 213, 217–220, 228, 229
    432 BC, 197, 201, 203–206, 209–211, 213, 215–219
    AD 140, 200, 218, 219
speculum metal, 98
sphere of the fixed stars, 25, 29, 35, 43, 44, 129–137, 178, 180, 196, 230–232
spyglass, 80, 83
Stobaeus, Johannes, 20, 22, 24, 25
Strabo of Amasia, 153, 169, 173
Strato of Lampsacus, 20–23, 42, 45, 50, 93, 141, 193
Suda, 221
Sufi, Abd al-Rahman al-, 50, 231
summary of
    On Sizes 1, 75
    On Sizes 2, 195
    hestiocentrism, 32
Sun, 25, 30, 32
    distance, 45, 51, 56, 58, 59, 61, 68, 73, 74, 76, 77, 79, 80, 82, 83, 85–88, 90–92, 94, 96, 97, 101–110, 112, 113, 116–121, 123–127, 130, 136, 138, 139, 147, 148, 151–156, 158–164, 167, 168, 171–175, 182, 184, 186, 187, 190, 196, 237, 239, 242, 247, 249, 252, 255, 257, 260
sunspots, 233

Swerdlow, Noel Mark, 153, 154, 157, 158, 165–167, 247, 253
Syene, 142
Syon House, 80, 85
syzygy, 93

Tannery, Paul, 222, 224, 225, 229
telescope, 79, 85–87, 92, 125
terminator, 51, 53, 63, 65, 66, 69, 80, 81, 85, 93, 106, 107, 114, 116, 118, 121, 126, 127
Thabit ibn Qurra, 13, 17, 64, 183
Theodoret of Cyrus, 20
Theodosius of Bithynia, 16
Theon of Alexandria, 48, 51
Theon of Smyrna, 30, 38, 165, 170, 171
Theophrastus of Eresus, 20, 21, 33
Thurston, Hugh, 199, 200, 203
Timagoras of Gela, 21
Toomer, Gerald James, 153, 165–170, 174, 247, 253, 257–259
Troad, 169

Valla, Giorgio, 13, 19
Vat. Gr. 191, 214, 215
Vat. Gr. 204, 17, 64
Vat. Gr. 218, 18
Vat. Gr. 381, 212
Vatican Library, 19, 212, 214
Venus, 25, 29, 30, 32, 37–39, 41, 139, 141, 233, 234
    dichotomy, 93
    transits, 87, 88, 174
Vettius Valens, 212, 214, 215, 229
Vitruvius Pollio, Marcus, 18, 38, 165, 216, 236

Voltaire, 13, 15, 39–41, 232

Wallis, John, 39
Wendelin, Godfrey, 85, 87, 88, 92, 104, 121, 125
Westfall, John Edward, 79, 92–96, 101, 104–106, 108, 110, 112–125
William of Moerbeke, 19

year, 191, 196–198, 201, 208, 212, 214, 217, 228, 229
    Almagest, 208, 216
    Aristarchus-Censorinus, 222, 223
    Aristarchus-Hipparchus A, 208, 210
    Aristarchus-Hipparchus B, 207
    Callippus, 206, 213, 221, 223–225, 229
    Meton, 203, 215
    Meton-Aristarchus, 211, 212, 217
    Meton-Hipparchus A, 197, 226
    Philolaus, 26
    saronic, 224, 229, 236
    sidereal, 198, 227
    tropical, 197, 198, 200, 214, 224, 229, 236
        135 BC, 197
        280 BC, 209
    VSOP2000, 197

Zeno of Elea, 39
Zeus, *see* Jupiter
Zeuxippus, 129, 130, 179
Zimmermann, Johan Jacob, 15
zodiac, 191, 193
Zu Chongzi, 153

www.ingramcontent.com/pod-product-compliance
Ingram Content Group UK Ltd.
Pitfield, Milton Keynes, MK11 3LW, UK
UKHW021822140426
5217IPUK00004B/42